2000
LLEWELLYN'S
MOON SIGN BOOK

AND GARDENING ALMANAC

With Lunar Forecasts by Gloria Star

Editor/Designer: Michael Fallon
Cover Design and Illust...
Special thanks to...

L...

St. ...

Table of Contents

How to Use This Book

A re you ready to harness the power of the Moon? With the *Moon Sign Book* you can do just that. The *Moon Sign Book* provides you with four essential tools to reap the benefits of the Moon. You can use these tools alone or in any combination to help you achieve success in 2000. The first tool is our unique, easy-to-use Astro Almanac: a list of the best dates in 2000 to begin important activities. The second tool is a complete how-to on using the Moon to fine-tune your timing. This takes the Astro Almanac one step further and teaches you *how* to choose the best dates for your activities. The third tool consists of insightful lunar astrological forecasts by astrologer Gloria Star. The fourth tool is a mass of informative articles on using the Moon in the home, business, garden, and everywhere you go. Read on to find out more about how to use each of these features.

The Astro Almanac

The simplest method for using the *Moon Sign Book* in lunar timing is to turn straight to the Astro Almanac, beginning on page 12. The Astro Almanac lists the best days to perform sixty different activities, based on the sign and phase of the Moon and on lunar aspects. All you need to do is find the particular activity that you are interested in, and the best dates for each month will be listed across the page. When working with the Moon's energies we consider two things—the inception, or beginning of an activity, and the desired outcome. We begin a project under a certain Moon sign and phase in order to achieve certain results. The results are influenced by the attributes of the sign and phase under which we started the project. Therefore, the Astro Almanac lists the best times to begin many activities.

The Moon Tables

The Astro Almanac is a general guide to the best days for the listed activities, but the Astro Almanac can't take everyone's special needs into account. For a more in-depth exploration of lunar timing, the *Moon Sign Book* provides the Moon Tables. Although we can provide generally favorable dates for everyone, not everyone will have the same goal for each activity that they start. Therefore, not everyone will want to start every activity at the same time. For example, let's say you decide to plant a flower garden. Which attributes would you like most in your flowers? Beauty? Then you may want to plant in Libra, because Libra is ruled by Venus, which in turn governs appearance. How about planting for quantity or abundance? Then you might try Cancer or Pisces, the two most fertile signs. What if you were going to be transporting the flowers somewhere, either in pots or as cut blooms? Then you might want to try Scorpio for sturdiness, or Taurus for hardiness. The Astro Almanac also does not take into account retrogrades, Moon void-of-course, or favorable and unfavorable days.

The procedure for using the Moon tables is more complex than simply consulting the Astro Almanac, but we encourage you to try it so that you can tailor the *Moon Sign Book* information to your needs and fully harness the potential of the Moon. The directions for using the tables to choose your own dates are in the section called "Using the Moon Tables," which begins on page 22. Be sure to read all of the directions, paying special attention to the information on the signs, Moon void-of-course, retrogrades, and favorable and unfavorable days. The sections titled "The Moon's Quarters & Signs" (page 52), "Retrogrades" (page 54), and the Moon Void-of-Course Table (page 56) are provided as supplementary material and should be read as well. These sections will give you a deeper understanding of how the process of lunar planning works, and give you background helpful in making use of the articles in this book.

Personal Lunar Forecasts

The third tool for working with the Moon is the Personal Lunar Forecasts section, written by Gloria Star. This section begins on page 375. Here Gloria tells you, based on your Moon sign, what's in store for you for 2000. This approach is different than that of other astrology books, including Llewellyn's *Sun Sign Book*, which make forecasts based on Sun sign. While the Sun in an astrological chart represents the basic essence or personality, the Moon represents the internal or private you—your feelings, emotions, and subconscious. Knowing what's in store for your Moon in 2000 can give you great insight for personal growth. If you already know your Moon sign, go ahead and turn to the corresponding section in the back of the book (forecasts begin on page 379). If you don't know your Moon sign, you can figure it out using the procedure outlined beginning on page 62.

Articles, Articles, Articles

Scattered throughout the *Moon Sign Book* are articles on using the Moon for activities from fishing to business. These articles are written by people who successfully use the Moon to enhance their daily lives, and are chosen to entertain you and enhance your knowledge of what the Moon can do for you. The articles can be found in the Home, Health, & Beauty section; the Leisure & Recreation section; the Business section; and the Gardening section. Check the table of contents for specific topics.

Some Final Notes

We get a number of letters and phone calls every year from readers asking the same types of questions. Most of these have to do with how to find certain information in the *Moon Sign Book* and how to use this information.

The best advice we can give is to read the *entire* introduction (pages 5–74), in particular the section on how to use the tables. We provide examples using the current Moon and aspect tables so that you can follow along and get familiar with the process. At first, using the Moon tables may seem confusing because there are several factors to take into account, but if you read the directions carefully and practice a little bit, you'll be a Moon sign pro in no time.

Remember, for quick reference for the best dates to begin an activity, turn to the Astro Almanac. To choose special dates for an activity that are tailor-made just for you, turn to "Using the Moon Tables" For insight into your personal Moon sign, check out Gloria Star's lunar forecasts. Finally, to learn about the many ways you can harness the power of the Moon, turn to the articles in the Home, Health & Beauty; Leisure & Recreation; Business & Legal; and Farm, Garden, & Weather sections.

Get ready to improve your life with the power of the Moon!

Important!

All times given in the *Moon Sign Book* are set in Eastern Standard Time (EST). You must adjust for your time zone. There is a time zone conversions chart on page 8 to assist you. You must also adjust for Daylight Saving Time where applicable.

Time Zone Conversions

World Time Zones

(Compared to Eastern Standard Time)

(R) EST—Used

(S) CST—Subtract 1 hour

(T) MST—Subtract 2 hours

(U) PST—Subtract 3 hours

(V) Subtract 4 hours

(V*) Subtract 4½ hours

(W) Subtract 5 hours

(X) Subtract 6 hours

(Y) Subtract 7 hours

(Q) Add 1 hour

(P) Add 2 hours

(P*) Add 2½ hours

(O) Add 3 hours

(N) Add 4 hours

(Z) Add 5 hours

(A) Add 6 hours

(B) Add 7 hours

(C) Add 8 hours

(C*) Add 8½ hours

(D) Add 9 hours

(D*) Add 9½ hours

(E) Add 10 hours

(E*) Add 10½ hours

(F) Add 11 hours

(F*) Add 11½ hours

(G) Add 12 hours

(H) Add 13 hours

(I) Add 14 hours

(I*) Add 14½ hours

(K) Add 15 hours

(K*) Add 15½ hours

(L) Add 16 hours

(L*) Add 16½ hours

(M) Add 17 hours

(M*) Add 17½ hours

Important!

All times given in the *Moon Sign Book* are set in Eastern Standard Time (EST). You must adjust for your time zone. You must also adjust for Daylight Saving Time where applicable.

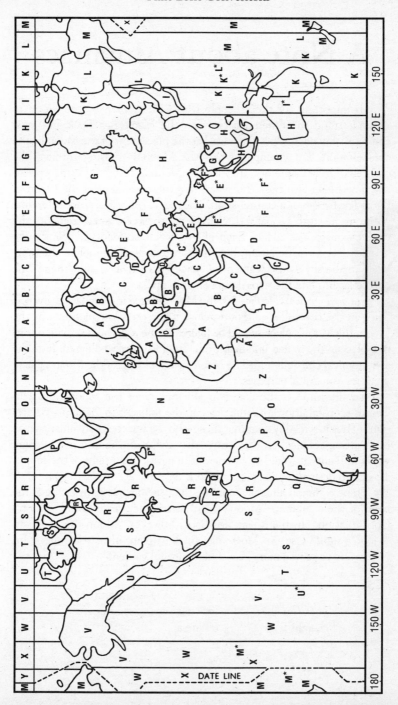

A Note about Almanacs

It is important for those people who wish to plan by the Moon to understand the difference between the *Moon Sign Book* and most common almanacs. Most almanacs list the placement of the Moon by the constellation. For example, when the Moon is passing through the constellation of Capricorn, they list the Moon as being in Capricorn.

The *Moon Sign Book*, however, lists the placement of the Moon in the zodiac by *sign*, not constellation. The zodiac is a belt of space extending out from the earth's equator. It is divided into twelve equal segments: the twelve signs of the zodiac. Each of the twelve segments happens to be named after a constellation, but the constellations are not in the same place in the sky as the segment of space (sign) named after them. The constellations and the signs do not "match up."

For *astronomical* calculations, the Moon's place in almanacs is given as being in the constellation. For *astrological* purposes, like planning by the Moon, the Moon's place should be figured in the zodiacal *sign*, which is its true place in the zodiac, and nearly one sign (30 degrees) different from the astronomical constellation. The *Moon Sign Book* figures the Moon's placement for *astrological* purposes.

To illustrate: If the common almanac gives the Moon's place in Taurus (constellation), its true place in the zodiac is in Gemini (zodiacal sign). Thus it is readily seen that those who use the common almanac may be planting seeds when they think that the Moon is in a fruitful sign, while in reality it would be in one of the most barren signs of the zodiac. To obtain desired results, planning must be done according to *sign*.

Some common almanacs confuse the issue further by inserting at the head of their columns "Moon's Sign" when they really mean "Moon's Constellation." In the *Moon Sign Book*, however, "Moon's sign" means "Moon's sign!" Use the *Moon Sign Book* to plan all of your important events and to grow a more beautiful, bountiful garden.

Using the Astro Almanac

Llewellyn's unique Astro Almanac (pages 12–21) is provided for quick reference. Use it to find the best dates for anything from asking for a raise to buying a car!

By reader request, we have included several new categories in the Astro Almanac relating to business, from hiring and firing staff to the best time to advertise on the internet. We hope you will find them useful. If you have suggestions for other activities to be added to the Astro Almanac, please write us at the address listed on the title page of this book.

The dates provided are determined from the sign and phase of the Moon and the aspects to the Moon. These are approximate dates only. We have removed dates that have long Moon void-of-course periods from the list (we did not do this before 1998). Although some of these dates may meet the criteria listed for your particular activity, the Moon void would nullify the positive influences of that day. We have not removed dates with short Moon voids, however, and we have not taken planetary retrogrades into account. To learn more about Moon void-of-course and planetary retrogrades, see pages 54–61.

This year we have also removed eclipse dates and days with lots of squares to the Moon. Like Moon voids, squares could nullify the "good" influences of a given day. Eclipses lend an unpredictable energy to a day, so we have removed eclipse dates so that you may begin your activities on the strongest footing possible.

Another thing to bear in mind when using the Astro Almanac is that sometimes the dates given may not be favorable for your Sun sign or for your particular interests. The Astro Almanac does not take personal factors into account, such as your Sun and Moon sign, your schedule, etc. That's why it is important for you to learn how to use the entire process to come up with the most beneficial dates for you. To do this, read the instructions under "Using the Moon Tables" (page 22). That way, you can get the most out of the power of the Moon!

Astro Almanac

Activity	Jan.	Feb.	Mar.	Apr.	May	Jun.	Jul.	Aug.	Sep.	Oct.	Nov.	Dec.
Advertise Sale	4, 31	1	27	22						4		25
Advertise New Venture	4, 18	1, 15	13, 27	22		3		23	19, 20	4		10, 25
Advertise in the News-paper	4, 6, 9, 11, 14, 16, 18, 22, 26, 31	1, 6, 11, 15, 16, 19, 25, 29	5, 9, 13, 18, 22, 27, 28	1, 2, 6, 7, 10, 11, 15, 22, 24, 28, 29	3, 4, 8, 12, 13, 22, 25, 27, 30	1, 3, 5, 8, 9, 13, 19, 23, 24, 28	2, 5, 7, 10, 17, 20, 22, 25, 26, 29, 30	2, 3, 8, 13, 18, 20, 23, 25, 27, 29, 31	3, 9, 10, 15, 19, 20, 23, 24, 28, 29	4, 7, 10, 12, 16, 19, 21, 23, 25, 27, 31	3, 5, 8, 12, 14, 17, 19, 21, 24, 29, 30	5, 10, 14, 15, 18, 19, 25, 27, 31
Advertise on TV, Radio, Internet	9, 18								10			
Apply for Job	6, 16, 25	29	10, 18	6, 15, 24	4, 12, 13, 22, 23	1, 19, 27, 28	5, 7, 26	3, 13	8, 27	16	1, 3, 12, 20, 30	18
Apply for Copy-rights, patents	4, 6, 9, 11, 16, 18, 22	1, 6, 11, 19, 29	5, 9, 13, 18, 27	1, 6, 10, 15, 24, 29	3, 4, 8, 12, 13, 22	1, 5, 19, 28	2, 5, 7, 10, 17, 22, 26, 30	3, 13, 18, 20, 23, 31	10, 15, 19, 23, 28	7, 12, 16, 25, 27	3, 8, 12, 17, 21, 30	5, 10, 14, 18, 25, 31

Astro Almanac

Activity	Jan.	Feb.	Mar.	Apr.	May	Jun.	Jul.	Aug.	Sep.	Oct.	Nov.	Dec.
Ask for Raise	4, 6, 9 13, 14, 16, 27	6, 7, 12, 16, 21, 27	5, 9, 13, 18, 27, 28	1–3, 7, 10–12, 15, 22, 24, 29	3, 7, 8, 12, 13, 22, 23, 28, 30	3, 5, 6, 8, 11, 13, 19, 22, 23, 27	2, 5–7, 10, 17, 20, 22, 25, 27, 29, 31	2, 3, 11, 18, 20, 23, 25, 26, 29–31	9, 10, 15, 19, 20, 23, 24, 28, 29	4, 5, 7, 12, 19, 20, 21, 23, 25, 27, 30	3, 4, 10, 17, 19, 24, 30	5, 10, 14, 18, 19, 24, 25, 31
Bid on Contracts	26		22					31	28, 29	25	21	19
Brewing	1, 2, 29	24–26	5, 6, 23, 24	2, 3, 19, 20, 29, 30	26–28	23, 24	1, 20, 21, 29, 30	16–18, 25, 26	14, 21, 22, 23	19, 20, 27	15, 16, 24, 25	12, 13, 21, 22
Buy Stocks	6	2, 29	28	6, 24, 25	4, 22	1, 19, 28	16, 26	13	9	6, 16	2, 3, 12, 30	27
Buy Animals	7, 9, 13, 14	6, 9, 10, 12	8, 9, 13	7, 10, 11	5–9	3–6, 8	2, 3, 5–7, 31	2–5, 30, 31	1, 2, 5, 28, 29	2–5, 30, 31	1–4, 26–30	1, 26–28, 30
Buy Antiques	1, 2, 6, 7, 19, 20, 28, 29	2–4, 16, 17, 24–26, 29	1, 2, 14, 15, 23, 28	10, 11, 19, 20, 24, 25, 26	8, 9, 16, 17, 18, 21–23	4, 5, 13, 14, 18, 19	1, 2, 10, 11, 15, 16, 29, 30	6–8, 11–13, 25, 26	3, 4, 7–9, 21–23, 30	1, 5, 6, 19, 20, 27–29	1–3, 15, 16, 24, 25, 28–30	12, 13, 21, 22, 26, 27

Astro Almanac

Activity	Jan.	Feb.	Mar.	Apr.	May	Jun.	Jul.	Aug.	Sep.	Oct.	Nov.	Dec.
Buy Appliances	3, 9, 13, 15, 18, 22, 26	5, 10, 14, 22, 27	3, 8, 12, 17, 21, 26, 31	4, 7, 9, 17, 22, 27	2, 4, 6, 15, 17, 20, 25, 29	2, 6, 11, 13, 16, 21, 26, 28, 30	8, 10, 13, 18, 23, 25, 27, 31	4, 9, 14, 19, 23	1, 3, 5, 10, 15, 20, 24, 28, 30	3, 8, 12, 15, 17, 25, 28, 30	4, 9, 13, 17, 22, 24, 26	1, 6, 11, 15, 19, 24, 29
Buy Cameras	2, 12, 17, 25, 30	9, 13, 21, 26	7, 11, 20, 25	3, 8, 16, 21	1, 5, 14, 18, 28	1, 10, 15, 25, 29	7, 12, 22, 26	3, 8, 18, 22, 31	4, 14, 19, 27	2, 11, 16, 24, 29	8, 12, 21, 25	5, 10, 18, 23
Buy a Car	3, 18	14, 27	12, 13, 26	9, 22	6, 20	2, 3, 16, 30	13, 27	9, 23	5, 20	3, 4, 17, 30, 31	13, 14, 26	11, 24, 25
Buy Electronics	3, 9, 13, 15, 18, 22, 26	5, 10, 14, 22, 25, 27	3, 8, 12, 17, 21, 26, 31	4, 7, 9, 17, 22, 27	2, 6, 15, 20, 25, 29	2, 6, 11, 13, 16, 21, 26, 28, 30	8, 10, 13, 18, 23, 25, 27, 31	4, 9, 14, 19, 23	1, 3, 5, 10, 15, 20, 24, 28, 30	3, 8, 12, 15, 17, 25, 28, 30	4, 9, 13, 17, 22, 24, 26	1, 6, 11, 15, 19, 24, 29
Buy a House		12	9, 10	6, 10	4, 8	4–7	2, 3, 31	13				
Buy Real Estate	13, 14	11	9	10	8	5, 11	2	13, 31	28	12	10	

Astro Almanac

Activity	Jan.	Feb.	Mar.	Apr.	May	Jun.	Jul.	Aug.	Sep.	Oct.	Nov.	Dec.
Canning	1, 2, 28, 29	24–26	5, 6, 23, 24	2, 3, 19, 20, 29, 30	26–28	23, 24	1, 20, 21, 29, 30	16–18, 25, 26	14, 21, 22, 23	19, 20, 27	15, 16, 24, 25	12, 13, 21, 22
Collect Money	13, 14, 27	2, 11, 29	8, 9, 28	10, 24	8, 22, 23	5, 11, 19	1, 2, 22, 30	5, 11, 13, 26, 31	15, 23, 28, 29	5, 12, 20, 25	3, 10, 21, 29, 30	19, 27
Cut Hair to Decrease Growth	24, 25	20, 21		4	1, 2, 29, 30	2, 25, 26, 29, 30	22–24, 27, 28	19, 20, 23, 24, 29	15, 16, 19, 20, 26, 27	17, 18, 23, 24	13, 14, 19, 24	16–18
Cut Hair to Increase Growth	11, 12	7, 8	7	10, 11	7, 8	4, 5	2					3, 31
Cut Hair for Thickness	18–20	16–19								13	10, 11	8, 9, 11
Cut Timber	3–6, 21–27, 30, 31	1–5, 20–23, 27–29	1–4, 20–22, 25–31	1, 4, 21–28	1–3, 19–25, 29–31	1, 2, 17–22, 25–30	17–19, 22–28	19–24, 27–29	15–20, 24–27	14–18, 21–26	12–14, 17–23	14–20, 23–25

Astro Almanac

Activity	Jan.	Feb.	Mar.	Apr.	May	Jun.	Jul.	Aug.	Sep.	Oct.	Nov.	Dec.
Dock or Dehorn Animals	4, 8, 9, 13, 30, 31	1–5, 28, 29	3, 4, 11–13, 30, 31	1, 8–12, 26, 27	6, 7, 9, 10	2–8, 29, 30	1–7, 27–31	1–7, 23–31	1–5, 20–30	2–4, 20–26, 29–31	3, 18–27	1, 18–24, 28, 29
End a Relation-ship	1–6, 21–31	1–5, 20–29	1–6, 20–31	1–4, 19–30	1–3, 19–31	1, 2, 17–30	1, 17–30	16–29	14–27	14–23, 27	12–20, 23–25	12–17, 21–25
Entertain	8–10, 17, 18, 21–23, 26, 27	5, 6, 14, 15, 18, 19, 22, 23	3, 4, 12, 13, 16, 17, 20–22, 30, 31	1, 8, 9, 12–14, 17, 18, 27, 28	6, 7, 10, 11, 14, 15, 24, 25	2, 3, 6, 7, 10, 11, 12, 20–22, 29, 30	3, 4, 8, 9, 17–19, 27, 28, 31	1, 4, 5, 14, 15, 23, 24, 27, 28, 31	1, 2, 10, 11, 19, 20, 24, 25, 28, 29	7–9, 17, 18, 21, 22, 25, 26	4, 5, 13, 14, 17, 18, 21–23	1, 2, 10, 11, 14, 15, 19, 20, 28–30
Extract Teeth	3, 4, 30, 31	27, 28	12, 13	8, 9	6, 7	2, 3, 8, 29, 30	5, 6, 7, 27, 28	2, 3, 23, 24, 29, 30	5, 26, 27	2, 3, 23, 24, 30, 31	19, 20, 26, 27	18, 23, 24, 25
Get a Perm	22	12		7, 12	3	6, 27	31	21	25			9
Fire Staff	1–6, 21–31	1–4, 20–23, 27–29	1–5, 20–31	1–4, 19–30	1–3, 19–31	1, 2, 17–30	1, 17–30	16–29	14–27	14–23, 27	12–20, 23–25	12–17, 21–25

Astro Almanac

Activity	Jan.	Feb.	Mar.	Apr.	May	Jun.	Jul.	Aug.	Sep.	Oct.	Nov.	Dec.
Hire Staff	9, 11, 14, 16, 18	6, 11, 16, 19	9, 13, 18	6, 7, 10, 11, 15	4, 8, 12, 13	3, 5, 8, 9, 13	2, 5, 7, 10	2, 3, 8, 13, 31	3, 9, 10, 28, 29	4, 7, 10, 12, 31	3, 5, 8, 29, 30	5, 10, 27, 31
Legal Matters	4	1, 6	27									
Marriage Ceremony	4, 6, 8, 9, 13, 16, 18, 30	1, 2, 5, 6, 10, 11, 27	3, 5, 6, 8–10, 13, 15, 18, 28, 30	1, 3, 4, 6, 7, 9, 10, 12, 15, 28, 29	3, 4, 7, 8, 12, 13, 27, 28	1, 2, 5, 6, 11, 27, 28	1, 2, 5–7, 10, 11, 26, 27, 30, 31	3–5, 9, 11, 13, 23, 24, 26, 27, 29–31	2, 4, 8, 10, 23, 25, 27–29	2, 5, 7, 8, 10, 12, 21, 22, 25, 27, 30	1, 3, 4, 6, 8, 10, 19–21, 24, 25, 29, 30	1, 5, 6, 9, 10, 18–20, 24, 25, 29, 31
Marry for Happiness	4, 13, 16, 22	12	13	7, 10, 12	3, 4, 8	1, 5, 6, 11, 27, 28	2, 26, 31	4, 26, 31	2, 23, 28, 29	16, 20–22, 25	12, 16, 17	9, 14, 15, 19
Marry for Longevity		12	10	7	4							
Mow Lawn for Less Growth	3–6, 21–31	1–5, 20–23, 27–29	1–4, 20–31	1, 4, 21–28	1–3, 19–25, 29–31	1, 2, 17–22, 25–30	17–19, 22–28	19–24, 27–29	15–20, 24–27	14–18, 21–27	12–14, 17–22	14–20, 23–25

Astro Almanac

Activity	Jan.	Feb.	Mar.	Apr.	May	Jun.	Jul.	Aug.	Sep.	Oct.	Nov.	Dec.
Neutering or Spaying Animals	6–10, 17	2–6, 14	1–4, 12, 28–31	1, 8, 9,12–14, 27, 28	6, 10, 11	2, 6, 7, 29, 30	3, 4, 27, 31	1, 11–13, 23, 24, 27, 28	7–11, 23–25	5–9, 21, 22	1–5, 28–30	1, 2, 26–30
Open a Business			18	6, 15	4, 12	9	6, 7, 16	3, 13, 30	9	6	2, 3, 30	27
Paint House	8, 22	12	3	7, 12, 28	3	6, 22, 27	31	21	10, 25		4	9, 29
Pour Concrete	8, 9, 21, 22	18, 19	18, 28	10, 15, 25, 30	8, 12, 22, 27	5, 9, 19, 24	2, 6, 16, 21, 30	3, 13, 18, 26, 30	9, 14, 23, 27	6, 11, 20, 24	2, 7, 16, 20, 30	5, 13, 17, 27
Remodel a Business			5, 18	15	12			13, 18	23			
Remodel a House	3, 21, 23, 31	18, 27	16, 25	12, 22	5, 10, 19	6, 15	3, 12, 31	8, 27	5, 25	2, 16, 21, 30	17, 26	14, 23

Astro Almanac

Activity	Jan.	Feb.	Mar.	Apr.	May	Jun.	Jul.	Aug.	Sep.	Oct.	Nov.	Dec.
Repair a Car	9, 22	5	17, 31	7	29	6	18, 25	14	10, 24	8, 15, 21	4, 17	1
Repair Electronics	9, 15, 22, 29	5, 12, 13, 25	3, 10, 17, 31	1, 7, 27	15, 24, 25, 29	6, 7, 13, 21, 22, 27, 28	18, 26	8, 14, 25	3, 10, 24, 30	8, 15, 21, 28	4, 12, 17	1, 15, 29, 30
Roofing	21, 22, 23	5	3, 4, 30, 31	1, 27, 28	3, 24, 25, 31	1, 20–22, 27, 28	17–19, 25, 26	21, 22, 27, 28	17, 18, 24, 25	14–16, 21, 22	12, 17, 18	14, 15
Seek Favors or Credit	4, 8, 9, 11, 22, 30	1, 5, 6, 7, 19, 27	3, 5, 6, 27	1, 3, 12, 22, 23, 28, 29	27, 28	6, 22, 24	17, 21, 31	9, 18, 27	10, 25	2, 7, 8, 10, 21, 22, 30	4, 6, 8, 17	1, 5, 14, 15, 24, 25, 29, 31
Set Fence Posts	6, 19, 22	4, 5, 20, 29	3, 14, 18, 28	10, 15, 25, 30	8, 12, 22, 27	5, 9, 19, 24	2, 6, 16, 21, 30	3, 13, 18, 26, 30	9, 14, 23, 27	6, 11, 20, 24	2, 7, 16, 20, 30	5, 13, 17, 27
Sign Contracts	9, 16, 22	6, 19		1, 6, 7, 28	3, 4, 25	1, 28	17, 25, 26	27	10, 24	7, 16, 21	5, 12, 17	14, 15

Astro Almanac

Activity	Jan.	Feb.	Mar.	Apr.	May	Jun.	Jul.	Aug.	Sep.	Oct.	Nov.	Dec.
Sell Items		12		7	3	27		21				9
Sell Real Estate	14, 19	11, 16	9, 14	10	8	5	2, 22, 30	26	15, 23	12, 20	16	13
Sports	4, 5, 14, 22	1, 11, 19, 27	9, 17, 27		29	7	22	27	6, 15, 16	12, 21	17, 26	14, 15, 25
Start Diet	3–5, 21–25, 30, 31	1, 5, 20, 21, 27, 28	3, 4, 25–27, 30, 31	1, 4, 21–23, 27, 28	1, 2, 19, 20, 24, 25, 29, 30	17, 20–22, 25, 26	17–19, 22–24	19, 20, 27–29	15, 16, 24–27	21–24	17–20	14–17, 23–25
Start House Building	6, 11, 19, 23	2, 7, 16, 20, 29	5, 14, 18, 28	2, 10, 15, 25, 30	8, 12, 22, 27	5, 9, 19, 24	2, 6, 16, 21, 30	3, 13, 18, 26, 30	9, 14, 23, 27	6, 11, 20, 24	2, 7, 16, 20, 30	5, 13, 17, 27
Start Savings Account		12		6, 7	4							9

Astro Almanac

Activity	Jan.	Feb.	Mar.	Apr.	May	Jun.	Jul.	Aug.	Sep.	Oct.	Nov.	Dec.
Stop a Bad Habit	23, 25	20, 21	18	15, 16	12	9	6, 7	3, 30	27	24	20	17, 18
Travel	3, 18, 22	14, 27	12, 13, 17, 26	9, 22	6, 20	3, 6, 16, 30	13, 27, 31	9, 23	5, 20, 24	3, 4, 21, 30	13, 17, 26	11, 24, 25
Visit Dentist	11, 19, 23	7, 16, 20	5, 14, 18	2, 10, 15, 30	8, 12, 27	5, 9, 24	2, 6, 21, 30	3, 18, 26, 30	14, 23, 27	11, 20, 24	7, 16, 20	5, 13, 17
Visit Physician			18, 19	15, 16	12, 13	8, 9	5, 6, 7	2, 3, 30				
Work with Consultants	4, 6, 9, 11, 14, 16, 18, 22, 26, 31	1, 6, 11, 15, 16, 25, 29	5, 9, 13, 18, 22, 27, 28	1, 2, 6, 7, 10, 11, 15, 22, 24, 28, 29	3, 4, 8, 12, 13, 22, 25, 27, 30	1, 3, 5, 8, 9, 13, 19, 23, 24, 28	2, 5, 7, 10, 17, 20, 22, 25, 26, 29, 30	2, 3, 8, 13, 18, 20, 23, 25, 27, 29, 31	3, 9, 10, 15, 19, 20, 23, 24, 28, 29	4, 7, 10, 12, 16, 19, 21, 23, 25, 27, 31	3, 5, 8, 12, 14, 17, 19, 21, 24, 29, 30	5, 10, 14, 15, 18, 19, 25, 27, 31
Write Letters	11, 31		5, 13	2, 22	13	3, 8, 23	5, 20	2, 29	20	4, 10, 23, 31	14, 19	5, 25, 31

Using the Moon Tables

Timing by the Moon

Timing your activities is one of the most important things you can do to ensure success. In many Eastern countries, timing by the planets is so important that practically no event takes place without first setting up a chart for it. Weddings have occurred in the middle of the night because that was when the influences were best. You may not want to take it that far, but you can still make use of the influences of the Moon whenever possible. It's easy and it works!

In the *Moon Sign Book* you will find the information you need to plan just about any activity: weddings, fishing, buying a car or house, cutting your hair, traveling, and more. Not all of the things you do will fall on favorable days, but we provide the guidelines you need to pick the best day out of the several from which you have to choose. The primary method in the *Moon Sign Book* for choosing your own dates is to use the Moon Tables that begin on page 28. Following are instructions for choosing the best dates for your activities using the *Moon Sign Book*, several examples, directions on how to read the Moon Tables themselves, and more advanced information on using the Favorable and Unfavorable Days Tables, Void-of-Course, and Retrograde information to choose the dates that are best for you personally. To enhance your understanding of the directions given below, we highly recommend that you read the sections of this book called "A Note about Almanacs" (page 10), "The Moon's Quarters & Signs" (page 52), "Retrogrades" (page 54), and "Moon Void-of-Course" (page 55). It is not essential that you read these before you try the examples below, but reading them will deepen your understanding of the date-choosing process.

The Four Basic Steps

Step One: Use the Directions for Choosing Dates

Look up the directions for choosing dates for the activity that you wish to begin. The directions are listed at the beginning of the following sections of this book: Home, Health, & Beauty; Leisure & Recreation; Business & Legal; and Farm, Garden, & Weather. Check the Table of Contents to see in what section the directions for your specific activity are listed. The activities contained in each section are listed in italics after the name of the section in the Table of Contents. For example, directions for choosing a good day for canning are listed in the Home, Health, & Beauty Section, and directions for choosing a good day to throw a party are in the Leisure Section. Read the directions for your activity, then go to step two.

Step Two: Check the Moon Tables

Next, turn to the Moon Tables that begin on page 28. In the Moon Tables section, there are two tables for each month of the year. Use the Moon Tables to determine what dates the Moon is in the phase and sign listed in the directions for your particular activity. The Moon Tables are the tables on the left-hand pages, and include the day, date, the sign the Moon is in, the element of that sign, the nature of the sign, the Moon's phase, and the times that it changes sign or phase.

If there is a time listed after a date, such as Sun. 4:32 pm on January 2, that time is the time when the Moon moves into the zodiac sign listed for that day (which in this case would be Sagittarius). Until then, the Moon is considered to be in the sign for the previous day (Scorpio in this example).

The abbreviation Full signifies Full Moon and New signifies New Moon. The times listed directly after the abbreviation are the times when the Moon changes sign. The times listed after the phase indicate when the Moon changes phase.

If you know the specific month you would like to begin your activity, turn directly to that month. When you begin choosing your own dates, you will be using the Moon's sign and phase information most often. All times are listed in Eastern Standard Time (EST). You need to adjust them according to your own time zone. (There is a time zone conversion map on page 8.)

When you have found some dates that meet the criteria for the correct Moon phase and sign for your activity, you may have completed the process. For certain simple activities, such as getting a haircut, the phase and sign information is all that is needed. For other activities, however, we need to meet further criteria in order to choose the best date. If the directions for your activity include information on certain lunar aspects, you should consult the Lunar Aspectarian. An example of this would be if the directions told you that you should not perform a certain activity when the Moon is square (Q) Jupiter.

Step Three: Turn to the Lunar Aspectarian

On the pages opposite the Moon Tables you will find the Lunar Aspectarian and the Favorable and Unfavorable Days Tables. The Lunar Aspectarian gives the aspects (or angles) of the Moon to the other planets. In a nutshell, it tells where the Moon is in relation to the other planets in the sky. Some placements of the Moon in relation to other planets are favorable, while others are not. To use the Lunar Aspectarian, which is the left half of this table, find the planet that the directions list as favorable for your activity, and run down the column to the date desired. For example, if you are planning surgery and the Health & Beauty section says that you should avoid aspects to Mars, you would look for Mars across the top and then run down that column looking for days where there are no aspects to Mars (as signified by empty boxes). If you want to find a favorable aspect (sextile [X] or trine [T]) to Mercury, run your finger down the column under Mercury until you find an X or T. Negative or adverse aspects (square or opposition) are signified by a Q or O. A conjunction, C, is sometimes good, sometimes bad, depending on the activity or planets.

Step Four: Use the Favorable and Unfavorable Days Tables

The Favorable and Unfavorable Days Tables are helpful in choosing your personal best dates, because they consider your Sun sign. All Sun signs are listed on the right-hand side of the Lunar Aspectarian table. Once you have determined which days meet the criteria for phase, sign, and aspects for your activity, you can find out if those days are positive for you. To find out if a day is positive for you, find your Sun sign and then look down the column. If it is marked F, it is very favorable. If it is marked f, it is slightly favorable. U means very unfavorable and u means slightly unfavorable. Once you have selected good dates for the activity you are about to begin, you can go straight to the examples section beginning on the next page. However, if you are up to the challenge and would like to learn how to fine-tune your selections even further, read on!

Step Five: Check for Moon Void-of-Course and Retrogrades

This last step is perhaps the most advanced portion of the procedure. It is generally considered a bad idea to make decisions, sign important papers, or start special activities during a Moon void-of-course period or during a planetary retrograde. Once you have chosen the best date for your activity based on steps one through four, you can check the Void-of-Course Table on page 56 to find out if any of the dates you have chosen have void periods. The Moon is said to be void-of-course after it has made its last aspect to a planet within a particular sign, but before it has moved into the next sign. Put simply, during the void-of-course period the Moon is "at rest," so activities initiated at this time generally don't come to fruition. You will notice that there are many void periods during the year, and it is nearly impossible to avoid all of them. Some people choose to ignore these altogether and do not take them into consideration when planning activities.

Next, you can check the Planetary Retrograde Table on page 54 to see what planets are retrograde during your chosen date(s). A planet is said to be retrograde when it appears to move backward in the sky as viewed from the Earth. Generally, the farther a planet is away from the Sun, the longer it can stay retrograde. Some planets will retrograde for several months at a time. Avoiding retrogrades is not as important in lunar planning as avoiding the Moon void-of-course, with the exception of the planet Mercury. Mercury rules thought and communication, so it is important not to sign papers, initiate important business or legal work, or make crucial decisions during these times. As with the Moon void-of-course, it is difficult to avoid all planetary retrogrades when beginning events, and you may choose to ignore this step of the process. Following are some examples using some or all of the steps outlined above.

Using What You've Learned
Example Number One

Let's say you need to make an appointment to have your hair cut. Your hair is thin and you would like it to look thicker. You look in the Table of Contents to find the section of the book with directions for hair care. You find that it is in the Home, Health, & Beauty section. Turning to that section you see that for thicker hair you should cut hair while the Moon is Full and in the sign of Taurus, Cancer, or Leo. You should avoid the Moon in Aries, Gemini, or Virgo. We'll say that it is the month of January. Look up January in the Moon Tables (page 28). The Full Moon falls on January 20 at 11:40 pm. The Moon moves into the sign of Leo that day at 10:58 pm, so this day meets both the phase and sign criteria.

Example Number Two

That was easy. Let's move on to a more difficult example using the sign and phase of the Moon. You want to buy a house for a permanent home. After checking the Table of Contents to see where the house purchasing instructions are, look in the Home, Health, & Beauty section under House. It says that you should buy a home when the Moon is in Taurus, Leo, Scorpio, or Aquarius (fixed signs). You need to get a loan, so you should also look in the Business & Legal section under Loans. Here it says that the third and fourth Moon quarters favor the borrower (you). You are going to buy the house in January. Look up January in the Moon Tables. The Moon is in the third quarter from January 20–28, and in the fourth quarter from January 28–February 5. The best days for obtaining a loan would be January 21–23 while the Moon is in Leo or January 27–29 while it is in Scorpio. Just match up the best signs and phases (quarters) to come up with the best dates. With all activities, be sure to check the Favorable and Unfavorable Days for your Sun sign in the table adjoining the Lunar Aspectarian. If there is a choice between several dates, pick the one most favorable for you (marked F under your Sun sign). Because buying a home is an important business decision, you may also wish to see if there are Moon voids or a Mercury retrograde during these dates.

Example Number Three

Now let's look at an example that uses signs, phases, and aspects. Our example this time is fixing your car. We will use January as the example month. Look in the Home, Health, & Beauty section under automobile repair. It says that the Moon should be in a fixed sign (Taurus, Leo, Scorpio, or Aquarius) in the first or second quarter and well aspected to Uranus. (Good aspects are sextiles and trines, marked X and T. Conjunctions are also usually considered good if they are not conjunctions to Mars, Saturn, or Neptune.) It also tells you to avoid negative aspects to Mars, Saturn, Uranus, Neptune, and Pluto. (Negative aspects are squares and oppositions, marked Q and O.) Look in the Moon Tables under January. You will see that the Moon is in the first and second quarters from January 6 to 20. The Moon is in Aquarius on January 7 at 5:53 pm through January 10 at 4:59 am. The Moon is in Taurus from January 14 at 7:38 pm through January 16 at 10:25 pm, and it moves into Leo from on January 20 at 10:58 pm. Although the Moon remains in Leo till January 23, the Full Moon occurs just a few minutes after it passes into Leo, beginning the third quarter and creating unfavorable conditions. Now, looking to the Lunar Aspectarian, we see that January 9 has a positive aspect to Uranus (conjunction), but that none of the other chosen dates do. In addition, there are no negative aspects to Mars, Saturn, Uranus, Neptune, and Pluto on January 9. There is a Moon void-of-course to avoid after 8:41 pm. Otherwise, if you wanted to fix your car in January, this would be the best date.

Use Common Sense!

Some activities depend on outside factors. Obviously, you can't go out and plant when there is a foot of snow on the ground. You should adjust to the conditions at hand. If the weather was bad during the first quarter when it was best to plant crops, do it during the second quarter while the Moon is in a fruitful sign. If the Moon is not in a fruitful sign during the first or second quarter, choose a day when it is in a semi-fruitful sign. The best advice is to choose either the sign or phase that is most favorable when the two don't coincide.

To summarize, in order to make the most of your activities, check with the *Moon Sign Book*. First, look up the activity under the proper heading, then look for the information given in the tables (the Moon Tables, Lunar Aspectarian, or Favorable and Unfavorable Days). Choose the best date considering the number of positive factors in effect. If most of the dates are favorable, there is no problem choosing the one that will fit your schedule. However, if there aren't any really good dates, pick the ones with the least number of negative influences.

Key of Abbreviations for the Moon Tables

X: sextile/positive

T: trine/positive

Q: square/negative

O: opposition/negative

C: conjunction/positive, negative, or neutral depending on planets involved; conjunctions to Mars, Saturn, or Neptune are sometimes negative.

F: very favorable

f: slightly favorable

U: very unfavorable

u: slightly unfavorable

Full: Full Moon

New: New Moon

January Moon Table

Date	Sign	Element	Nature	Phase
1 Sat.	Scorpio	Water	Fruitful	4th
2 Sun. 4:32 pm	Sagittarius	Fire	Barren	4th
3 Mon.	Sagittarius	Fire	Barren	4th
4 Tue.	Sagittarius	Fire	Barren	4th
5 Wed. 5:24 am	Capricorn	Earth	Semi-fruit	4th
6 Thu.	Capricorn	Earth	Semi-fruit	New 1:14 pm
7 Fri. 5:53 pm	Aquarius	Air	Barren	1st
8 Sat.	Aquarius	Air	Barren	1st
9 Sun.	Aquarius	Air	Barren	1st
10 Mon. 4:59 am	Pisces	Water	Fruitful	1st
11 Tue.	Pisces	Water	Fruitful	1st
12 Wed. 1:48 pm	Aries	Fire	Barren	1st
13 Thu.	Aries	Fire	Barren	1st
14 Fri. 7:38 pm	Taurus	Earth	Semi-fruit	2nd 8:34 am
15 Sat.	Taurus	Earth	Semi-fruit	2nd
16 Sun. 10:25 pm	Gemini	Air	Barren	2nd
17 Mon.	Gemini	Air	Barren	2nd
18 Tue. 11:01 pm	Cancer	Water	Fruitful	2nd
19 Wed.	Cancer	Water	Fruitful	2nd
20 Thu. 10:58 pm	Leo	Fire	Barren	Full 11:40 pm
21 Fri.	Leo	Fire	Barren	3rd
22 Sat.	Leo	Fire	Barren	3rd
23 Sun. 12:07 am	Virgo	Earth	Barren	3rd
24 Mon.	Virgo	Earth	Barren	3rd
25 Tue. 4:09 am	Libra	Air	Semi-fruit	3rd
26 Wed.	Libra	Air	Semi-fruit	3rd
27 Thu. 12:01 pm	Scorpio	Water	Fruitful	3rd
28 Fri.	Scorpio	Water	Fruitful	4th 2:57 am
29 Sat. 11:17 pm	Sagittarius	Fire	Barren	4th
30 Sun.	Sagittarius	Fire	Barren	4th
31 Mon.	Sagittarius	Fire	Barren	4th

January

Lunar Aspectarian   Favorable and Unfavorable Days

Date	Sun	Mercury	Venus	Mars	Jupiter	Saturn	Uranus	Neptune	Pluto	Aries	Taurus	Gemini	Cancer	Leo	Virgo	Libra	Scorpio	Sagittarius	Capricorn	Aquarius	Pisces
1	X					0	Q				U		f	u	f		F		f	u	f
2			C	Q				X			U		f	u	f		F		f	u	f
3							X		C	f		U		f	u	f		F		f	u
4					T					f		U		f	u	f		F		f	u
5				X						f		U		f	u	f		F		f	u
6	C	C				T				u	f		U		f	u	f		F		f
7					Q					u	f		U		f	u	f		F		f
8			X		Q			C	X	f	u	f		U		f	u	f		F	
9				X			C			f	u	f		U		f	u	f		F	
10			C							f	u	f		U		f	u	f		F	
11	X	X	Q			X			Q		f	u	f		U		f	u	f		F
12								X			f	u	f		U		f	u	f		F
13			T				X		T	F		f	u	f		U		f	u	f	
14	Q	Q			C					F		f	u	f		U		f	u	f	
15				X		C	Q	Q			F		f	u	f		U		f	u	f
16	T	T									F		f	u	f		U		f	u	f
17			Q					T	0	f		F		f	u	f		U		f	u
18		0		X		T				f		F		f	u	f		U		f	u
19			T			X				u	f		F		f	u	f		U		f
20	0			Q						u	f		F		f	u	f		U		f
21		0				Q		0	T	f	u	f		F		f	u	f		U	
22			T	T		0				f	u	f		F		f	u	f		U	
23						T			Q	f	u	f		F		f	u	f		U	
24				0							f	u	f		F		f	u	f		U
25	T		Q				T				f	u	f		F		f	u	f		U
26		T				T		X		U		f	u	f		F		f	u	f	
27			X		0		Q			U		f	u	f		F		f	u	f	
28	Q	Q				0	Q				U		f	u	f		F		f	u	f
29				T							U		f	u	f		F		f	u	f
30	X							X		f		U		f	u	f		F		f	u
31		X		Q			X		C	f		U		f	u	f		F		f	u

February Moon Table

Date	Sign	Element	Nature	Phase
1 Tue. 12:10 pm	Capricorn	Earth	Semi-fruit	4th
2 Wed.	Capricorn	Earth	Semi-fruit	4th
3 Thu.	Capricorn	Earth	Semi-fruit	4th
4 Fri. 12:31 am	Aquarius	Air	Barren	4th
5 Sat.	Aquarius	Air	Barren	New 8:03 am
6 Sun. 11:02 am	Pisces	Water	Fruitful	1st
7 Mon.	Pisces	Water	Fruitful	1st
8 Tue. 7:17 pm	Aries	Fire	Barren	1st
9 Wed.	Aries	Fire	Barren	1st
10 Thu.	Aries	Fire	Barren	1st
11 Fri. 1:21 am	Taurus	Earth	Semi-fruit	1st
12 Sat.	Taurus	Earth	Semi-fruit	2nd 6:21 pm
13 Sun. 5:23 am	Gemini	Air	Barren	2nd
14 Mon.	Gemini	Air	Barren	2nd
15 Tue. 7:45 am	Cancer	Water	Fruitful	2nd
16 Wed.	Cancer	Water	Fruitful	2nd
17 Thu. 9:11 am	Leo	Fire	Barren	2nd
18 Fri.	Leo	Fire	Barren	2nd
19 Sat. 10:53 am	Virgo	Earth	Barren	Full 11:27 am
20 Sun.	Virgo	Earth	Barren	3rd
21 Mon. 2:21 pm	Libra	Air	Semi-fruit	3rd
22 Tue.	Libra	Air	Semi-fruit	3rd
23 Wed. 8:58 pm	Scorpio	Water	Fruitful	3rd
24 Thu.	Scorpio	Water	Fruitful	3rd
25 Fri.	Scorpio	Water	Fruitful	3rd
26 Sat. 7:10 am	Sagittarius	Fire	Barren	4th 10:53 pm
27 Sun.	Sagittarius	Fire	Barren	4th
28 Mon. 7:45 pm	Capricorn	Earth	Semi-fruit	4th
29 Tue.	Capricorn	Earth	Semi-fruit	4th

February

Lunar Aspectarian   Favorable and Unfavorable Days

Date	Sun	Mercury	Venus	Mars	Jupiter	Saturn	Uranus	Neptune	Pluto	Aries	Taurus	Gemini	Cancer	Leo	Virgo	Libra	Scorpio	Sagittarius	Capricorn	Aquarius	Pisces
1						T				f		U		f	u	f		F		f	u
2			C			T				u	f		U		f	u	f		F		f
3				X	Q					u	f		U		f	u	f		F		f
4						Q		C		u	f		U		f	u	f		F		f
5	C						C		X	f	u	f		U		f	u	f		F	
6		C		X						f	u	f		U		f	u	f		F	
7			X			X			Q		f	u	f		U		f	u	f		F
8			C								f	u	f		U		f	u	f		F
9								X	T	F		f	u	f		U		f	u	f	
10	X		Q				X			F		f	u	f		U		f	u	f	
11		X			C	C		Q			F		f	u	f		U		f	u	f
12	Q		T				Q				F		f	u	f		U		f	u	f
13				X				T			F		f	u	f		U		f	u	f
14		Q				T			0	f		F		f	u	f		U		f	u
15	T			Q	X					f		F		f	u	f		U		f	u
16		T				X				u	f		F		f	u	f		U		f
17			0	T	Q			0		u	f		F		f	u	f		U		f
18						Q	0		T	f	u	f		F		f	u	f		U	
19	0				T					f	u	f		F		f	u	f		U	
20		0				T			Q		f	u	f		F		f	u	f		U
21			T				T				f	u	f		F		f	u	f		U
22			0			T			X	U		f	u	f		F		f	u	f	
23						0				U		f	u	f		F		f	u	f	
24	T		Q			0		Q			U		f	u	f		F		f	u	f
25		T				Q					U		f	u	f		F		f	u	f
26	Q						X				U		f	u	f		F		f	u	f
27		Q	X	T			X		C	f		U		f	u	f		F		f	u
28										f		U		f	u	f		F		f	u
29	X	X		Q	T	T				u	f		U		f	u	f		F		f

March Moon Table

Date	Sign	Element	Nature	Phase
1 Wed.	Capricorn	Earth	Semi-fruit	4th
2 Thu. 8:14 am	Aquarius	Air	Barren	4th
3 Fri.	Aquarius	Air	Barren	4th
4 Sat. 6:30 pm	Pisces	Water	Fruitful	4th
5 Sun.	Pisces	Water	Fruitful	4th
6 Mon.	Pisces	Water	Fruitful	New 12:17 am
7 Tue. 1:54 am	Aries	Fire	Barren	1st
8 Wed.	Aries	Fire	Barren	1st
9 Thu. 7:01 am	Taurus	Earth	Semi-fruit	1st
10 Fri.	Taurus	Earth	Semi-fruit	1st
11 Sat. 10:46 am	Gemini	Air	Barren	1st
12 Sun.	Gemini	Air	Barren	1st
13 Mon. 1:51 pm	Cancer	Water	Fruitful	2nd 1:59 am
14 Tue.	Cancer	Water	Fruitful	2nd
15 Wed. 4:43 pm	Leo	Fire	Barren	2nd
16 Thu.	Leo	Fire	Barren	2nd
17 Fri. 7:48 pm	Virgo	Earth	Barren	2nd
18 Sat.	Virgo	Earth	Barren	2nd
19 Sun. 11:57 pm	Libra	Air	Semi-fruit	Full 11:44 pm
20 Mon.	Libra	Air	Semi-fruit	3rd
21 Tue.	Libra	Air	Semi-fruit	3rd
22 Wed. 6:17 am	Scorpio	Water	Fruitful	3rd
23 Thu.	Scorpio	Water	Fruitful	3rd
24 Fri. 3:43 pm	Sagittarius	Fire	Barren	3rd
25 Sat.	Sagittarius	Fire	Barren	3rd
26 Sun.	Sagittarius	Fire	Barren	3rd
27 Mon. 3:51 am	Capricorn	Earth	Semi-fruit	4th 7:21 pm
28 Tue.	Capricorn	Earth	Semi-fruit	4th
29 Wed. 4:34 pm	Aquarius	Air	Barren	4th
30 Thu.	Aquarius	Air	Barren	4th
31 Fri.	Aquarius	Air	Barren	4th

March

Lunar Aspectarian   Favorable and Unfavorable Days

Date	Sun	Mercury	Venus	Mars	Jupiter	Saturn	Uranus	Neptune	Pluto	Aries	Taurus	Gemini	Cancer	Leo	Virgo	Libra	Scorpio	Sagittarius	Capricorn	Aquarius	Pisces
1										u	f		U		f	u	f		F		f
2						Q		C		u	f		U		f	u	f		F		f
3			C	X		Q	C		X	f	u	f		U		f	u	f		F	
4										f	u	f		U		f	u	f		F	
5		C			X	X			Q		f	u	f		U		f	u	f		F
6	C										f	u	f		U		f	u	f		F
7							X			F		f	u	f		U		f	u	f	
8			X	C			X		T	F		f	u	f		U		f	u	f	
9		X			C			Q		F		f	u	f		U		f	u	f	
10	X				C	Q					F		f	u	f		U		f	u	f
11		Q	Q					T			F		f	u	f		U		f	u	f
12						T			0	f		F		f	u	f		U		f	u
13	Q	T	T	X	X					f		F		f	u	f		U		f	u
14						X				u	f		F		f	u	f		U		f
15	T			Q						u	f		F		f	u	f		U		f
16					Q	Q		0	T	f	u	f		F		f	u	f		U	
17			T					0		f	u	f		F		f	u	f		U	
18		0	0		T	T			Q		f	u	f		F		f	u	f		U
19	0										f	u	f		F		f	u	f		U
20							T	X	U		f	u	f		F		f	u	f	
21						T			U		f	u	f		F		f	u	f	
22		T		0	0			Q	U		f	u	f		F		f	u	f	
23			T		0	Q				U		f	u	f		F		f	u	f
24										U		f	u	f		F		f	u	f
25	T	Q	Q				X	C	f		U		f	u	f		F		f	u
26						X			f		U		f	u	f		F		f	u
27	Q			T	T				f		U		f	u	f		F		f	u
28		X	X		T				u	f		U		f	u	f		F		f
29									u	f		U		f	u	f		F		f
30	X			Q	Q	Q		C	X	f	u	f		U		f	u	f		F	
31						C				f	u	f		U		f	u	f		F	

April Moon Table

Date	Sign	Element	Nature	Phase
1 Sat. 3:12 am	Pisces	Water	Fruitful	4th
2 Sun.	Pisces	Water	Fruitful	4th
3 Mon. 10:22 am	Aries	Fire	Barren	4th
4 Tue.	Aries	Fire	Barren	New 1:12 pm
5 Wed. 2:29 pm	Taurus	Earth	Semi-fruit	1st
6 Thu.	Taurus	Earth	Semi-fruit	1st
7 Fri. 4:58 pm	Gemini	Air	Barren	1st
8 Sat.	Gemini	Air	Barren	1st
9 Sun. 7:16 pm	Cancer	Water	Fruitful	1st
10 Mon.	Cancer	Water	Fruitful	1st
11 Tue. 10:16 pm	Leo	Fire	Barren	2nd 8:30 am
12 Wed.	Leo	Fire	Barren	2nd
13 Thu.	Leo	Fire	Barren	2nd
14 Fri. 2:19 am	Virgo	Earth	Barren	2nd
15 Sat.	Virgo	Earth	Barren	2nd
16 Sun. 7:36 am	Libra	Air	Semi-fruit	2nd
17 Mon.	Libra	Air	Semi-fruit	2nd
18 Tue. 2:35 pm	Scorpio	Water	Fruitful	Full 12:41 pm
19 Wed.	Scorpio	Water	Fruitful	3rd
20 Thu. 11:58 pm	Sagittarius	Fire	Barren	3rd
21 Fri.	Sagittarius	Fire	Barren	3rd
22 Sat.	Sagittarius	Fire	Barren	3rd
23 Sun. 11:47 am	Capricorn	Earth	Semi-fruit	3rd
24 Mon.	Capricorn	Earth	Semi-fruit	3rd
25 Tue.	Capricorn	Earth	Semi-fruit	3rd
26 Wed. 12:42 am	Aquarius	Air	Barren	4th 2:30 pm
27 Thu.	Aquarius	Air	Barren	4th
28 Fri. 12:06 pm	Pisces	Water	Fruitful	4th
29 Sat.	Pisces	Water	Fruitful	4th
30 Sun. 7:54 pm	Aries	Fire	Barren	4th

April

Lunar Aspectarian   Favorable and Unfavorable Days

Date	Sun	Mercury	Venus	Mars	Jupiter	Saturn	Uranus	Neptune	Pluto	Aries	Taurus	Gemini	Cancer	Leo	Virgo	Libra	Scorpio	Sagittarius	Capricorn	Aquarius	Pisces
1				X	X					f	u	f		U		f	u	f		F	
2		C				X			Q	f	u	f			U		f	u	f		F
3			C					X		f	u	f			U		f	u	f		F
4	C					X			T	F		f	u	f		U		f	u	f	
5										F		f	u	f		U		f	u	f	
6				C	C	C		Q			F		f	u	f		U		f	u	f
7		X	X				Q				F		f	u	f		U		f	u	f
8								T	0	f		F		f	u	f		U		f	u
9	X	Q					T			f		F		f	u	f		U		f	u
10			Q	X	X	X				u	f		F		f	u	f		U		f
11	Q	T								u	f		F		f	u	f		U		f
12			T		Q			0	T	f	u	f		F		f	u	f		U	
13	T			Q		Q	0			f	u	f		F		f	u	f		U	
14										f	u	f		F		f	u	f		U	
15				T	T	T			Q		f	u	f		F		f	u	f		U
16		0						T			f	u	f		F		f	u	f		U
17			0					T	X	U		f	u	f		F		f	u	f	
18	0									U		f	u	f		F		f	u	f	
19				0				Q			U		f	u	f		F		f	u	f
20			0		0	Q					U		f	u	f		F		f	u	f
21								X		f		U		f	u	f		F		f	u
22		T	T				X		C	f		U		f	u	f		F		f	u
23	T									f		U		f	u	f		F		f	u
24					T					u	f		U		f	u	f		F		f
25		Q	Q	T		T				u	f		U		f	u	f		F		f
26	Q							C		u	f		U		f	u	f		F		f
27				Q	Q	C			X	f	u	f		U		f	u	f		F	
28		X	X	Q						f	u	f		U		f	u	f		F	
29	X				X				Q		f	u	f		U		f	u	f		F
30				X		X					f	u	f		U		f	u	f		F

May Moon Table

Date	Sign	Element	Nature	Phase
1 Mon.	Aries	Fire	Barren	4th
2 Tue. 11:54 pm	Taurus	Earth	Semi-fruit	4th
3 Wed.	Taurus	Earth	Semi-fruit	New 11:12 pm
4 Thu.	Taurus	Earth	Semi-fruit	1st
5 Fri. 1:23 am	Gemini	Air	Barren	1st
6 Sat.	Gemini	Air	Barren	1st
7 Sun. 2:14 am	Cancer	Water	Fruitful	1st
8 Mon.	Cancer	Water	Fruitful	1st
9 Tue. 4:01 am	Leo	Fire	Barren	1st
10 Wed.	Leo	Fire	Barren	2nd 3:00 pm
11 Thu. 7:41 am	Virgo	Earth	Barren	2nd
12 Fri.	Virgo	Earth	Barren	2nd
13 Sat. 1:27 pm	Libra	Air	Semi-fruit	2nd
14 Sun.	Libra	Air	Semi-fruit	2nd
15 Mon. 9:16 pm	Scorpio	Water	Fruitful	2nd
16 Tue.	Scorpio	Water	Fruitful	2nd
17 Wed.	Scorpio	Water	Fruitful	2nd
18 Thu. 7:09 am	Sagittarius	Fire	Barren	Full 2:34 am
19 Fri.	Sagittarius	Fire	Barren	3rd
20 Sat. 7:01 pm	Capricorn	Earth	Semi-fruit	3rd
21 Sun.	Capricorn	Earth	Semi-fruit	3rd
22 Mon.	Capricorn	Earth	Semi-fruit	3rd
23 Tue. 8:00 am	Aquarius	Air	Barren	3rd
24 Wed.	Aquarius	Air	Barren	3rd
25 Thu. 8:07 pm	Pisces	Water	Fruitful	3rd
26 Fri.	Pisces	Water	Fruitful	4th 6:55 am
27 Sat.	Pisces	Water	Fruitful	4th
28 Sun. 5:08 am	Aries	Fire	Barren	4th
29 Mon.	Aries	Fire	Barren	4th
30 Tue. 10:02 am	Taurus	Earth	Semi-fruit	4th
31 Wed.	Taurus	Earth	Semi-fruit	4th

May

Lunar Aspectarian   Favorable and Unfavorable Days

Date	Sun	Mercury	Venus	Mars	Jupiter	Saturn	Uranus	Neptune	Pluto	Aries	Taurus	Gemini	Cancer	Leo	Virgo	Libra	Scorpio	Sagittarius	Capricorn	Aquarius	Pisces
1								X	T	F		f	u	f		U		f	u	f	
2							X			F		f	u	f		U		f	u	f	
3	C	C	C					Q			F		f	u	f		U		f	u	f
4					C	C	Q				F		f	u	f		U		f	u	f
5			C					T	0		F		f	u	f		U		f	u	f
6							T			f		F		f	u	f		U		f	u
7			X							f		F		f	u	f		U		f	u
8	X	X			X	X				u	f		F		f	u	f		U		f
9			Q	X				0		u	f		F		f	u	f		U		f
10	Q	Q			Q	Q	0		T	f	u	f		F		f	u	f		U	
11				Q						f	u	f		F		f	u	f		U	
12			T		T	T			Q		f	u	f		F		f	u	f		U
13	T	T									f	u	f		F		f	u	f		U
14				T				T	X	U		f	u	f		F		f	u	f	
15								T		U		f	u	f		F		f	u	f	
16								Q			U		f	u	f		F		f	u	f
17		0			0	0	Q				U		f	u	f		F		f	u	f
18	0							X			U		f	u	f		F		f	u	f
19		0		0					C	f		U		f	u	f		F		f	u
20							X			f		U		f	u	f		F		f	u
21										u	f		U		f	u	f		F		f
22					T	T				u	f		U		f	u	f		F		f
23	T		T						C	u	f		U		f	u	f		F		f
24				T					X	f	u	f		U		f	u	f		F	
25		T	Q		Q	Q	C			f	u	f		U		f	u	f		F	
26	Q								Q		f	u	f		U		f	u	f		F
27		Q		Q	X	X					f	u	f		U		f	u	f		F
28	X		X					X			f	u	f		U		f	u	f		F
29				X		X			T	F		f	u	f		U		f	u	f	
30		X							Q	F		f	u	f		U		f	u	f	
31								Q			F		f	u	f		U		f	u	f

June Moon Table

Date	Sign	Element	Nature	Phase
1 Thu. 11:34 am	Gemini	Air	Barren	4th
2 Fri.	Gemini	Air	Barren	New 7:14 am
3 Sat. 11:30 am	Cancer	Water	Fruitful	1st
4 Sun.	Cancer	Water	Fruitful	1st
5 Mon. 11:45 am	Leo	Fire	Barren	1st
6 Tue.	Leo	Fire	Barren	1st
7 Wed. 1:57 pm	Virgo	Earth	Barren	1st
8 Thu.	Virgo	Earth	Barren	2nd 10:29 pm
9 Fri. 6:58 pm	Libra	Air	Semi-fruit	2nd
10 Sat.	Libra	Air	Semi-fruit	2nd
11 Sun.	Libra	Air	Semi-fruit	2nd
12 Mon. 2:55 am	Scorpio	Water	Fruitful	2nd
13 Tue.	Scorpio	Water	Fruitful	2nd
14 Wed. 1:18 pm	Sagittarius	Fire	Barren	2nd
15 Thu.	Sagittarius	Fire	Barren	2nd
16 Fri.	Sagittarius	Fire	Barren	Full 5:27 pm
17 Sat. 1:26 am	Capricorn	Earth	Semi-fruit	3rd
18 Sun.	Capricorn	Earth	Semi-fruit	3rd
19 Mon. 2:26 pm	Aquarius	Air	Barren	3rd
20 Tue.	Aquarius	Air	Barren	3rd
21 Wed.	Aquarius	Air	Barren	3rd
22 Thu. 2:52 am	Pisces	Water	Fruitful	3rd
23 Fri.	Pisces	Water	Fruitful	3rd
24 Sat. 12:55 pm	Aries	Fire	Barren	4th 8:00 pm
25 Sun.	Aries	Fire	Barren	4th
26 Mon. 7:19 pm	Taurus	Earth	Semi-fruit	4th
27 Tue.	Taurus	Earth	Semi-fruit	4th
28 Wed. 9:59 pm	Gemini	Air	Barren	4th
29 Thu.	Gemini	Air	Barren	4th
30 Fri. 10:09 pm	Cancer	Water	Fruitful	4th

June

Lunar Aspectarian

Favorable and Unfavorable Days

Date	Sun	Mercury	Venus	Mars	Jupiter	Saturn	Uranus	Neptune	Pluto	Aries	Taurus	Gemini	Cancer	Leo	Virgo	Libra	Scorpio	Sagittarius	Capricorn	Aquarius	Pisces
1					C	C		T			F		f	u	f		U		f	u	f
2	C		C	C			T		0	f		F		f	u	f		U		f	u
3		C								f		F		f	u	f		U		f	u
4										u	f		F		f	u	f		U		f
5					X	X		0		u	f		F		f	u	f		U		f
6	X		X				0		T	f	u	f		F			f	u	f	U	
7				X	Q	Q				f	u	f		F			f	u	f	U	
8	Q	X	Q						Q		f	u	f		F		f	u	f		U
9				Q	T	T					f	u	f		F		f	u	f		U
10		Q						T	X	U		f	u	f		F		f	u	f	
11	T		T	T			T			U		f	u	f		F		f	u	f	
12								Q		U		f	u	f		F		f	u	f	
13		T						Q				U	f	u	f		F		f	u	f
14					0	0						U	f	u	f		F		f	u	f
15								X	C	f		U	f	u	f			F		f	u
16	0		0				X			f		U	f	u	f			F		f	u
17				0						f		U	f	u	f			F		f	u
18		0								u	f		U		f	u	f	F			f
19					T	T				u	f		U		f	u	f	F			f
20								C	X	f	u	f		U	f	u	f		F		
21					Q	Q	C			f	u	f		U	f	u	f		F		
22	T		T	T						f	u	f		U	f	u	f		F		
23		T							Q		f	u	f	U		f	u	f				F
24	Q			Q	X	X					f	u	f	U		f	u	f			F
25			Q					X	T	F		f	u	f	U		f	u	f		
26		Q					X			F		f	u	f	U		f	u	f		
27	X		X	X				Q			F		f	u	f		U		f	u	f
28		X			C	C	Q				F		f	u	f		U		f	u	f
29								T	0	f		F		f	u	f		U		f	u
30							T			f		F		f	u	f		U		f	u

July Moon Table

Date	Sign	Element	Nature	Phase
1 Sat.	Cancer	Water	Fruitful	New 2:20 pm
2 Sun. 9:38 pm	Leo	Fire	Barren	1st
3 Mon.	Leo	Fire	Barren	1st
4 Tue. 10:19 pm	Virgo	Earth	Barren	1st
5 Wed.	Virgo	Earth	Barren	1st
6 Thu.	Virgo	Earth	Barren	1st
7 Fri. 1:47 am	Libra	Air	Semi-fruit	1st
8 Sat.	Libra	Air	Semi-fruit	2nd 7:53 am
9 Sun. 8:48 am	Scorpio	Water	Fruitful	2nd
10 Mon.	Scorpio	Water	Fruitful	2nd
11 Tue. 7:06 pm	Sagittarius	Fire	Barren	2nd
12 Wed.	Sagittarius	Fire	Barren	2nd
13 Thu.	Sagittarius	Fire	Barren	2nd
14 Fri. 7:27 am	Capricorn	Earth	Semi-fruit	2nd
15 Sat.	Capricorn	Earth	Semi-fruit	2nd
16 Sun. 8:27 pm	Aquarius	Air	Barren	Full 8:55 am
17 Mon.	Aquarius	Air	Barren	3rd
18 Tue.	Aquarius	Air	Barren	3rd
19 Wed. 8:44 am	Pisces	Water	Fruitful	3rd
20 Thu.	Pisces	Water	Fruitful	3rd
21 Fri. 7:09 pm	Aries	Fire	Barren	3rd
22 Sat.	Aries	Fire	Barren	3rd
23 Sun.	Aries	Fire	Barren	3rd
24 Mon. 2:44 am	Taurus	Earth	Semi-fruit	4th 6:02 am
25 Tue.	Taurus	Earth	Semi-fruit	4th
26 Wed. 7:01 am	Gemini	Air	Barren	4th
27 Thu.	Gemini	Air	Barren	4th
28 Fri. 8:30 am	Cancer	Water	Fruitful	4th
29 Sat.	Cancer	Water	Fruitful	4th
30 Sun. 8:23 am	Leo	Fire	Barren	New 9:25 pm
31 Mon.	Leo	Fire	Barren	1st

July

Lunar Aspectarian   Favorable and Unfavorable Days

Date	Sun	Mercury	Venus	Mars	Jupiter	Saturn	Uranus	Neptune	Pluto	Aries	Taurus	Gemini	Cancer	Leo	Virgo	Libra	Scorpio	Sagittarius	Capricorn	Aquarius	Pisces
1	C		C	C						u	f		F		f	u	f		U		f
2		C			X	X				u	f		F		f	u	f		U		f
3								0	T	f	u	f		F		f	u	f		U	
4					Q	Q	0			f	u	f		F		f	u	f		U	
5	X	X		X					Q		f	u	f		F		f	u	f		U
6			X			T					f	u	f		F		f	u	f		U
7					T			T	X		f	u	f		F		f	u	f		U
8	Q	Q	Q	Q			T			U		f	u	f		F		f	u	f	
9								Q		U		f	u	f		F		f	u	f	
10	T	T		T			Q				U		f	u	f		F		f	u	f
11			T		0	0					U		f	u	f		F		f	u	f
12							X	C	f		U		f	u	f		F		f	u
13						X				f		U		f	u	f		F		f	u
14										f		U		f	u	f		F		f	u
15		0		0						u	f		U		f	u	f		F		f
16	0					T				u	f		U		f	u	f		F		f
17			0		T			C	X	f	u	f		U		f	u	f		F	
18							C			f	u	f		U		f	u	f		F	
19					Q	Q				f	u	f		U		f	u	f		F	
20		T							Q		f	u	f		U		f	u	f		F
21	T			T		X					f	u	f		U		f	u	f		F
22		Q	T		X			X	T	F		f	u	f		U		f	u	f	
23			Q			X				F		f	u	f		U		f	u	f	
24	Q							Q		F		f	u	f		U		f	u	f	
25		X	Q				Q				F		f	u	f		U		f	u	f
26	X			X	C	C		T			F		f	u	f		U		f	u	f
27			X			T			0	f		F		f	u	f		U		f	u
28										f		F		f	u	f		U		f	u
29		C								u	f		F		f	u	f		U		f
30	C			C	X	X		0		u	f		F		f	u	f		U		f
31			C				0		T	f	u	f		F		f	u	f		U	

August Moon Table

Date	Sign	Element	Nature	Phase
1 Tue. 8:27 am	Virgo	Earth	Barren	1st
2 Wed.	Virgo	Earth	Barren	1st
3 Thu. 10:31 am	Libra	Air	Semi-fruit	1st
4 Fri.	Libra	Air	Semi-fruit	1st
5 Sat. 4:04 pm	Scorpio	Water	Fruitful	1st
6 Sun.	Scorpio	Water	Fruitful	2nd 8:02 pm
7 Mon.	Scorpio	Water	Fruitful	2nd
8 Tue. 1:30 am	Sagittarius	Fire	Barren	2nd
9 Wed.	Sagittarius	Fire	Barren	2nd
10 Thu. 1:44 pm	Capricorn	Earth	Semi-fruit	2nd
11 Fri.	Capricorn	Earth	Semi-fruit	2nd
12 Sat.	Capricorn	Earth	Semi-fruit	2nd
13 Sun. 2:43 am	Aquarius	Air	Barren	2nd
14 Mon.	Aquarius	Air	Barren	2nd
15 Tue. 2:41 pm	Pisces	Water	Fruitful	Full 12:13 am
16 Wed.	Pisces	Water	Fruitful	3rd
17 Thu.	Pisces	Water	Fruitful	3rd
18 Fri. 12:44 am	Aries	Fire	Barren	3rd
19 Sat.	Aries	Fire	Barren	3rd
20 Sun. 8:31 am	Taurus	Earth	Semi-fruit	3rd
21 Mon.	Taurus	Earth	Semi-fruit	3rd
22 Tue. 1:55 pm	Gemini	Air	Barren	4th 1:51 pm
23 Wed.	Gemini	Air	Barren	4th
24 Thu. 4:59 pm	Cancer	Water	Fruitful	4th
25 Fri.	Cancer	Water	Fruitful	4th
26 Sat. 6:17 pm	Leo	Fire	Barren	4th
27 Sun.	Leo	Fire	Barren	4th
28 Mon. 6:55 pm	Virgo	Earth	Barren	4th
29 Tue.	Virgo	Earth	Barren	New 5:19 am
30 Wed. 8:33 pm	Libra	Air	Semi-fruit	1st
31 Thu.	Libra	Air	Semi-fruit	1st

August

Lunar Aspectarian   Favorable and Unfavorable Days

Date	Sun	Mercury	Venus	Mars	Jupiter	Saturn	Uranus	Neptune	Pluto	Aries	Taurus	Gemini	Cancer	Leo	Virgo	Libra	Scorpio	Sagittarius	Capricorn	Aquarius	Pisces
1					Q	Q					f	u	f		F		f	u	f		U
2		X							Q		f	u	f		F		f	u	f		U
3				X	T	T		T			f	u	f		F		f	u	f		U
4	X						T		X	U		f	u	f		F		f	u	f	
5		Q	X	Q						U		f	u	f		F		f	u	f	
6	Q							Q			U		f	u	f		F		f	u	f
7						Q					U		f	u	f		F		f	u	f
8		T	Q	T	O	O		X	C		U		f	u	f		F		f	u	f
9	T						X			f		U		f	u	f		F		f	u
10										f		U		f	u	f		F		f	u
11			T							u	f		U		f	u	f		F		f
12										u	f		U		f	u	f		F		f
13			O		T	T		C	X	u	f		U		f	u	f		F		f
14		O					C			f	u	f		U		f	u	f		F	
15	O				Q					f	u	f		U		f	u	f		F	
16			O	Q					Q		f	u	f		U		f	u	f		F
17											f	u	f		U		f	u	f		F
18				T	X	X		X	T		f	u	f		U		f	u	f		F
19							X			F		f	u	f		U		f	u	f	
20	T	T						Q		F		f	u	f		U		f	u	f	
21			T	Q			Q				F		f	u	f		U		f	u	f
22	Q	Q				C		T			F		f	u	f		U		f	u	f
23				X	C	T			O	f		F		f	u	f		U		f	u
24	X		Q							f		F		f	u	f		U		f	u
25		X								u	f		F		f	u	f		U		f
26			X			X				u	f		F		f	u	f		U		f
27				C	X		O	O	T	f	u	f		F		f	u	f		U	
28					Q					f	u	f		F		f	u	f		U	
29	C	C			Q				Q		f	u	f		F		f	u	f		U
30			C			T					f	u	f		F		f	u	f		U
31					T			T	X	U		f	u	f		F		f	u	f	

September Moon Table

Date	Sign	Element	Nature	Phase
1 Fri.	Libra	Air	Semi-fruit	1st
2 Sat. 12:55 am	Scorpio	Water	Fruitful	1st
3 Sun.	Scorpio	Water	Fruitful	1st
4 Mon. 9:08 am	Sagittarius	Fire	Barren	1st
5 Tue.	Sagittarius	Fire	Barren	2nd 11:27 am
6 Wed. 8:47 pm	Capricorn	Earth	Semi-fruit	2nd
7 Thu.	Capricorn	Earth	Semi-fruit	2nd
8 Fri.	Capricorn	Earth	Semi-fruit	2nd
9 Sat. 9:44 am	Aquarius	Air	Barren	2nd
10 Sun.	Aquarius	Air	Barren	2nd
11 Mon. 9:34 pm	Pisces	Water	Fruitful	2nd
12 Tue.	Pisces	Water	Fruitful	2nd
13 Wed.	Pisces	Water	Fruitful	Full 2:37 pm
14 Thu. 7:00 am	Aries	Fire	Barren	3rd
15 Fri.	Aries	Fire	Barren	3rd
16 Sat. 2:05 pm	Taurus	Earth	Semi-fruit	3rd
17 Sun.	Taurus	Earth	Semi-fruit	3rd
18 Mon. 7:22 pm	Gemini	Air	Barren	3rd
19 Tue.	Gemini	Air	Barren	3rd
20 Wed. 11:16 pm	Cancer	Water	Fruitful	4th 8:28 pm
21 Thu.	Cancer	Water	Fruitful	4th
22 Fri.	Cancer	Water	Fruitful	4th
23 Sat. 2:00 am	Leo	Fire	Barren	4th
24 Sun.	Leo	Fire	Barren	4th
25 Mon. 4:02 am	Virgo	Earth	Barren	4th
26 Tue.	Virgo	Earth	Barren	4th
27 Wed. 6:22 am	Libra	Air	Semi-fruit	New 2:53 pm
28 Thu.	Libra	Air	Semi-fruit	1st
29 Fri. 10:30 am	Scorpio	Water	Fruitful	1st
30 Sat.	Scorpio	Water	Fruitful	1st

September

Lunar Aspectarian   Favorable and Unfavorable Days

Date	Sun	Mercury	Venus	Mars	Jupiter	Saturn	Uranus	Neptune	Pluto	Aries	Taurus	Gemini	Cancer	Leo	Virgo	Libra	Scorpio	Sagittarius	Capricorn	Aquarius	Pisces
1				X				T		U		f	u	f		F		f	u	f	
2	X							Q		U		f	u	f		F		f	u	f	
3		X		Q			Q				U		f	u	f		F		f	u	f
4			X			0		X			U		f	u	f		F		f	u	f
5	Q				0		X		C	f		U		f	u	f		F		f	u
6		Q		T						f		U		f	u	f		F		f	u
7			Q							u	f		U		f	u	f		F		f
8	T									u	f		U		f	u	f		F		f
9		T					T	C		u	f		U		f	u	f		F		f
10			T		T		C	X		f	u	f		U		f	u	f		F	
11				0		Q				f	u	f		U		f	u	f		F	
12					Q				Q		f	u	f		U		f	u	f		F
13	0										f	u	f		U		f	u	f		F
14						X		X			f	u	f		U		f	u	f		F
15		0	0		X		X		T	F		f	u	f		U		f	u	f	
16			T					Q		F		f	u	f		U		f	u	f	
17								Q			F		f	u	f		U		f	u	f
18	T			Q		C					F		f	u	f		U		f	u	f
19				C				T	0	f		F		f	u	f		U		f	u
20	Q	T	T					T		f		F		f	u	f		U		f	u
21				X						u	f		F		f	u	f		U		f
22		Q	Q							u	f		F		f	u	f		U		f
23	X				X	X		0	T	u	f		F		f	u	f		U		f
24		X						0		f	u	f		F		f	u	f		U	
25		X	C	Q	Q				Q	f	u	f		F		f	u	f		U	
26											f	u	f		F		f	u	f		U
27	C					T		T			f	u	f		F		f	u	f		U
28					T	T			X	U		f	u	f		F		f	u	f	
29		C	C					Q		U		f	u	f		F		f	u	f	
30				X				Q			U		f	u	f		F		f	u	f

October Moon Table

Date	Sign	Element	Nature	Phase
1 Sun. 5:50 pm	Sagittarius	Fire	Barren	1st
2 Mon.	Sagittarius	Fire	Barren	1st
3 Tue.	Sagittarius	Fire	Barren	1st
4 Wed. 4:42 am	Capricorn	Earth	Semi-fruit	1st
5 Thu.	Capricorn	Earth	Semi-fruit	2nd 5:59 am
6 Fri. 5:33 pm	Aquarius	Air	Barren	2nd
7 Sat.	Aquarius	Air	Barren	2nd
8 Sun.	Aquarius	Air	Barren	2nd
9 Mon. 5:36 am	Pisces	Water	Fruitful	2nd
10 Tue.	Pisces	Water	Fruitful	2nd
11 Wed. 2:51 pm	Aries	Fire	Barren	2nd
12 Thu.	Aries	Fire	Barren	2nd
13 Fri. 9:06 pm	Taurus	Earth	Semi-fruit	Full 3:53 am
14 Sat.	Taurus	Earth	Semi-fruit	3rd
15 Sun.	Taurus	Earth	Semi-fruit	3rd
16 Mon. 1:19 am	Gemini	Air	Barren	3rd
17 Tue.	Gemini	Air	Barren	3rd
18 Wed. 4:37 am	Cancer	Water	Fruitful	3rd
19 Thu.	Cancer	Water	Fruitful	3rd
20 Fri. 7:42 am	Leo	Fire	Barren	4th 2:59 am
21 Sat.	Leo	Fire	Barren	4th
22 Sun. 10:52 am	Virgo	Earth	Barren	4th
23 Mon.	Virgo	Earth	Barren	4th
24 Tue. 2:30 pm	Libra	Air	Semi-fruit	4th
25 Wed.	Libra	Air	Semi-fruit	4th
26 Thu. 7:23 pm	Scorpio	Water	Fruitful	4th
27 Fri.	Scorpio	Water	Fruitful	New 2:58 am
28 Sat.	Scorpio	Water	Fruitful	1st
29 Sun. 2:40 am	Sagittarius	Fire	Barren	1st
30 Mon.	Sagittarius	Fire	Barren	1st
31 Tue. 1:01 pm	Capricorn	Earth	Semi-fruit	1st

October

Lunar Aspectarian   Favorable and Unfavorable Days

Date	Sun	Mercury	Venus	Mars	Jupiter	Saturn	Uranus	Neptune	Pluto	Aries	Taurus	Gemini	Cancer	Leo	Virgo	Libra	Scorpio	Sagittarius	Capricorn	Aquarius	Pisces
1						0					U		f	u	f		F		f	u	f
2	X			Q	0			X	C	f		U		f	u	f		F		f	u
3							X			f		U		f	u	f		F		f	u
4		X								f		U		f	u	f		F		f	u
5	Q		X	T						u	f		U		f	u	f		F		f
6						T				u	f		U		f	u	f		F		f
7		Q			T			C	X	f	u	f		U		f	u	f		F	
8	T		Q					C		f	u	f		U		f	u	f		F	
9						Q				f	u	f		U		f	u	f		F	
10		T	T	0	Q				Q		f	u	f		U		f	u	f		F
11						X		X			f	u	f		U		f	u	f		F
12					X			X	T	F		f	u	f		U		f	u	f	
13	0									F		f	u	f		U		f	u	f	
14		0						Q			F		f	u	f		U		f	u	f
15			0	T				Q			F		f	u	f		U		f	u	f
16					C	C		T	0		F		f	u	f		U		f	u	f
17	T			Q			T			f		F		f	u	f		U		f	u
18										f		F		f	u	f		U		f	u
19		T		X						u	f		F		f	u	f		U		f
20	Q		T			X		0		u	f		F		f	u	f		U		f
21		Q			X		0		T	f	u	f		F		f	u	f		U	
22	X		Q		Q					f	u	f		F		f	u	f		U	
23		X			Q				Q		f	u	f		F		f	u	f		U
24				C		T		T			f	u	f		F		f	u	f		U
25			X		T		T	X		U		f	u	f		F		f	u	f	
26										U		f	u	f		F		f	u	f	
27	C	C						Q			U		f	u	f		F		f	u	f
28				X			Q				U		f	u	f		F		f	u	f
29					0	0		X			U		f	u	f		F		f	u	f
30			C				X		C	f		U		f	u	f		F		f	u
31		X		Q						f		U		f	u	f		F		f	u

November Moon Table

Date	Sign	Element	Nature	Phase
1 Wed.	Capricorn	Earth	Semi-fruit	1st
2 Thu.	Capricorn	Earth	Semi-fruit	1st
3 Fri. 1:41 am	Aquarius	Air	Barren	1st
4 Sat.	Aquarius	Air	Barren	2nd 2:27 am
5 Sun. 2:13 pm	Pisces	Water	Fruitful	2nd
6 Mon.	Pisces	Water	Fruitful	2nd
7 Tue.	Pisces	Water	Fruitful	2nd
8 Wed. 12:02 am	Aries	Fire	Barren	2nd
9 Thu.	Aries	Fire	Barren	2nd
10 Fri. 6:12 am	Taurus	Earth	Semi-fruit	2nd
11 Sat.	Taurus	Earth	Semi-fruit	Full 4:15 pm
12 Sun. 9:27 am	Gemini	Air	Barren	3rd
13 Mon.	Gemini	Air	Barren	3rd
14 Tue. 11:21 am	Cancer	Water	Fruitful	3rd
15 Wed.	Cancer	Water	Fruitful	3rd
16 Thu. 1:19 pm	Leo	Fire	Barren	3rd
17 Fri.	Leo	Fire	Barren	3rd
18 Sat. 4:15 pm	Virgo	Earth	Barren	4th 10:24 am
19 Sun.	Virgo	Earth	Barren	4th
20 Mon. 8:35 pm	Libra	Air	Semi-fruit	4th
21 Tue.	Libra	Air	Semi-fruit	4th
22 Wed.	Libra	Air	Semi-fruit	4th
23 Thu. 2:33 am	Scorpio	Water	Fruitful	4th
24 Fri.	Scorpio	Water	Fruitful	4th
25 Sat. 10:33 am	Sagittarius	Fire	Barren	New 6:11 pm
26 Sun.	Sagittarius	Fire	Barren	1st
27 Mon. 8:57 pm	Capricorn	Earth	Semi-fruit	1st
28 Tue.	Capricorn	Earth	Semi-fruit	1st
29 Wed.	Capricorn	Earth	Semi-fruit	1st
30 Thu. 9:26 am	Aquarius	Air	Barren	1st

November

Lunar Aspectarian Favorable and Unfavorable Days

Date	Sun	Mercury	Venus	Mars	Jupiter	Saturn	Uranus	Neptune	Pluto	Aries	Taurus	Gemini	Cancer	Leo	Virgo	Libra	Scorpio	Sagittarius	Capricorn	Aquarius	Pisces
1	X									u	f		U		f	u	f		F		f
2							T			u	f		U		f	u	f		F		f
3		Q			T	T		C		u	f		U		f	u	f		F		f
4	Q		X				C		X	f	u	f		U		f	u	f		F	
5		T				Q				f	u	f		U		f	u	f		F	
6	T				Q				Q		f	u	f		U		f	u	f		F
7			Q			X					f	u	f		U		f	u	f		F
8				0	X			X	T		f	u	f		U		f	u	f		F
9							X			F		f	u	f		U		f	u	f	
10		0	T					Q		F		f	u	f		U		f	u	f	
11	0						Q				F		f	u	f		U		f	u	f
12				T	C	C		T			F		f	u	f		U		f	u	f
13							T		0	f		F		f	u	f		U		f	u
14		T	0	Q						f		F		f	u	f		U		f	u
15										u	f		F		f	u	f		U		f
16	T	Q				X		0		u	f		F		f	u	f		U		f
17				X	X		0		T	f	u	f		F		f	u	f		U	
18	Q					Q				f	u	f		F		f	u	f		U	
19		X	T		Q				Q		f	u	f		F		f	u	f		U
20	X					T					f	u	f		F		f	u	f		U
21			Q	C	T			T	X	U		f	u	f		F		f	u	f	
22						T				U		f	u	f		F		f	u	f	
23								Q		U		f	u	f		F		f	u	f	
24		C	X			Q					U		f	u	f		F		f	u	f
25	C				0	0		X			U		f	u	f		F		f	u	f
26				X			X		C	f		U		f	u	f		F		f	u
27										f		U		f	u	f		F		f	u
28										u	f		U		f	u	f		F		f
29		X	C	Q						u	f		U		f	u	f		F		f
30					T	T		C		u	f		U		f	u	f		F		f

December Moon Table

Date	Sign	Element	Nature	Phase
1 Fri.	Aquarius	Air	Barren	1st
2 Sat. 10:23 pm	Pisces	Water	Fruitful	1st
3 Sun.	Pisces	Water	Fruitful	2nd 10:55 pm
4 Mon.	Pisces	Water	Fruitful	2nd
5 Tue. 9:17 am	Aries	Fire	Barren	2nd
6 Wed.	Aries	Fire	Barren	2nd
7 Thu. 4:26 pm	Taurus	Earth	Semi-fruit	2nd
8 Fri.	Taurus	Earth	Semi-fruit	2nd
9 Sat. 7:50 pm	Gemini	Air	Barren	2nd
10 Sun.	Gemini	Air	Barren	2nd
11 Mon. 8:48 pm	Cancer	Water	Fruitful	Full 4:03 am
12 Tue.	Cancer	Water	Fruitful	3rd
13 Wed. 9:09 pm	Leo	Fire	Barren	3rd
14 Thu.	Leo	Fire	Barren	3rd
15 Fri. 10:30 pm	Virgo	Earth	Barren	3rd
16 Sat.	Virgo	Earth	Barren	3rd
17 Sun.	Virgo	Earth	Barren	4th 7:41 pm
18 Mon. 2:01 am	Libra	Air	Semi-fruit	4th
19 Tue.	Libra	Air	Semi-fruit	4th
20 Wed. 8:12 am	Scorpio	Water	Fruitful	4th
21 Thu.	Scorpio	Water	Fruitful	4th
22 Fri. 4:57 pm	Sagittarius	Fire	Barren	4th
23 Sat.	Sagittarius	Fire	Barren	4th
24 Sun.	Sagittarius	Fire	Barren	4th
25 Mon. 3:54 am	Capricorn	Earth	Semi-fruit	New 12:22 pm
26 Tue.	Capricorn	Earth	Semi-fruit	1st
27 Wed. 4:25 pm	Aquarius	Air	Barren	1st
28 Thu.	Aquarius	Air	Barren	1st
29 Fri.	Aquarius	Air	Barren	1st
30 Sat. 5:27 am	Pisces	Water	Fruitful	1st
31 Sun.	Pisces	Water	Fruitful	1st

December

Lunar Aspectarian   Favorable and Unfavorable Days

Date	Sun	Mercury	Venus	Mars	Jupiter	Saturn	Uranus	Neptune	Pluto	Aries	Taurus	Gemini	Cancer	Leo	Virgo	Libra	Scorpio	Sagittarius	Capricorn	Aquarius	Pisces
1	X			T				C	X	f	u	f		U		f	u	f		F	
2		Q				Q				f	u	f		U		f	u	f		F	
3	Q				Q				Q		f	u	f		U		f	u	f		F
4											f	u	f		U		f	u	f		F
5		T	X		X	X		X			f	u	f		U		f	u	f		F
6	T			O			X		T	F		f	u	f		U		f	u	f	
7			Q							F		f	u	f		U		f	u	f	
8							Q	Q			F		f	u	f		U		f	u	f
9			T			C					F		f	u	f		U		f	u	f
10		O			C			T	O	f		F		f	u	f		U		f	u
11	O			T			T			f		F		f	u	f		U		f	u
12										u	f		F		f	u	f		U		f
13				Q	X					u	f		F		f	u	f		U		f
14			O	X				O	T	f	u	f		F		f	u	f		U	
15	T	T		X		Q	O			f	u	f		F		f	u	f		U	
16					Q				Q		f	u	f		F		f	u	f		U
17	Q	Q				T					f	u	f		F		f	u	f		U
18					T			T			f	u	f		F		f	u	f		U
19		X	T				T		X	U		f	u	f		F		f	u	f	
20	X			C				Q		U		f	u	f		F		f	u	f	
21			Q				Q				U		f	u	f		F		f	u	f
22					O	O					U		f	u	f		F		f	u	f
23							X		C	f		U		f	u	f		F		f	u
24			X				X			f		U		f	u	f		F		f	u
25	C	C		X						f		U		f	u	f		F		f	u
26										u	f		U		f	u	f		F		f
27				Q	T	T				u	f		U		f	u	f		F		f
28								C	X	f	u	f		U		f	u	f		F	
29			C				Q	C		f	u	f		U		f	u	f		F	
30				T	Q					f	u	f		U		f	u	f		F	
31	X	X							Q		f	u	f		U		f	u	f		F

The Moon's Quarters & Signs

Everyone has seen the Moon wax and wane through a period of approximately twenty-nine and a half days. This circuit from New Moon to Full Moon and back again is called the lunation cycle. The cycle is divided into parts, called quarters or phases. There are several methods by which this can be done, and the system used in the *Moon Sign Book* may not correspond to those used in other almanacs.

The Quarters

First Quarter

The first quarter begins at the New Moon, when the Sun and Moon are in the same place, or conjunct. (This means that the Sun and Moon are in the same degree of the same sign.) The Moon is not visible at first, since it rises at the same time as the Sun. The **New Moon** is the time of new beginnings, beginnings of projects that favor growth, externalization of activities, and the expansion of ideas. The first quarter is the time of germination, emergence, beginnings, and outwardly directed activity.

Second Quarter

The second quarter begins halfway between the New Moon and the Full Moon when the Sun and Moon are at right angles (90 degrees). This half Moon rises around noon and sets around midnight, so it can be seen in the western sky during the first half of the night. The second quarter is the time of growth, development, and articulation of things that already exist.

Third Quarter

The third quarter begins at the Full Moon, when the Sun is opposite the Moon and its full light can shine on the full sphere of the Moon. The round Moon can be seen rising in the east at sunset, and then rising a little later each evening. The **Full Moon** stands for illumination, fulfillment, culmination, completion, drawing inward, unrest, emotional expressions, and hasty actions leading to failure. The third quarter is a time of maturity, fruition, and the assumption of the full form of expression.

Fourth Quarter

The fourth quarter begins about halfway between the Full Moon and New Moon, when the Sun and Moon are again at 90 degrees, or square. This decreasing Moon rises at midnight, and can be seen in the east during the last half of the night, reaching the overhead position just about as the Sun rises. The fourth quarter is a time of disintegration, drawing back for reorganization, and reflection.

The Signs

Moon in Aries is good for starting things but lacks staying power. Things occur rapidly but also quickly pass.

With Moon in Taurus, things begun last the longest and tend to increase in value. Things begun now become habitual and hard to alter.

Moon in Gemini is an inconsistent position for the Moon, characterized by a lot of talk. Things begun now are easily changed by outside influence.

Moon in Cancer stimulates emotional rapport between people. It pinpoints need, and supports growth and nurturance.

Moon in Leo accents showmanship, being seen, drama, recreation, and happy pursuits. It may be concerned with praise and subject to flattery.

Moon in Virgo favors accomplishment of details and commands from higher up while discouraging independent thinking.

Moon in Libra increases self-awareness. It favors self-examination and interaction with others, but discourages spontaneous initiative.

Moon in Scorpio increases awareness of psychic power. It precipitates psychic crises and ends connections thoroughly.

Moon in Sagittarius encourages expansionary flights of imagination and confidence in the flow of life.

Moon in Capricorn increases awareness of the need for structure, discipline, and organization. Institutional activities are favored.

Moon in Aquarius favors activities that are unique and individualistic, concern for humanitarian issues, society as a whole, and improvements that can be made.

During Moon in Pisces, energy withdraws from the surface of life, hibernates within, secretly reorganizing and realigning for a new day.

Retrogrades

When the planets cross the sky, they occasionally appear to move backward as seen from Earth. When a planet turns "backward" it is said to be *retrograde*. When it turns forward again, it is said to go *direct*. The point at which the movement changes from one direction to another is called a *station*.

When a planet is retrograde, its expression is delayed or out of kilter with the normal progression of events. Generally, it can be said that whatever is planned during this period will be delayed, but usually it will come to fruition when the retrograde is over. Of course, this only applies to activities ruled by the planet that is retrograde. Mercury retrogrades are easy to follow.

Mercury Retrograde

Mercury rules informal communications—reading, writing, speaking, and short errands. Whenever Mercury goes retrograde, personal communications get fouled up or misunderstood. The general rule is *when Mercury is retrograde, avoid informal means of communication*.

Planetary Retrogrades for 2000 (EST)

Planet	Begin		End	
Saturn	8/29/99	7:09 pm	1/11/00	11:59 pm
Mercury	2/21/00	7:47 am	3/14/00	3:40 pm
Pluto	3/15/00	6:50 pm	8/20/00	5:42 pm
Neptune	5/8/00	7:31 am	10/15/00	9:13 am
Uranus	5/25/00	3:22 am	10/26/00	10:24 am
Mercury	6/23/00	3:32 am	7/17/00	8:20 am
Saturn	9/12/00	6:35 am	1/24/01	7:25 pm
Jupiter	9/29/00	7:53 am	1/28/01	3:39 am
Mercury	10/18/00	8:42 am	11/07/00	9:26 pm

Moon Void-of-Course

Kim Rogers-Gallagher

The Moon makes a loop around the Earth in about twenty-eight days, moving through each of the signs in two-and-a-half days (or so). As she passes through the thirty degrees of each sign, she "visits" with the planets in numerical order by forming angles or aspects with them. Because she moves one degree in just two to two-and-a-half hours, her influence on each planet lasts only a few hours, then she moves along. As she approaches the late degrees of the sign she's passing through, she eventually reaches the planet that's in the highest degree of any sign, and forms what will be her final aspect before leaving the sign. From this point until she actually enters the new sign, she is referred to as void-of-course, or void.

Think of it this way: The Moon is the emotional "tone" of the day, carrying feelings with her particular to the sign she's "wearing" at the moment. After she has contacted each of the planets, she symbolically "rests" before changing her costume, so her instinct is temporarily on hold. It's during this time that many people feel "fuzzy" or "vague"—scattered, even. Plans or decisions we make now will usually not pan out. Without the instinctual "knowing" the Moon provides as she touches each planet, we tend to be unrealistic or exercise poor judgment. The traditional definition of the void Moon is that "nothing will come of this," and it seems to be true. Actions initiated under a void Moon are often wasted, irrelevant, or incorrect—usually because information is hidden or missing, or has been overlooked.

Although it's not a good time to initiate plans, routine tasks seem to go along just fine. However, this period is really ideal for what the Moon does best: reflection. It's at this time that we can assimilate what the world has tossed at us over the past few days.

On the lighter side, remember that there are other good uses for the void Moon. This is the time period when the universe seems to be most open to loopholes. It's a great time to make plans you don't want to fulfill or schedule things you don't want to do. See the table on pages 56–61 for a schedule of the 2000 void-of-course Moons.

Moon Void-of-Course

Last Aspect		Moon Enters New Sign		
Date	Time	Date	Sign	Time
		January		
2	2:28 pm	2	Sagittarius	4:32 pm
4	8:06 pm	5	Capricorn	5:24 am
7	9:00 am	7	Aquarius	5:53 pm
9	8:41 pm	10	Pisces	4:59 am
11	9:23 pm	12	Aries	1:48 pm
14	12:47 pm	14	Taurus	7:38 pm
16	4:50 pm	16	Gemini	10:25 pm
18	5:21 pm	18	Cancer	11:01 pm
20	5:36 pm	20	Leo	10:58 pm
22	8:30 pm	23	Virgo	12:07 am
24	2:48 am	25	Libra	4:09 am
27	6:59 am	27	Scorpio	12:01 pm
29	2:11 am	29	Sagittarius	11:17 pm
		February		
1	8:08 am	1	Capricorn	12:10 pm
3	9:16 pm	4	Aquarius	12:31 am
6	8:34 am	6	Pisces	11:02 am
8	2:46 pm	8	Aries	7:17 pm
11	12:19 am	11	Taurus	1:21 am
12	6:22 pm	13	Gemini	5:23 am
15	12:51 am	15	Cancer	7:45 am
17	7:51 am	17	Leo	9:11 am
18	2:05 pm	19	Virgo	10:53 am
20	3:59 pm	21	Libra	2:21 pm
22	10:15 pm	23	Scorpio	8:58 pm
25	7:18 am	26	Sagittarius	7:10 am
27	7:28 pm	28	Capricorn	7:45 pm
29	11:38 pm	2 (March)	Aquarius	8:14 am
		March		
3	8:12 pm	4	Pisces	6:30 pm

Moon Void-of-Course

Last Aspect		Moon Enters New Sign		
Date	Time	Date	Sign	Time
6	12:17 am	7	Aries	1:54 am
8	9:34 pm	9	Taurus	7:01 am
11	6:31 am	11	Gemini	10:46 am
13	1:59 am	13	Cancer	1:51 pm
15	8:43 am	15	Leo	4:43 pm
17	1:07 pm	17	Virgo	7:48 pm
19	11:44 pm	19	Libra	11:57 pm
22	5:26 am	22	Scorpio	6:17 am
23	6:53 pm	24	Sagittarius	3:43 pm
26	6:26 am	27	Capricorn	3:51 am
28	6:43 pm	29	Aquarius	4:34 pm
31	7:19 am	1 (April)	Pisces	3:12 am
April				
3	2:44 am	3	Aries	10:22 am
4	9:04 pm	5	Taurus	2:29 pm
7	3:24 am	7	Gemini	4:58 pm
9	11:01 am	9	Cancer	7:16 pm
11	7:45 pm	11	Leo	10:16 pm
13	4:14 pm	14	Virgo	2:19 am
15	8:45 am	16	Libra	7:36 am
18	12:41 pm	18	Scorpio	2:35 pm
20	5:36 am	20	Sagittarius	11:58 pm
22	4:25 pm	23	Capricorn	11:47 am
25	1:12 pm	26	Aquarius	12:42 am
28	5:44 am	28	Pisces	12:06 pm
30	4:13 pm	30	Aries	7:54 pm
May				
2	7:59 am	2	Taurus	11:54 pm
4	10:07 am	5	Gemini	1:23 am
6	11:01 am	7	Cancer	2:14 am
8	11:31 am	9	Leo	4:01 am

Moon Void-of-Course

Last Aspect		Moon Enters New Sign		
Date	Time	Date	Sign	Time
10	7:11 pm	11	Virgo	7:41 am
13	10:57 am	13	Libra	1:27 pm
15	3:55 am	15	Scorpio	9:16 pm
18	2:34 am	18	Sagittarius	7:09 am
20	12:30 am	20	Capricorn	7:01 pm
23	2:31 am	23	Aquarius	8:00 am
25	4:56 am	25	Pisces	8:07 pm
27	11:17 pm	28	Aries	5:08 am
29	6:15 pm	30	Taurus	10:02 am
June				
1	1:08 am	1	Gemini	11:34 am
2	9:03 pm	3	Cancer	11:30 am
5	2:48 am	5	Leo	11:45 am
7	5:22 am	7	Virgo	1:57 pm
9	10:48 am	9	Libra	6:58 pm
11	9:14 pm	12	Scorpio	2:55 am
14	6:31 am	14	Sagittarius	1:18 pm
16	8:50 pm	17	Capricorn	1:26 am
19	9:46 am	19	Aquarius	2:26 pm
21	11:25 pm	22	Pisces	2:52 am
24	10:40 am	24	Aries	12:55 pm
26	2:23 am	26	Taurus	7:19 pm
28	9:34 pm	28	Gemini	9:59 pm
30	6:47 am	30	Cancer	10:09 pm
July				
2	4:36 pm	2	Leo	9:38 pm
4	5:26 pm	4	Virgo	10:19 pm
6	8:57 pm	7	Libra	1:47 am
8	11:10 pm	9	Scorpio	8:48 am
11	3:29 pm	11	Sagittarius	7:06 pm
13	11:02 am	14	Capricorn	7:27 am

Moon Void-of-Course

Last Aspect		Moon Enters New Sign		
Date	Time	Date	Sign	Time
16	4:48 pm	16	Aquarius	8:27 pm
19	5:37 am	19	Pisces	8:44 am
21	6:08 pm	21	Aries	7:09 pm
23	5:11 pm	24	Taurus	2:44 am
26	5:20 am	26	Gemini	7:01 am
27	3:18 pm	28	Cancer	8:30 am
30	7:18 am	30	Leo	8:23 am
August				
1	7:34 am	1	Virgo	8:27 am
3	9:50 am	3	Libra	10:31 am
5	1:56 pm	5	Scorpio	4:04 pm
8	1:17 am	8	Sagittarius	1:30 am
9	3:15 pm	10	Capricorn	1:44 pm
11	1:02 am	13	Aquarius	2:43 am
15	12:13 am	15	Pisces	2:41 pm
16	2:58 pm	18	Aries	12:44 am
20	4:14 am	20	Taurus	8:31 am
22	1:51 pm	22	Gemini	1:55 pm
24	2:57 am	24	Cancer	4:59 pm
26	9:10 am	26	Leo	6:17 pm
27	11:44 pm	28	Virgo	6:55 pm
30	8:21 pm	30	Libra	8:33 pm
September				
1	7:23 am	2	Scorpio	12:55 am
3	8:34 pm	4	Sagittarius	9:08 am
6	5:26 pm	6	Capricorn	8:47 pm
8	5:27 am	9	Aquarius	9:44 am
11	3:09 pm	11	Pisces	9:34 pm
13	2:37 pm	14	Aries	7:00 am
16	1:50 pm	16	Taurus	2:05 pm
18	12:31 pm	18	Gemini	7:22 pm

Moon Void-of-Course

Last Aspect		Moon Enters New Sign		
Date	Time	Date	Sign	Time
20	8:28 pm	20	Cancer	11:16 pm
22	10:58 pm	23	Leo	2:00 am
24	8:33 pm	25	Virgo	4:02 am
25	10:44 pm	27	Libra	6:22 am
28	11:57 am	29	Scorpio	10:30 am
30	5:42 pm	1 (October)	Sagittarius	5:50 pm
October				
3	3:03 am	4	Capricorn	4:42 am
5	7:34 am	6	Aquarius	5:33 pm
8	3:55 am	9	Pisces	5:36 am
10	8:09 pm	11	Aries	2:51 pm
13	3:53 am	13	Taurus	9:06 pm
16	1:17 am	16	Gemini	1:19 am
17	8:01 pm	18	Cancer	4:37 am
20	7:15 am	20	Leo	7:42 am
22	10:12 am	22	Virgo	10:52 am
24	1:34 pm	24	Libra	2:30 pm
25	8:03 pm	26	Scorpio	7:23 pm
29	1:04 am	29	Sagittarius	2:40 am
31	8:43 am	31	Capricorn	1:01 pm
November				
3	12:37 am	3	Aquarius	1:41 am
5	11:26 am	5	Pisces	2:13 pm
7	9:04 pm	8	Aries	12:02 am
10	12:07 am	10	Taurus	6:12 am
12	6:12 am	12	Gemini	9:27 am
13	1:51 pm	14	Cancer	11:21 am
16	9:30 am	16	Leo	1:19 pm
18	12:03 pm	18	Virgo	4:15 pm
20	6:45 pm	20	Libra	8:35 pm
22	3:18 am	23	Scorpio	2:33 am

Moon Void-of-Course

Last Aspect		Moon Enters New Sign		
Date	Time	Date	Sign	Time
25	4:52 am	25	Sagittarius	10:33 am
26	7:57 pm	27	Capricorn	8:57 pm
30	2:34 am	30	Aquarius	9:26 am
		December		
2	7:51 pm	2	Pisces	10:23 pm
5	2:26 am	5	Aries	9:17 am
7	3:22 pm	7	Taurus	4:26 pm
9	1:00 pm	9	Gemini	7:50 pm
11	9:15 am	11	Cancer	8:48 pm
13	2:03 pm	13	Leo	9:09 pm
15	2:57 pm	15	Virgo	10:30 pm
17	7:41 pm	18	Libra	2:01 am
20	6:07 am	20	Scorpio	8:12 am
22	7:28 am	22	Sagittarius	4:57 pm
24	5:03 am	25	Capricorn	3:54 am
27	5:52 am	27	Aquarius	4:25 pm
29	6:47 pm	30	Pisces	5:27 am

Find Your Moon Sign

E very year we give tables for the position of the Moon during that
year, but it is more complicated to provide tables for the Moon's
position in any given year because of its continuous movement. However, the problem was solved by Grant Lewi in *Astrology for the Millions*
(available from Llewellyn Publications).

Grant Lewi's System

1. Find your birth year in the Natal Moon Tables (pages 65–74).

2. Run down the left-hand column and see if your date is there.

3. If your date is in the left-hand column, run over this line until you
   come to the column under your birth year. Here you will find a number.
   This is your base number. Write it down, and go directly to the direction
   under the heading "What to Do with Your Base Number" on page 63.

4. If your birth date is not in the left-hand column, get a pencil and paper.
   Your birth date falls between two numbers in the left-hand column.
   Look at the date closest after your birth date; run across this line to your
   birth year. Write down the number you find there, and label it "top
   number." Having done this, write directly beneath it on your piece of
   paper the number printed just above it in the table. Label this "bottom
   number." Subtract the bottom number from the top number. If the top
   number is smaller, add 360 and subtract. The result is your difference.

5. Go back to the left-hand column and find the date before your birth
   date. Determine the number of days between this date and your birth
   date. Write this down and label it "intervening days."

6. Note which group your difference (found at 4, above) falls in. If your
   difference was 80–87, your daily motion was 12 degrees. If your difference
   was 88–94, your daily motion was 13 degrees. If your difference was
   95–101, your daily motion was 14 degrees. If your difference is 102–
   106, your daily motion is 15 degrees. *Note: If you were born in a leap*

year and use the difference between February 26 and March 5, then the daily motion is slightly different. If you fall into this category and your difference is 94–99, your daily motion is 12 degrees. If your difference is 100–108, your daily motion is 13 degrees. If your difference is 109–115, your daily motion is 14 degrees. If your difference is 115–122, your daily motion is 15 degrees.

7. Write down the "daily motion" corresponding to your place in the proper table of difference above. Multiply daily motion by the number labeled "intervening days" (found at step 5).

8. Add the result of step 7 to your bottom number (under step 4). This is your base number. If it is more than 360, subtract 360 from it and call the result your base number.

What to Do with Your Base Number

Turn to the Table of Base Numbers on page 64 and locate your base number in it. At the top of the column you will find the sign your Moon was in. In the far left-hand column you will find the degree the Moon occupied at: 7 am of your birth date if you were born under Eastern Standard Time (EST); 6 am of your birth date if you were born under Central Standard Time (CST); 5 am of your birth date if you were born under Mountain Standard Time (MST); or 4 am of your birth date if you were born under Pacific Standard Time (PST).

If you don't know the hour of your birth, accept this as your Moon's sign and degree. If you do know the hour of your birth, get the exact degree as follows:

If you were born after 7 am Eastern Standard Time (6 am Central Standard Time, etc.), determine the number of hours after the time that you were born. Divide this by two, rounding up if necessary. Add this to your base number, and the result in the table will be the exact degree and sign of the Moon on the year, month, date, and hour of your birth.

If you were born before 7 am Eastern Standard Time (6 am Central Standard Time, etc.), determine the number of hours before the time that you were born. Divide this by two. Subtract this from your base number, and the result in the table will be the exact degree and sign of the Moon on the year, month, date, and hour of your birth.

Table of Base Numbers

	♈ (13)	♉ (14)	♊ (15)	♋ (16)	♌ (17)	♍ (18)	♎ (19)	♏ (20)	♐ (21)	♑ (22)	♒ (23)	♓ (24)
0°	0	30	60	90	120	150	180	210	240	270	300	330
1°	1	31	61	91	121	151	181	211	241	271	301	331
2°	2	32	62	92	122	152	182	212	242	272	302	332
3°	3	33	63	93	123	153	183	213	243	273	303	333
4°	4	34	64	94	124	154	184	214	244	274	304	334
5°	5	35	65	95	125	155	185	215	245	275	305	335
6°	6	36	66	96	126	156	186	216	246	276	306	336
7°	7	37	67	97	127	157	187	217	247	277	307	337
8°	8	38	68	98	128	158	188	218	248	278	308	338
9°	9	39	69	99	129	159	189	219	249	279	309	339
10°	10	40	70	100	130	160	190	220	250	280	310	340
11°	11	41	71	101	131	161	191	221	251	281	311	341
12°	12	42	72	102	132	162	192	222	252	282	312	342
13°	13	43	73	103	133	163	193	223	253	283	313	343
14°	14	44	74	104	134	164	194	224	254	284	314	344
15°	15	45	75	105	135	165	195	225	255	285	315	345
16°	16	46	76	106	136	166	196	226	256	286	316	346
17°	17	47	77	107	137	167	197	227	257	287	317	347
18°	18	48	78	108	138	168	198	228	258	288	318	248
19°	19	49	79	109	139	169	199	229	259	289	319	349
20°	20	50	80	110	140	170	200	230	260	290	320	350
21°	21	51	81	111	141	171	201	231	261	291	321	351
22°	22	52	82	112	142	172	202	232	262	292	322	352
23°	23	53	83	113	143	173	203	233	263	293	323	353
24°	24	54	84	114	144	174	204	234	264	294	324	354
25°	25	55	85	115	145	175	205	235	265	295	325	355
26°	26	56	86	116	146	176	206	236	266	296	326	356
27°	27	57	87	117	147	177	207	237	267	297	327	357
28°	28	58	88	118	148	178	208	238	268	298	328	358
29°	29	59	89	119	149	179	209	239	269	299	329	359

Month	Date	1901	1902	1903	1904	1905	1906	1907	1908	1909	1910
Jan.	1	55	188	308	76	227	358	119	246	39	168
Jan.	8	149	272	37	179	319	82	208	350	129	252
Jan.	15	234	2	141	270	43	174	311	81	213	346
Jan.	22	327	101	234	353	138	273	44	164	309	84
Jan.	29	66	196	317	84	238	6	128	255	50	175
Feb.	5	158	280	46	188	328	90	219	359	138	259
Feb.	12	241	12	149	279	51	184	319	90	221	356
Feb.	19	335	111	242	2	146	283	52	173	317	94
Feb.	26	76	204	326	92	248	13	136	264	60	184
Mar.	5	166	288	57	211	336	98	229	21	147	267
Mar.	12	249	22	157	300	60	194	328	110	230	5
Mar.	19	344	121	250	24	154	293	60	195	325	105
Mar.	26	86	212	334	116	258	22	144	288	69	192
Apr.	2	175	296	68	219	345	106	240	29	155	276
Apr.	9	258	31	167	309	69	202	338	118	240	13
Apr.	16	352	132	258	33	163	304	68	204	334	115
Apr.	23	96	220	342	127	267	31	152	299	77	201
Apr.	30	184	304	78	227	354	114	250	38	164	285
May	7	267	40	177	317	78	210	348	126	249	21
May	14	1	142	266	42	172	313	76	212	344	124
May	21	104	229	350	138	275	40	160	310	85	210
May	28	193	313	87	236	2	123	259	47	172	294
Jun.	4	277	48	187	324	88	219	358	134	258	30
Jun.	11	11	151	275	50	182	322	85	220	355	132
Jun.	18	112	238	359	149	283	48	169	320	93	218
Jun.	25	201	322	96	245	11	133	267	57	180	304
Jul.	2	286	57	197	333	97	228	8	142	267	40
Jul.	9	21	160	283	58	193	330	94	228	6	140
Jul.	16	121	247	7	159	291	57	178	330	102	226
Jul.	23	209	332	105	255	18	143	276	66	188	314
Jul.	30	295	66	206	341	105	239	17	151	275	51
Aug.	6	32	168	292	66	204	338	103	237	17	148
Aug.	13	130	255	17	168	301	65	188	339	111	234
Aug.	20	217	341	113	265	27	152	285	76	197	323
Aug.	27	303	77	215	350	113	250	25	160	283	62
Sep.	3	43	176	301	75	215	346	111	246	27	157
Sep.	10	139	263	27	176	310	73	198	347	121	242
Sep.	17	225	350	123	274	35	161	294	85	205	331
Sep.	24	311	88	223	358	122	261	33	169	292	73
Oct.	1	53	185	309	85	224	355	119	256	35	166
Oct.	8	149	271	36	185	320	81	207	356	93	250
Oct.	15	233	359	133	283	44	169	305	93	214	339
Oct.	22	319	99	231	7	130	271	42	177	301	83
Oct.	29	62	194	317	95	233	5	127	266	44	176
Nov.	5	158	279	45	193	329	89	216	5	139	259
Nov.	12	242	6	144	291	53	177	316	101	223	347
Nov.	19	328	109	239	15	140	280	50	185	311	91
Nov.	26	70	203	325	105	241	14	135	276	52	185
Dec.	3	168	288	54	203	338	98	224	15	148	268
Dec.	10	251	14	155	299	61	185	327	109	231	356
Dec.	17	338	118	248	23	150	289	59	193	322	99
Dec.	24	78	213	333	115	249	23	143	286	61	194
Dec.	31	176	296	61	213	346	107	232	26	155	277

2000 Moon Sign Book

Month	Date	1911	1912	1913	1914	1915	1916	1917	1918	1919	1920
Jan.	1	289	57	211	337	100	228	23	147	270	39
Jan.	8	20	162	299	61	192	332	110	231	5	143
Jan.	15	122	251	23	158	293	61	193	329	103	231
Jan.	22	214	335	120	256	23	145	290	68	193	316
Jan.	29	298	66	221	345	108	237	32	155	278	49
Feb.	5	31	170	308	69	203	340	118	239	16	150
Feb.	12	130	260	32	167	302	70	203	338	113	239
Feb.	19	222	344	128	266	31	154	298	78	201	325
Feb.	26	306	75	231	353	116	248	41	164	286	60
Mar.	5	42	192	317	77	214	2	127	248	26	172
Mar.	12	140	280	41	176	311	89	212	346	123	259
Mar.	19	230	5	136	276	39	176	308	87	209	346
Mar.	26	314	100	239	2	124	273	49	173	294	85
Apr.	2	52	200	326	86	223	10	135	257	35	181
Apr.	9	150	288	51	184	321	97	222	355	133	267
Apr.	16	238	14	146	286	48	184	318	96	218	355
Apr.	23	322	111	247	11	132	284	57	181	303	96
Apr.	30	61	208	334	96	232	19	143	267	43	190
May	7	160	296	60	192	331	105	231	4	142	275
May	14	246	22	156	294	56	192	329	104	227	3
May	21	331	122	255	20	141	294	66	190	312	105
May	28	69	218	342	106	240	29	151	277	51	200
Jun.	4	170	304	69	202	341	114	240	14	151	284
Jun.	11	255	30	167	302	65	200	340	112	235	11
Jun.	18	340	132	264	28	151	304	74	198	322	114
Jun.	25	78	228	350	115	249	39	159	286	60	209
Jul.	2	179	312	78	212	349	122	248	25	159	293
Jul.	9	264	39	178	310	74	209	350	120	244	20
Jul.	16	349	141	273	36	161	312	84	206	332	123
Jul.	23	87	237	358	125	258	48	168	295	70	218
Jul.	30	187	321	86	223	357	131	256	36	167	302
Aug.	6	272	48	188	319	82	219	360	129	252	31
Aug.	13	359	150	282	44	171	320	93	214	342	131
Aug.	20	96	246	6	133	268	57	177	303	81	226
Aug.	27	195	330	94	234	5	140	265	46	175	310
Sep.	3	281	57	198	328	90	229	9	138	260	41
Sep.	10	9	158	292	52	180	329	102	222	351	140
Sep.	17	107	255	15	141	279	65	186	312	91	234
Sep.	24	203	339	103	244	13	149	274	56	184	319
Oct.	1	288	68	206	337	98	240	17	148	268	52
Oct.	8	18	167	301	61	189	338	111	231	360	150
Oct.	15	118	263	24	149	290	73.	195	320	102	242
Oct.	22	212	347	113	254	22	157	284	65	193	326
Oct.	29	296	78	214	346	106	250	25	157	276	61
Nov.	5	26	177	309	70	197	348	119	240	7	161
Nov.	12	129	271	33	158	300	81	203	329	112	250
Nov.	19	221	355	123	262	31	164	295	73	202	334
Nov.	26	305	88	223	355	115	259	34	165	285	70
Dec.	3	34	187	317	79	205	359	127	249	16	171
Dec.	10	138	279	41	168	310	89	211	340	120	259
Dec.	17	230	3	134	270	40	172	305	81	211	343
Dec.	24	313	97	232	3	124	267	44	173	294	78
Dec.	31	42	198	325	87	214	9	135	257	25	181

Month	Date	1921	1922	1923	1924	1925	1926	1927	1928	1929	1930
Jan.	1	194	317	80	211	5	127	250	23	176	297
Jan.	8	280	41	177	313	90	211	349	123	260	22
Jan.	15	4	141	275	41	175	312	86	211	346	123
Jan.	22	101	239	3	127	272	51	172	297	83	222
Jan.	29	203	325	88	222	13	135	258	34	184	306
Feb.	5	289	49	188	321	99	220	359	131	269	31
Feb.	12	14	149	284	49	185	320	95	219	356	131
Feb.	19	110	249	11	135	281	60	181	305	93	230
Feb.	26	211	334	96	233	21	144	266	45	191	314
Mar.	5	297	58	197	343	107	230	8	153	276	41
Mar.	12	23	157	294	69	194	328	105	238	6	140
Mar.	19	119	258	19	157	292	68	190	327	104	238
Mar.	26	219	343	104	258	29	153	275	70	200	323
Apr.	2	305	68	205	352	115	240	16	163	284	51
Apr.	9	33	166	304	77	204	337	114	247	14	149
Apr.	16	130	266	28	164	303	76	198	335	115	246
Apr.	23	227	351	114	268	38	161	285	79	208	331
Apr.	30	313	78	214	1	123	250	25	172	292	61
May	7	42	176	313	85	212	348	123	256	23	160
May	14	141	274	37	173	314	84	207	344	125	254
May	21	236	359	123	277	47	169	295	88	217	339
May	28	321	88	222	11	131	259	34	181	301	70
Jun.	4	50	186	321	94	220	358	131	264	31	171
Jun.	11	152	282	45	182	324	93	215	354	135	263
Jun.	18	245	7	134	285	56	177	305	96	226	347
Jun.	25	330	97	232	20	139	268	44	190	310	78
Jul.	2	58	197	329	103	229	9	139	273	40	181
Jul.	9	162	291	54	192	333	101	223	4	144	272
Jul.	16	254	15	144	294	65	185	315	104	236	355
Jul.	23	338	106	242	28	148	276	54	198	319	87
Jul.	30	67	208	337	112	238	20	147	282	49	191
Aug.	6	171	300	62	202	341	110	231	15	152	281
Aug.	13	264	24	153	302	74	194	324	114	244	4
Aug.	20	347	114	253	36	157	285	65	206	328	95
Aug.	27	76	218	346	120	248	29	156	290	59	200
Sep.	3	179	309	70	213	350	119	239	25	161	290
Sep.	10	273	32	162	312	83	203	332	124	252	13
Sep.	17	356	122	264	44	166	293	75	214	337	105
Sep.	24	86	227	354	128	258	38	165	298	70	208
Oct.	1	187	318	78	223	358	128	248	35	169	298
Oct.	8	281	41	170	322	91	212	340	134	260	23
Oct.	15	5	132	274	52	175	303	85	222	345	115
Oct.	22	97	235	3	136	269	46	174	306	81	216
Oct.	29	196	327	87	232	7	137	257	44	179	307
Nov.	5	289	50	178	332	99	221	349	144	268	31
Nov.	12	13	142	283	61	183	313	93	231	353	126
Nov.	19	107	243	12	144	279	54	183	315	91	225
Nov.	26	206	335	96	241	17	145	266	52	189	314
Dec.	3	297	59	187	343	107	230	359	154	276	39
Dec.	10	21	152	291	70	191	324	101	240	1	137
Dec.	17	117	252	21	153	289	63	191	324	99	234
Dec.	24	216	343	105	249	28	152	275	60	199	322
Dec.	31	305	67	197	352	115	237	9	162	285	47

Month	Date	1931	1932	1933	1934	1935	1936	1937	1938	1939	1940
Jan.	1	60	196	346	107	231	8	156	277	41	181
Jan.	8	162	294	70	193	333	104	240	4	144	275
Jan.	15	257	20	158	294	68	190	329	104	239	360
Jan.	22	342	108	255	32	152	278	67	202	323	88
Jan.	29	68	207	353	116	239	19	163	286	49	191
Feb.	5	171	302	78	203	342	113	248	14	153	284
Feb.	12	267	28	168	302	78	198	339	113	248	8
Feb.	19	351	116	266	40	161	286	78	210	332	96
Feb.	26	77	217	1	124	248	29	171	294	59	200
Mar.	5	179	324	86	213	350	135	256	25	161	306
Mar.	12	276	48	176	311	86	218	347	123	256	29
Mar.	19	360	137	277	48	170	308	89	218	340	119
Mar.	26	86	241	10	132	258	52	180	302	69	223
Apr.	2	187	334	94	223	358	144	264	34	169	315
Apr.	9	285	57	185	321	95	227	355	133	264	38
Apr.	16	9	146	287	56	178	317	99	226	349	128
Apr.	23	96	250	18	140	268	61	189	310	80	231
Apr.	30	196	343	102	232	7	153	273	43	179	323
May	7	293	66	193	332	103	237	4	144	272	47
May	14	17	155	297	64	187	327	108	235	357	139
May	21	107	258	28	148	278	69	198	318	90	239
May	28	205	351	111	241	17	161	282	51	189	331
Jun.	4	301	75	201	343	111	245	13	154	280	55
Jun.	11	25	165	306	73	195	337	117	244	5	150
Jun.	18	117	267	37	157	288	78	207	327	99	248
Jun.	25	215	360	120	249	28	169	291	60	200	339
Jul.	2	309	84	211	353	119	254	23	164	289	64
Jul.	9	33	176	315	82	203	348	125	253	13	160
Jul.	16	126	276	46	165	297	87	216	336	108	258
Jul.	23	226	8	130	258	38	177	300	69	210	347
Jul.	30	317	92	221	2	128	262	33	173	298	72
Aug.	6	41	187	323	91	211	359	133	261	21	170
Aug.	13	135	285	54	175	305	97	224	346	116	268
Aug.	20	237	16	138	267	49	185	308	78	220	355
Aug.	27	326	100	232	10	136	270	44	181	307	80
Sep.	3	49	197	331	100	220	8	142	270	31	179
Sep.	10	143	295	62	184	314	107	232	355	125	278
Sep.	17	247	24	147	277	58	194	317	89	228	4
Sep.	24	335	108	243	18	145	278	55	189	316	88
Oct.	1	58	206	341	108	229	17	152	278	40	188
Oct.	8	151	306	70	193	322	117	240	4	134	288
Oct.	15	256	32	155	287	66	203	324	100	236	13
Oct.	22	344	116	253	27	154	287	64	198	324	98
Oct.	29	68	214	350	116	239	25	162	286	49	196
Nov.	5	161	316	78	201	332	126	248	12	145	297
Nov.	12	264	41	162	298	74	212	333	111	244	22
Nov.	19	353	125	262	36	162	296	73	207	332	108
Nov.	26	77	222	0	124	248	33	172	294	58	205
Dec.	3	171	325	87	209	343	135	257	19	156	305
Dec.	10	272	50	171	309	82	220	341	120	253	30
Dec.	17	1	135	271	45	170	306	81	217	340	118
Dec.	24	86	231	10	132	256	43	181	302	66	214
Dec.	31	182	333	95	217	354	142	265	27	167	313

Month	Date	1941	1942	1943	1944	1945	1946	1947	1948	1949	1950
Jan.	1	325	88	211	353	135	258	22	165	305	68
Jan.	8	50	176	315	85	219	348	126	256	29	160
Jan.	15	141	276	50	169	312	87	220	340	123	258
Jan.	22	239	12	133	258	52	182	303	69	224	352
Jan.	29	333	96	221	2	143	266	32	174	314	75
Feb.	5	57	186	323	95	227	358	134	265	37	170
Feb.	12	150	285	58	178	320	96	228	349	131	268
Feb.	19	250	20	142	267	62	190	312	78	234	359
Feb.	26	342	104	231	11	152	274	43	182	323	83
Mar.	5	65	196	331	116	236	8	142	286	46	179
Mar.	12	158	295	66	199	328	107	236	10	139	279
Mar.	19	261	28	150	290	72	198	320	102	243	8
Mar.	26	351	112	242	34	161	281	53	204	332	91
Apr.	2	74	205	340	125	244	16	152	294	55	187
Apr.	9	166	306	74	208	337	117	244	19	148	289
Apr.	16	270	36	158	300	81	206	328	112	252	17
Apr.	23	360	120	252	42	170	290	63	212	340	100
Apr.	30	83	214	350	133	254	25	162	302	64	195
May	7	174	316	82	217	346	127	252	27	158	299
May	14	279	45	166	311	90	215	336	123	260	26
May	21	9	128	261	50	179	299	72	221	349	110
May	28	92	222	1	141	263	33	173	310	73	204
Jun.	4	184	326	91	226	356	137	261	36	168	307
Jun.	11	287	54	174	322	98	224	344	134	268	34
Jun.	18	17	137	270	60	187	308	81	231	357	119
Jun.	25	102	231	11	149	272	42	183	318	82	213
Jul.	2	194	335	99	234	7	145	269	44	179	316
Jul.	9	296	63	183	332	106	233	353	144	277	43
Jul.	16	25	147	279	70	195	318	89	241	5	129
Jul.	23	110	240	21	157	280	52	192	327	91	224
Jul.	30	205	343	108	242	18	153	278	52	190	324
Aug.	6	304	71	192	341	115	241	3	153	286	51
Aug.	13	33	156	287	80	203	327	98	251	13	138
Aug.	20	119	250	30	165	289	63	201	336	99	235
Aug.	27	216	351	117	250	28	162	287	61	200	332
Sep.	3	314	80	201	350	125	249	13	161	296	59
Sep.	10	41	165	296	90	211	336	108	260	21	146
Sep.	17	127	261	39	174	297	74	209	345	107	246
Sep.	24	226	359	126	259	38	170	295	70	209	341
Oct.	1	323	88	211	358	135	257	22	170	306	67
Oct.	8	49	174	306	99	220	344	118	269	30	154
Oct.	15	135	272	47	183	305	84	217	353	116	256
Oct.	22	236	8	134	269	47	180	303	80	217	351
Oct.	29	333	95	220	7	144	265	31	179	315	75
Nov.	5	58	181	317	107	229	352	129	277	39	162
Nov.	12	143	283	55	192	314	94	225	1	125	265
Nov.	19	244	18	141	279	55	189	311	90	225	0
Nov.	26	343	104	229	16	153	274	39	189	323	84
Dec.	3	67	189	328	115	237	360	140	284	47	171
Dec.	10	153	292	64	200	324	103	234	9	136	274
Dec.	17	252	28	149	289	63	199	319	100	234	9
Dec.	24	351	112	237	27	161	282	47	199	331	93
Dec.	31	76	198	338	123	246	9	150	293	55	180

Month	Date	1951	1952	1953	1954	1955	1956	1957	1958	1959	1960
Jan.	1	194	336	115	238	6	147	285	47	178	317
Jan.	8	297	67	199	331	107	237	9	143	278	47
Jan.	15	30	150	294	70	200	320	104	242	9	131
Jan.	22	114	240	35	161	284	51	207	331	94	223
Jan.	29	204	344	124	245	17	155	294	55	189	325
Feb.	5	305	76	207	341	116	246	18	152	287	56
Feb.	12	38	159	302	80	208	330	112	252	17	140
Feb.	19	122	249	45	169	292	61	216	340	102	233
Feb.	26	215	352	133	253	27	163	303	63	199	333
Mar.	5	314	96	216	350	125	266	27	161	297	75
Mar.	12	46	180	310	91	216	351	121	262	25	161
Mar.	19	130	274	54	178	300	86	224	349	110	259
Mar.	26	225	14	142	262	37	185	312	72	208	356
Apr.	2	324	104	226	358	135	274	37	169	307	83
Apr.	9	54	189	319	100	224	360	131	271	34	170
Apr.	16	138	285	62	187	308	97	232	357	118	269
Apr.	23	235	23	150	271	46	194	320	82	217	5
Apr.	30	334	112	235	6	146	282	48	177	317	91
May	7	62	197	330	109	232	8	142	279	42	177
May	14	146	296	70	196	316	107	240	6	127	279
May	21	243	32	158	280	54	204	328	91	225	15
May	28	344	120	244	15	155	290	55	187	326	100
Jun.	4	71	205	341	117	241	16	153	288	51	186
Jun.	11	155	306	79	204	325	117	249	14	137	288
Jun.	18	252	42	166	290	63	214	336	101	234	25
Jun.	25	354	128	253	26	164	298	63	198	335	109
Jul.	2	80	214	351	125	250	24	164	296	60	195
Jul.	9	164	315	88	212	335	126	259	22	147	297
Jul.	16	260	52	174	299	72	223	344	110	243	34
Jul.	23	3	137	261	37	173	307	71	209	343	118
Jul.	30	89	222	2	134	258	33	174	304	68	205
Aug.	6	174	324	97	220	345	134	268	30	156	305
Aug.	13	270	62	182	308	82	232	353	118	254	42
Aug.	20	11	146	269	48	181	316	79	220	351	126
Aug.	27	97	232	11	143	267	43	183	314	76	215
Sep.	3	184	332	107	228	355	143	278	38	166	314
Sep.	10	280	71	191	316	92	241	2	127	265	50
Sep.	17	19	155	278	58	189	325	88	230	359	135
Sep.	24	105	242	20	152	274	54	191	323	84	225
Oct.	1	193	341	116	237	4	152	287	47	174	324
Oct.	8	291	79	200	324	103	249	11	135	276	58
Oct.	15	27	163	287	68	198	333	98	239	8	143
Oct.	22	113	252	28	162	282	64	199	332	92	235
Oct.	29	201	350	125	245	12	162	295	56	182	334
Nov.	5	302	87	209	333	114	256	19	144	286	66
Nov.	12	36	171	297	76	207	341	109	247	17	150
Nov.	19	121	262	37	171	291	73	208	341	101	244
Nov.	26	209	0	133	254	20	173	303	65	190	345
Dec.	3	312	95	217	342	124	265	27	154	295	75
Dec.	10	45	179	307	84	216	348	119	255	27	158
Dec.	17	129	271	46	180	299	82	218	350	110	252
Dec.	24	217	11	141	263	28	184	311	73	199	355
Dec.	31	321	103	225	352	132	273	35	164	303	84

Month	Date	1961	1962	1963	1964	1965	1966	1967	1968	1969	1970
Jan.	1	96	217	350	128	266	27	163	298	76	197
Jan.	8	179	315	89	217	350	126	260	27	161	297
Jan.	15	275	54	179	302	86	225	349	112	257	36
Jan.	22	18	141	264	35	189	311	74	207	359	122
Jan.	29	105	225	1	136	275	35	173	306	85	206
Feb.	5	188	323	99	225	360	134	270	35	171	305
Feb.	12	284	64	187	310	95	235	357	121	267	45
Feb.	19	26	150	272	46	197	320	81	218	7	130
Feb.	26	113	234	11	144	283	45	182	315	93	216
Mar.	5	198	331	109	245	9	142	280	54	180	313
Mar.	12	293	73	195	332	105	244	5	142	277	54
Mar.	19	34	159	280	71	205	329	90	243	15	139
Mar.	26	122	243	19	167	291	54	190	338	101	226
Apr.	2	208	340	119	253	18	151	290	63	189	323
Apr.	9	303	82	204	340	116	252	14	150	288	62
Apr.	16	42	167	288	81	213	337	99	253	23	147
Apr.	23	130	253	28	176	299	64	198	347	109	235
Apr.	30	216	349	128	261	27	161	298	71	197	333
May	7	314	90	213	348	127	260	23	158	299	70
May	14	51	176	298	91	222	345	109	262	32	155
May	21	137	263	36	186	307	74	207	357	117	245
May	28	225	359	137	270	35	172	307	80	205	344
Jun.	4	325	98	222	357	137	268	31	168	309	78
Jun.	11	60	184	308	99	231	353	119	270	42	163
Jun.	18	146	272	45	195	315	82	217	6	126	253
Jun.	25	233	10	145	279	43	183	315	89	214	355
Jul.	2	336	106	230	6	147	276	40	178	318	87
Jul.	9	70	191	318	108	241	1	129	279	51	171
Jul.	16	154	281	56	204	324	91	227	14	135	261
Jul.	23	241	21	153	288	52	193	323	98	223	5
Jul.	30	345	115	238	16	156	286	47	188	327	97
Aug.	6	79	200	327	116	250	10	138	288	60	180
Aug.	13	163	289	66	212	333	99	238	22	144	270
Aug.	20	250	32	161	296	61	203	331	106	233	14
Aug.	27	353	124	246	27	164	295	55	199	335	106
Sep.	3	88	208	336	126	259	19	147	297	68	189
Sep.	10	172	297	77	220	342	108	249	30	152	279
Sep.	17	260	41	170	304	72	212	340	114	244	23
Sep.	24	1	134	254	37	172	304	64	208	344	115
Oct.	1	97	217	344	136	267	28	155	308	76	198
Oct.	8	180	306	88	228	351	117	259	38	161	289
Oct.	15	270	50	179	312	82	220	350	122	254	31
Oct.	22	10	143	262	47	182	313	73	217	353	123
Oct.	29	105	226	352	146	275	37	163	318	84	207
Nov.	5	189	315	97	237	359	127	268	47	168	299
Nov.	12	281	58	188	320	93	228	359	130	264	39
Nov.	19	19	151	271	55	191	321	82	225	3	131
Nov.	26	113	235	1	157	282	45	172	328	92	215
Dec.	3	197	326	105	245	7	138	276	55	176	310
Dec.	10	291	66	197	328	102	237	7	139	273	48
Dec.	17	30	159	280	63	202	329	91	234	13	139
Dec.	24	121	243	11	167	291	53	183	337	101	223
Dec.	31	204	336	113	254	14	149	284	64	184	320

Month	Date	1971	1972	1973	1974	1975	1976	1977	1978	1979	1980
Jan.	1	335	109	246	8	147	279	56	179	318	90
Jan.	8	71	197	332	108	243	6	144	278	54	176
Jan.	15	158	283	69	207	328	93	240	18	139	263
Jan.	22	244	20	169	292	54	192	339	102	224	4
Jan.	29	344	117	255	17	156	288	64	188	327	99
Feb.	5	81	204	342	116	253	14	153	287	63	184
Feb.	12	167	291	79	216	337	101	251	26	147	271
Feb.	19	252	31	177	300	62	203	347	110	233	14
Feb.	26	353	126	263	27	164	297	72	199	334	109
Mar.	5	91	224	351	124	262	34	162	296	72	204
Mar.	12	176	312	90	224	346	122	262	34	156	203
Mar.	19	261	55	185	309	72	226	356	118	243	37
Mar.	26	1	149	270	37	172	320	80	208	343	130
Apr.	2	100	233	360	134	270	43	170	307	80	213
Apr.	9	184	320	101	232	355	131	273	42	164	302
Apr.	16	271	64	194	317	82	235	5	126	254	46
Apr.	23	9	158	278	47	181	329	88	217	352	139
Apr.	30	109	242	8	145	278	52	178	318	88	222
May	7	193	329	111	240	3	141	282	50	173	312
May	14	281	73	203	324	92	243	14	134	264	54
May	21	19	167	287	55	191	337	97	226	3	147
May	28	117	251	16	156	286	61	187	328	96	231
Jun.	4	201	339	120	249	11	151	291	59	180	323
Jun.	11	291	81	213	333	102	252	23	143	273	63
Jun.	18	29	176	296	64	201	346	106	234	13	155
Jun.	25	125	260	25	167	295	69	196	338	105	239
Jul.	2	209	349	129	258	19	162	299	68	188	334
Jul.	9	300	90	222	341	111	261	32	152	282	72
Jul.	16	40	184	305	72	212	354	115	243	24	163
Jul.	23	133	268	35	176	303	78	206	347	114	248
Jul.	30	217	0	137	267	27	172	308	77	197	344
Aug.	6	309	99	230	350	120	271	40	161	290	83
Aug.	13	51	192	314	81	223	2	124	252	34	171
Aug.	20	142	276	45	185	312	86	217	356	123	256
Aug.	27	225	10	146	276	36	182	317.	86	206	353
Sep.	3	317	109	238	360	128	281	48	170	299	93
Sep.	10	61	200	322	90	232	10	132	262	43	180
Sep.	17	151	284	56	193	321	94	228	4	132	264
Sep.	24	234	20	155	284	45	191	326	94	215	2
Oct.	1	325	120	246	9	136	291	56	179	308	103
Oct.	8	70	208	330	101	241	19	140	273	51	189
Oct.	15	160	292	66	202	330	102	238	12	140	273
Oct.	22	243	28	165	292	54	199	336	102	225	10
Oct.	29	334	130	254	17	146	301	64	187	318	112
Nov.	5	79	217	338	112	249	27	148	284	59	197
Nov.	12	169	300	76	210	339	111	247	21	148	282
Nov.	19	253	36	175	300	63	207	347	110	234	18
Nov.	26	344	139	262	25	156	310	73	195	329	120
Dec.	3	87	226	346	122	257	36	157	294	67	206
Dec.	10	177	310	84	220	347	121	255	31	156	292
Dec.	17	261	45	185	308	72	216	356	118	242	28
Dec.	24	355	148	271	33	167	318	81	203	340	128
Dec.	31	95	235	355	132	265	44	166	303	76	214

Month	Date	1981	1982	1983	1984	1985	1986	1987	1988	1989	1990
Jan.	1	226	350	129	260	36	162	300	71	205	333
Jan.	8	315	89	225	346	126	260	36	156	297	72
Jan.	15	53	188	309	73	225	358	119	243	37	168
Jan.	22	149	272	35	176	319	82	206	348	129	252
Jan.	29	234	0	137	270	43	172	308	81	213	343
Feb.	5	324	98	234	354	135	270	44	164	306	82
Feb.	12	64	196	317	81	236	6	128	252	48	175
Feb.	19	157	280	45	185	328	90	217	356	138	260
Feb.	26	242	10	145	279	51	182	316	90	222	353
Mar.	5	332	108	242	15	143	280	52	185	313	93
Mar.	12	74	204	326	104	246	14	136	275	57	184
Mar.	19	166	288	55	208	337	97	227	19	147	268
Mar.	26	250	20	154	300	60	191	326	111	230	1
Apr.	2	340	119	250	24	151	291	60	194	322	103
Apr.	9	84	212	334	114	255	22	144	286	66	192
Apr.	16	175	296	66	216	346	106	237	27	156	276
Apr.	23	259	28	164	309	69	199	336	119	240	9
Apr.	30	349	130	258	33	160	302	68	203	331	113
May	7	93	221	342	124	264	31	152	297	75	201
May	14	184	304	75	225	355	114	246	36	165	285
May	21	268	36	175	317	78	207	347	127	249	18
May	28	358	140	266	41	170	311	76	211	341	122
Jun.	4	102	230	350	135	272	40	160	307	83	210
Jun.	11	193	313	84	234	3	123	255	45	173	294
Jun.	18	277	45	185	325	87	216	357	135	258	27
Jun.	25	8	149	275	49	180	320	85	219	352	130
Jul.	2	110	239	359	146	281	49	169	317	92	219
Jul.	9	201	322	93	244	11	133	263	55	181	304
Jul.	16	286	54	196	333	96	225	7	143	266	37
Jul.	23	19	158	284	57	191	328	94	227	3	138
Jul.	30	119	248	7	155	290	57	178	327	101	227
Aug.	6	210	331	101	254	19	142	272	66	189	313
Aug.	13	294	64	205	341	104	236	16	152	274	48
Aug.	20	30	166	293	66	202	337	103	236	13	147
Aug.	27	128	256	17	164	299	65	187	335	111	235
Sep.	3	218	340	110	264	27	151	281	75	197	321
Sep.	10	302	75	214	350	112	247	24	160	282	59
Sep.	17	40	174	302	74	212	345	112	245	23	156
Sep.	24	138	264	26	172	309	73	197	343	121	243
Oct.	1	226	349	119	274	36	159	292	84	206	329
Oct.	8	310	86	222	359	120	258	32	169	291	70
Oct.	15	50	183	310	84	220	354	120	255	31	165
Oct.	22	148	272	35	181	319	81	206	352	130	251
Oct.	29	234	357	130	282	44	167	303	92	214	337
Nov.	5	318	96	230	8	129	268	40	178	300	79
Nov.	12	58	193	318	93	229	4	128	265	39	175
Nov.	19	158	280	44	190	329	90	214	2	139	260
Nov.	26	243	5	141	290	53	175	314	100	223	345
Dec.	3	327	106	238	16	139	277	49	185	310	88
Dec.	10	66	203	326	103	237	14	136	274	48	185
Dec.	17	167	288	52	200	337	98	222	12	147	269
Dec.	24	252	13	152	298	62	184	324	108	232	355
Dec.	31	337	114	248	24	149	285	59	193	320	96

Month	Date	1991	1992	1993	1994	1995	1996	1997	1998	1999	2000
Jan.	1	111	242	15	145	281	53	185	317	92	223
Jan.	8	206	326	108	244	16	136	279	56	186	307
Jan.	15	289	54	210	337	99	225	21	147	270	37
Jan.	22	18	158	299	61	190	329	110	231	2	140
Jan.	29	119	252	23	155	290	62	193	326	101	232
Feb.	5	214	335	116	254	24	145	287	66	193	315
Feb.	12	298	63	220	345	108	235	31	155	278	47
Feb.	19	29	166	308	69	201	337	119	239	12	148
Feb.	26	128	260	32	164	299	70	202	335	111	240
Mar.	5	222	356	124	265	32	166	295	76	201	337
Mar.	12	306	87	229	354	116	259	39	164	285	72
Mar.	19	39	189	317	77	211	360	128	248	22	170
Mar.	26	138	280	41	172	310	90	212	343	121	260
Apr.	2	230	5	133	275	40	175	305	86	210	345
Apr.	9	314	98	237	3	123	270	47	173	294	83
Apr.	16	49	198	326	86	220	9	136	257	31	180
Apr.	23	148	288	50	180	320	98	221	351	132	268
Apr.	30	238	13	143	284	48	183	315	95	218	353
May	7	322	109	245	12	132	281	55	182	302	93
May	14	57	207	335	95	228	18	144	267	39	190
May	21	158	296	59	189	330	106	230	1	141	276
May	28	247	21	154	292	57	191	326	103	227	1
Jun.	4	330	119	253	21	141	291	64	190	311	102
Jun.	11	66	217	343	105	236	28	152	276	48	199
Jun.	18	168	304	68	199	340	114	238	11	150	285
Jun.	25	256	29	165	300	66	199	337	111	236	10
Jul.	2	339	129	262	29	150	300	73	198	321	111
Jul.	9	74	227	351	114	245	38	160	285	57	209
Jul.	16	177	313	76	210	348	123	246	22	158	293
Jul.	23	265	38	175	309	75	208	347	120	245	19
Jul.	30	349	137	272	37	160	308	83	206	331	119
Aug.	6	83	237	359	123	255	48	169	293	67	218
Aug.	13	186	322	84	221	356	132	254	33	166	302
Aug.	20	273	47	185	318	83	218	356	129	253	29
Aug.	27	358	146	282	45	169	317	93	214	340	128
Sep.	3	93	246	7	131	265	56	177	301	78	226
Sep.	10	194	331	92	231	4	141	263	43	174	311
Sep.	17	281	56	194	327	91	228	5	138	261	39
Sep.	24	8	154	292	53	178	326	102	223	349	137
Oct.	1	104	254	16	139	276	64	186	310	89	234
Oct.	8	202	339	101	241	13	149	273	53	183	319
Oct.	15	289	66	202	337	99	238	13	148	269	49
Oct.	22	16	164	301	61	187	336	111	231	357	148
Oct.	29	115	262	25	148	287	72	195	318	100	242
Nov.	5	211	347	111	250	22	157	283	61	193	326
Nov.	12	297	76	211	346	107	247	22	157	277	58
Nov.	19	29	24	174	309	70	194	346	119	240	5	159
Nov.	26	126	270	33	156	297	80	203	328	109	251
Dec.	3	220	355	121	258	31	165	293	69	202	334
Dec.	10	305	85	220	355	115	256	31	165	286	67
Dec.	17	32	185	317	79	203	357	127	249	13	169
Dec.	24	135	278	41	166	306	89	211	338	117	260
Dec.	31	230	3	131	266	41	173	303	78	211	343

HOME,
HEALTH, &
BEAUTY

Home, Health, & Beauty

How to Choose the Best Dates

Automobile Purchase

The Moon is helpful when in favorable aspect to Mercury and Uranus and in the signs corresponding to travel (Gemini or Sagittarius) or to reliable purchases (Aquarius, Taurus, Leo, and Scorpio).

Automobile Repair

The Moon should be in favorable aspect to Uranus and in the signs of Taurus, Leo, Aquarius, or Virgo. The first and second quarters are best. Avoid any unfavorable aspects between the Moon and Mars, Saturn, Uranus, Neptune, or Pluto.

Baking

Baking should be done when the Moon is in Cancer. Bakers who have experimented say that dough rises higher and bread is lighter during the increase of the Moon (first or second quarter). If it is not possible to bake under the sign of Cancer, try Aries, Libra, or Capricorn.

Beauty Care

For beauty treatments, skin care, and massage, the Moon should be in Taurus, Cancer, Leo, Libra, or Aquarius, and sextile, trine, or conjunct (X, T, C) Venus or Jupiter.

Plastic surgery should be done in the increase of the Moon, when the Moon is not square or opposite (Q or O) Mars. Nor should the Moon be in the sign ruling the area to be operated on. Avoid days when the Moon is square or opposite Saturn or the Sun.

Fingernails should be cut when the Moon is not in any aspect with Mercury or Jupiter. Saturn and Mars must not be marked Q or O because this makes the nails grow slowly or thin and weak. The Moon should be in Aries, Taurus, Cancer, or Leo. For toenails, the Moon should not be in Gemini or Pisces. Corns are best cut in the third or fourth quarter.

Brewing

It is best to brew during the Full Moon and the fourth quarter. Plan to have the Moon in Cancer, Scorpio, or Pisces.

Building

Turning the first sod for the foundation of a home or laying the cornerstone for a building marks the beginning of the building. Excavate, lay foundations, and pour cement when the Moon is Full and in Taurus, Leo, or Aquarius. Saturn should be aspected, but not Mars.

Canning

Can fruits and vegetables when the Moon is in either the third or fourth quarter, and when it is in Cancer or Pisces. For preserves and jellies, use the same quarters but see that the Moon is in Cancer, Pisces, or Taurus.

Cement and Concrete

Pour cement and concrete during the Full Moon in the fixed signs of Taurus, Leo, or Aquarius.

Dental Work

Pick a day that is marked favorable for your Sun sign. Mars should be marked X, T, or C and Saturn, Uranus, and Jupiter should not be marked Q or O. Teeth are best removed during the increase of the Moon in the first or second quarter in Gemini, Virgo, Sagittarius, Capricorn, or Pisces. Avoid the Full Moon! The day should be favorable for your lunar cycle, and Mars and Saturn should be marked C, T, or X. Fillings should be done in the third or fourth quarters in the signs of Taurus, Leo, Scorpio, or Aquarius. The same applies for plates.

Dieting

Weight gain occurs more readily when the Moon is in a water sign (Cancer, Scorpio, Pisces). Experience has shown that weight may be lost if a diet is started when the Moon is decreasing in light (third or fourth quarter) and when it is in Aries, Leo, Virgo, Sagittarius, or Aquarius. The lunar cycle should be favorable on the day you wish to begin your diet.

Dressmaking

Design, cut, repair, or make clothes in Taurus, Leo, or Libra on a day marked favorable for your Sun sign. First and second quarters are best. Venus, Jupiter, and Mercury should be aspected, but avoid Mars or Saturn aspects. William Lilly wrote in 1676, "Make no new clothes, or first put them on when the Moon is in Scorpio or afflicted by Mars, for they will be apt to be torn and quickly worn out."

Eyeglasses

Eyes should be tested and glasses fitted on a day marked favorable for your Sun sign and on a day that falls during your favorable lunar cycle. Mars should not be in aspect with the Moon. The same applies for any treatment of the eyes, which should also be started during the increase of the Moon (first or second quarter).

Fence Posts and Poles

Set the posts or poles when the Moon is in the third or fourth quarters. The fixed signs Taurus, Leo, and Aquarius are best for this.

Habits

To end any habit, start on a day when the Moon is in the third or fourth quarter and in a barren sign. Gemini, Leo, or Virgo are the best times, although Aries and Capricorn may be suitable as well. Make sure your lunar cycle is favorable. Avoid lunar aspects to Mars or Jupiter. Aspects to Neptune or Saturn are helpful. These rules apply to smoking.

Hair Care

Haircuts are best when the Moon is in a mutable (Gemini, Sagittarius, Pisces) or earthy sign (Taurus, Capricorn), well-placed and aspected, but not in Virgo, which is barren. For faster growth, hair should be cut when the Moon is in Cancer or Pisces in the first or second quarter. To make hair grow thicker, cut it when the Moon is Full or in opposition to the Sun (marked O in the Lunar Aspectarian) in the signs of Taurus, Cancer, or Leo up to and at, but not after, the Full Moon. However, if you want your hair to grow more slowly, the Moon should be in Aries, Gemini, or Virgo in the third or fourth quarter, with Saturn square or opposite the Moon.

Permanents, straightening, and hair coloring will take well if the Moon is in Taurus or Leo and Venus is marked T or X. You should avoid doing your hair if Mars is marked Q or O, especially if heat is to be used. For permanents, a trine to Jupiter is helpful. The Moon also should be in the first quarter, and check the lunar cycle for a favorable day in relation to your Sun sign.

Health

Diagnosis is more likely to be successful when the Moon is in a cardinal sign (Aries, Cancer, Libra, Capricorn), and less so when in a mutable sign (Gemini, Sagittarius, Pisces, Virgo). Begin a recuperation program when the Moon is in a cardinal or fixed sign and the day is favorable to your sign. Enter hospitals at these times. For surgery, see Surgical Procedures. Buy medicines when the Moon is in Virgo or Scorpio.

House Furnishings

Days when Saturn is aspected make things wear longer and tend to a more conservative purchase. Saturn days are good for buying, and Jupiter days are good for selling.

House Purchasing

If you desire a permanent home, buy when the Moon is in Taurus, Leo, Scorpio, Aquarius, or Cancer, preferably when the Moon is New. If you're buying for speculation and a quick turnover, be certain that the Moon is not in a fixed sign, but in Aries, Cancer, or Libra.

Lost Articles

Search for lost articles during the first quarter and when your Sun sign is marked favorable. Also check to see that the planet ruling the lost item is trine, sextile, or conjunct the Moon. The Moon governs household utensils, Mercury letters and books, and Venus clothing, jewelry, and money.

Marriage

The best time for marriage to take place is during the increase of the Moon, just past the first quarter, but not under the Full Moon. Good signs for the Moon to be in are Taurus, Cancer, Leo, and Libra. The Moon in Taurus produces the most steadfast marriages, but if the partners later want to separate they may have a difficult time. Avoid Aries, Gemini, Virgo, Scorpio, and Aquarius. Make sure that the Moon is well aspected (X or T), especially to Venus or Jupiter. Avoid aspects to Mars, Uranus, or Pluto.

Moving

Make sure that Mars is not aspected to the Moon. Move on a day favorable to your Sun sign, or when the Moon is conjunct, sextile, or trine the Sun.

Mowing the Lawn

Mow the lawn in the first or second quarter to increase growth. If you wish to retard growth, mow in the third or fourth quarter.

Painting

The best time to paint buildings is during the decrease of the Moon (third and fourth quarter). If the weather is hot, do the painting while the Moon is in Taurus; if the weather is cold, paint while the Moon is in Leo. Another good sign for painting is Aquarius. By painting in the fourth quarter, the wood is drier and the paint will penetrate; when painting around the New Moon the wood is damp and the paint is subject to scalding when hot weather hits it. It is not advisable to paint while the Moon is in a water sign if the temperature is below seventy degrees, as it is apt to creep, check, or run.

Pets

Take home new pets when the date is favorable to your Sun sign, or the Moon is well aspected by the Sun, Venus, Jupiter, Uranus, or Neptune. Avoid days when the Moon is badly aspected (Q or O) by the Sun, Mars, Saturn, Uranus, Neptune, or Pluto. Train pets starting when the Moon is in Taurus. Neuter them in any sign but Virgo, Libra, Scorpio, or Sagittarius. Avoid the week before and after the Full Moon. Declaw cats in the dark of the Moon. Avoid the week before and after the Full Moon and the sign of Pisces. When selecting a new pet it is good to have the Moon well aspected by the planet that rules the animal. Cats are ruled by the Sun, dogs by Mercury, birds by Venus, horses by Jupiter, and fish by Neptune.

Predetermining Sex

Count from the last day of menstruation to the day of its next beginning, and divide the interval between the two dates into halves. Pregnancy in the first half produces females, but copulation should take place when the Moon is in a feminine sign. Pregnancy in the latter half, up to within three days of the beginning of menstruation, produces males, but copulation should take place when the Moon is in a masculine sign. This three-day period to the end of the first half of the next period again produces females.

Romance

The same principles hold true for starting a relationship as for marriage. However, since there is less control of when a romance starts, it is sometimes necessary to study it after the fact. Romances begun under an increasing Moon are more likely to be permanent, or at least satisfying. Those started on the waning Moon will more readily transform the participants. The general tone of the relationship can be guessed from the sign the Moon is in. For instance, romances begun when the Moon is in Aries may be impulsive and quick to burn out. Those begun in Capricorn will take greater effort to bring them to a desirable conclusion, but they may be very rewarding. Good aspects between the Moon and Venus are excellent influences. Avoid Mars, Uranus, and Pluto aspects. Ending relationships is facilitated by a decreasing Moon, particularly in the fourth quarter. This causes the least pain and attachment.

Sauerkraut

The best tasting sauerkraut is made just after the Full Moon in a fruitful sign (Cancer, Scorpio, or Pisces).

Shingling

Shingling should be done in the decrease of the Moon (third or fourth quarter) when it is in a fixed sign (Taurus, Leo, Scorpio, or Aquarius). If shingles are laid during the New Moon, they have a tendency to curl at the edges.

Surgical Procedures

The flow of blood, like the ocean tides, appears to be related to the Moon's phases. *Time* magazine (June, 6, 1960, page 74) reported that out of 1,000 tonsillectomy case histories analyzed by Dr. Edson J. Andrews, only eighteen percent of associated hemorrhaging occurred in the fourth and first quarters. Thus, an astrological rule: To reduce hemorrhage after a surgical procedure, plan to have the surgery within one week before or after the New Moon. Avoid surgery within one week before or after the Full Moon. Operate in the increase of the Moon if possible.

Also select a date when the Moon is not in the sign governing the part of the body involved in the operation. The further removed the Moon sign from the sign ruling the afflicted part of the body, the better for healing. The signs and the body parts they rule are as follows: Aries, head; Taurus, neck and throat; Gemini, lungs, nerves, arms, shoulders, hands, and fingers; Cancer, breast, chest, and stomach; Leo, heart, spine, and back; Virgo, nervous system and intestines; Libra, kidneys; Scorpio,

Zodiac Signs & Their Corresponding Body Parts

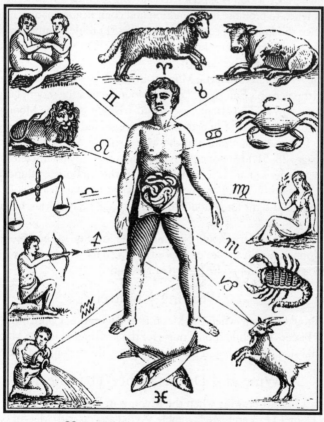

♈	= Aries	♎	= Libra
♉	= Taurus	♏	= Scorpio
♊	= Gemini	♐	= Sagittarius
♋	= Cancer	♑	= Capricorn
♌	= Leo	♒	= Aquarius
♍	= Virgo	♓	= Pisces

reproductive organs; Sagittarius, thighs, hips, and liver; Capricorn, knees, bones, and teeth; Aquarius, circulatory system, shins, and ankles; and Pisces, feet.

For successful operations, there should be no lunar aspects to Mars, and favorable aspects to Venus and Jupiter should be present. Do not operate when the Moon is applying to (moving toward) any aspect of Mars, which tends to promote inflammation and complications after the operation. See the Lunar Aspectarian (pages 28–51) to determine which days have Mars aspects. There should be good aspects to Venus and Jupiter (X or T in the Lunar Aspectarian).

Never operate when the Moon is in the same sign as at the patient's birth (the person's Sun sign). Let the Moon be in a fixed sign, but not in the same sign as the patient's ascendant (rising sign). The Moon should be free of all manner of impediment. There should be no Q or O aspects in the Lunar Aspectarian, and the Moon should not be void-of-course. (See pages 28–51.) Do not cut a nerve when Mercury is afflicted (marked Q or O in the Lunar Aspectarian). When the Moon is conjunct or opposed the Sun (C or O) or when it is opposed by Mars (O), avoid amputations. Good signs for abdominal operations are Sagittarius, Capricorn, or Aquarius.

Cosmetic surgery should be done in the increase of the Moon, when the Moon is not in square or opposition to Mars. Avoid days when the Moon is square or opposite Saturn or the Sun.

Weaning Children

This should be done when the Moon is in Sagittarius, Capricorn, Aquarius, or Pisces. The child should nurse for the last time when the Moon is in a fruitful sign. Venus should then be trine, sextile, or conjunct the Moon.

Wine and Drinks Other Than Beer

It is best to start brewing when the Moon is in Pisces or Taurus. Good aspects (X or T) to Venus are favorable. Avoid aspects with Mars or Saturn.

There's No Place Like Home

What Your Moon Can Tell You about Your Perfect Home

Skye Alexander

T he average American changes residence once every five years, and many of us are even more mobile than that. So if you're like millions of other people on the go, you may be thinking about relocating, buying a home, or just moving to a different apartment in another part of town. Before you pay a big down payment and security deposit, however, you might do well to take a look at the Moon's position in your birth chart.

When it comes to issues of home and family, the Moon's position is very significant. The sign and house in which it is placed, along with the aspects your Moon makes to other planets in your chart, will show the type of home that is perfect for you. In order for you to feel happy, comfortable, and secure in your home, the qualities described by your Moon must be met. The following interpretations of the twelve Moon signs reveal what kind of residence will suit your unique lifestyle.

Moon in Aries

You enjoy lots of excitement in your domestic life. Living in an isolated, rural setting would quickly bore you; you prefer to be "where the action is." Cleanliness, order, and picture-perfect decorating aren't very important to you. You believe a home should be lived in, and your house or apartment is more likely to resemble a locker room than a museum. You probably have an interest in sports, so your ideal home might contain an exercise room, a place to hang a basketball hoop, and plenty of space to store your bike, skis, and baseball equipment.

Moon in Taurus

You enjoy creature comforts and want a home that is as luxurious as your finances will allow. You'll stretch your budget to the limit to get the place and furnishings you want. As a natural interior decorator, your eye for design lets you turn any home into a show place. Because you love cooking and eating, a large kitchen is essential. Your ideal home will also feature a nice yard with a place to plant flowers and a vegetable garden, and maybe even a greenhouse.

Moon in Gemini

You need plenty of space to accommodate your hobbies, such as: a sewing room, basement workshop, painting studio, home office, and an attic for storing all your ongoing projects. You love to read and play games, so your ideal home might include a library or study with floor-to-ceiling bookcases, as well as  an alcove where you can set up a game table. Because you like to work with your hands and fancy yourself a handyman (or handywoman), you won't be turned off by a place that needs some fixing up, especially if it's a bargain.

Moon in Cancer

It is extremely important for you to own your own home and to establish yourself in a secure community. You love children and need plenty of space for your family, even if your children are grown and visit only occasionally. Many of your activities center around your home, and you enjoy cooking for family and friends. Therefore, your ideal home will have a large, well-designed kitchen, perhaps with an adjoining family room, and a dining room big enough for huge gatherings. You also appreciate antiques and would probably be most content in an older home or an apartment with great woodwork and details.

Moon in Leo

You think of your home as your castle. Nothing is too grand or ostentatious for you—you want people to be impressed when they see where you live. Your ideal home is something out of *Lifestyles of the Rich and Famous* or *Gone with the Wind* with plenty of gilt, sunlight, lush furnishings, bright colors, and dramatic details. For these things you're willing to pay top dollar. However, if your finances don't quite equal your fantasies, you have the creativity and flair to turn even an ordinary house or apartment into a palace.

Moon in Virgo

Your tastes are rather modest. You prefer a home that's practical and energy-efficient over one that's palatial. Cleanliness and orderliness are important, and your dwelling is always neat. Your ideal home is one that has plenty of storage space so you have a place for everything. Although you appreciate a building that has been well-maintained by previous residents, you're quite capable when it comes to repairs and are a good candidate for handyman specials. A location that is conveniently located relative to work, shopping, and schools means more to you than being in the fashionable part of town.

Moon in Libra

Your tastes are refined, conservative, and classic. You prefer a house that is neat and pretty over one that has character. Like Virgo-Moon people, you appreciate cleanliness and order, and you need a harmonious environment to be happy. You have an eye for beauty and can use your artistic sensibility to make your home a picture of grace and charm. However, you're not much for doing-it-yourself and prefer to avoid the handyman specials. A quiet, respectable neighborhood is an important factor in your choice of the perfect place to live.

Moon in Scorpio

You enjoy your privacy and tend to withdraw to your home when you want to be alone. Not much for entertaining, you don't require room to socialize. Household plumbing, though, is very important, and you won't be happy unless you have at least two bathrooms—maybe with a Jacuzzi. A safe, secluded location, particularly one near water, would suit you best. Your ideal home will provide a sense of security and enclose you in a dark, protective, cave-like environment.

Moon in Sagittarius

You are sociable and on the go. Friends, family, neighbors, and business associates feel welcome in your home for holiday gatherings, Saturday night card games, kids' birthday parties, and company cook outs. Your ideal home will have plenty of space for all your activities with a large den for drinks and pool and a big yard for volleyball and Frisbee. Like Moon-in-Aries people, you need storage room for athletic gear. You don't mind driving, so a location that is some distance from your workplace isn't a problem. However, your need to be surrounded by activity might make it difficult for you to live in an isolated, rural spot.

Moon in Capricorn

You appreciate tradition and history. Antique homes in older communities—especially the wealthier ones—appeal to you most. You want to own your home and are willing to work hard to save for the perfect property. You are concerned with your image in the community and want an address that connotes status, stability, and respect. Practical and responsible, you hate being in debt and won't assume a financial burden that you can't handle. You conscientiously maintain your residence and expect your neighbors to do the same.

Moon in Aquarius

You want a residence that is unique or distinctive. No "cookie-cutter" ranch houses for you! Your ideal home might be an artist's loft with huge windows and brightly-colored exposed pipes, or a converted lighthouse. Sunshine is important, too, and you enjoy locations high above the streets—perhaps in an ultra modern condo on the twenty-fourth floor overlooking the city. Since you feel compelled to put your stamp of individuality on your home, you will probably start renovations as soon as you move in—even if the place is brand new!

Moon in Pisces

You are reclusive and private about your domestic life. Your home is a retreat where you can withdraw from the stress of the outside world and relax. You rarely entertain more than a few close friends at a time, so your social areas are of less importance than your private ones. Bedrooms and baths are your special interests. Your ideal home will have a large, restful bathroom with a Jacuzzi and perhaps a swimming pool! You also love animals, and want a place where your pets can romp about safely. A quiet, secluded location, especially one near water, would appeal to you most.

Millennium Fever–
Stress and Your Moon Sign

Gretchen Lawlor

Straddling the Millennium and looking out over the year 2000, our general mood is one of anxious anticipation. We ask: What lies ahead for us in these new times?

While times look to be challenging in the days to come, they have not been easy of late. The global economy, which we now irrevocably rely on for our well-being, is threatened, and environmental and weather pattern disruptions are occurring more and more frequently. Furthermore, we are hearing a good deal about the Y2K problem—a simple computer system flaw that threatens to interrupt many of our basic services— power, food distribution, transportation, public services—for weeks or longer. To be sure, the astrological aspects that will occur during the change of Millennium seem to confirm that these concerns are real.

These are profoundly stressful times, and at such times we become particularly aware of our fragility. Each person has unique ways of responding to increased stress. Some faint at the sight of blood; some leave the room at first sign of tears. And yet stress is an integral part of a rich life—it is part of the ongoing, ever-present ebb and flow of our lives, and it challenges us to grow. We need to understand the place of stress in our lives and work with it rather than crumble under the weight of it.

How we work with stress indicates how we interact with life—with our relationships, our work, and our various commitments. Stress itself is not necessarily a bad thing. It is stimulating, an inspiration for change, and a great awakener. Our response to stress determines whether we will be healthy or unhealthy, energized or exhausted.

It is also important to remember that each of us can rise to the challenge of even the most difficult of circumstances. If we are aware of our own special talents, and able to utilize these talents, a stressful occasion may become an opportunity for the finest in us to emerge. Your natal chart—the astrological map that points to what gifts emerge in you when under pressure—contains indicators of what stresses you are particularly sensitive to, how you respond when you are overloaded, and what will help you cope under difficult circumstances.

In your natal chart, the Moon is the best indicator of what your instinctive response to stress is likely to be. Understanding the qualities of your Moon sign helps you know what to expect of yourself when the going gets tough. By focusing on what your Moon indicates, you increase your ability to cope with stress, and by knowing the strengths and weaknesses of your Moon sign and aspects, you can make healthy choices. You will know where to balance or adjust to circumstances, and when and where to look for support to help you through a particularly difficult time.

Let's look at a few examples. The Moon in Scorpio, which is at its very best in emergencies, has been known to provoke a crisis in order to get more of that passionate interaction it craves. On the other hand, people with Moon in Taurus seldom are ruffled by anything; their very presence can be a significant calming influence. Meanwhile, Moon in Gemini people make good traffic directors, and compassionate Moon in Pisces people respond well when anyone is in deep crisis, though they sometimes lose their grip on reality if they don't get regular time alone.

The Moon in your chart symbolizes a basic way of being that comes instinctively to you. It reflects your deepest emotional patterns and what you need to feel nourished. The Moon, by sign and aspect, shows how you are rooted in life, in your family, in your homes and larger community. It speaks of how you form attachments and what you need to feel secure and protect yourself. Let's look now at all twelve Moon signs, their response patterns to stress, and their special needs during times of stress.

A Look at Your Moon Signs Under Stress
Moon in Aries Under Stress

With a Moon in Aries, you are at your very best when times call for an immediate response. As a fire sign, your reaction to an emergency is most likely to be "Let's go!" Headstrong and future-oriented, you will enjoy the sense of adventure and innovation that Uranus and Neptune in Aquarius bring to this change in Millennium.

You enjoy trying new things, taking the initiative, and starting new projects. Never mind that you tend to lose interest after the thrill of adventure wears off. When this occurs, you may do well to hand things over to one of the more persevering Moon signs, such as Taurus or Virgo, and throw yourself into the next quest. One challenging new project after another will present itself, and your leadership skills will be needed.

Aries is the first sign of the zodiacal year, wakening from the winter's slumbers with zeal and energy. Stepping into a fresh new century is bliss for a person with their Moon in Aries. Your adventurous imagination and courageous spirit will carry timid souls forward into the unknown.

As Aries is a lone wolf, you do find it difficult to work cooperatively, preferring to be a self-sufficient catalyst. With the social signs of Aquarius and Sagittarius dominating the larger picture, however, times will require collaboration and community building, and Aries Moon will be frustrated. You have charm, vivacity, and verve, but not much patience or ability for self-reflection. For you, the biggest challenge during this time of transformation will be finding companions who appreciate and support your enthusiastic, zealous sense of adventure.

The Aries Moon won't succumb to depression or inaction for long. You are more likely to rage or indulge in compulsive or purposeless hyperactivity; these things are soothing for you in times of high stress. To harness the Aries Moon energy, you may create an exercise regime or strenuous routine, like chopping wood or pitting yourself against the elements—anything that helps you to discharge that excess energy.

Moon in Taurus Under Stress

Steady, dependable, and unflappable under pressure, the Moon in Taurus is a calming influence and source of peace in the midst of even the biggest storms. People with their Moon in Taurus are pragmatic and able to apply good common sense during times of stress. They can rise to most any emergency and are known for their perseverance and determination.

The flip side of this gift is a tendency for a Taurus Moon to become entrenched. Under pressure from Uranus and Neptune in Aquarius, Taurus Moon people need to accept that there are some areas of life that are ready for change. While the Taurus Moon prefers to change only gradually, this may not be an option during the Millennial reorganization.

Change doesn't have to be earth-shattering, and for the Taurus Moon it must be gradual. Fortunately, you can use your potent will and abundant resourcefulness to bring about practical improvements in general conditions. It is best when the Taurus Moon designs plans with some built-in flexibility and remains unattached to the outcome of any efforts.

Moon in Gemini Under Stress

Gemini is an air sign, and its strength lies in its incredible adaptability. If you have your Moon in Gemini, you tend to be restless and prone to seek out mental stimulation and conversation. Fortunately, this new era will be full of new ideas and challenges at every turn!

The Gemini Moon has an ability to disengage emotionally from uncomfortable situations. You are able to feel comfortable in noisy, chaotic, and emotionally charged settings, and you know how to tend to yourself and are capable of moving unscathed through events that might cripple another soul.

Under extreme stress, the Gemini Moon goes into a hyper-alert state and becomes excessively talkative in a superficial, overly intellectual way. You like to talk about how you feel but prefer not to have to feel your emotions for any period of time. You would rather move on than confront murky problems. When stressed, Moon in Gemini looks for distractions, such as compulsive reading or chatting on the Internet.

It can be hard at times for Gemini Moons to complete projects. Your adaptability sometimes becomes a liability in difficult circumstances. You find it hard to stay focused on multiple projects. Furthermore, being among the most independent Moon signs in the zodiac, dependency is not easy for you. With Uranus and Neptune in community-minded Aquarius for the next seven years, there will be many situations where you will be asked to join forces with others. You can gain by believing in a common ideal and taking a responsible role among a group. With your gifts at handling communications, at teaching others, and with overseeing the movement of people, information, and materials, you would be a valuable addition to any group.

In the end, writing is an excellent release for a stressed-out Moon in Gemini. A journal can help you sort out your problems and act as the germ for a book. Meditation also helps calm the overworked logical self and allows for the flow of wise instinct and intuition.

Moon in Cancer Under Stress

With the Moon in Cancer, you have deep concern and sympathy for others. You get great satisfaction from caring for others and are able to make any situation feel like home. You appreciate everyone's need to feel safe and protected, and you place a very high value on a comfortable, stable home environment. Overall, any worry about the Millennium may affect you more through your loved ones than it affects you.

Cancer Moon is able to recreate a caring sense of sanctuary even in difficult circumstances. In any group setting, your job is to make people feel they belong and to create an informal feeling of warmth and relaxation. You have an instinctive maternal streak, no matter your gender or age, and cooking for others may be soothing in difficult times.

On the other hand, you tend to hide your own emotions behind a shell of reserve. In the rapidly changing times ahead, you will have to learn to be more forthcoming and immediate if you want to feel more connected with others.

Finally, Cancer Moon is particularly affected by experiences from early years. Under stress, you pine for the good old days and attempt to recreate them. The outer planets that are currently moving into group and community-minded Aquarius suggest that everything Cancer values may go through vast changes. In these rapidly changing times, you are as apt to find close companionship with new friends as with family, though your loyalty will require you to weave both worlds together.

Moon in Leo Under Stress

Like Taurus and Scorpio, Leo is a fixed sign and can be quite stubborn about embracing change. For Moon in Leo, emotional well-being hinges on being appreciated. It is extremely painful for you to lose face or the respect of others. You may make many mistakes under the challenges of this new and unprecedented era, but so will we all! For Leo Moon, however, it is important to cultivate a sense of self-acceptance and even of humor about these natural and even inevitable errors of judgment.

Under stress, the Leo Moon is likely to become entrenched in old ways and susceptible to feelings of loss and disappointment. You have been known to create high drama to get attention, though seldom about what really concerns you. You like to put on a good front, hiding your feelings so no one will know how you really feel.

Leo likes to be in charge, and you can become dictatorial and self-righteous if thwarted in your desire to lead and inspire. With the collective focus of Leo's opposite sign, Aquarius, so strong in the next few years, Leo will be challenged by the necessity to function as part of a group. You have the capacity to motivate others through your own enthusiasm; however, it is important to cultivate detachment about your place in the group. You may often have to pass the baton to others more qualified than you in order to get the group through those challenges.

In order to cope with stress, find an area of expertise and hone your skills in this area. Pluto in Sagittarius is in good aspect to your Leo Moon over the next few years, and this will help you stay focused. Pay attention to chances for developing and celebrating your innate gifts.

In this Millennium shift, the personal challenge of the Moon in Leo has most to do with gaining inner respect rather than outer authority. Ultimately, you will be called to contribute your considerable gifts of passionate enthusiasm and graciousness to your larger community.

Moon in Virgo Under Stress

Virgo is an earth sign, and those of you with Moon in Virgo have an instinctive practicality that emerges in stressful times. You are attuned to the laws of nature and understand what is needed to care for your own and others' well-being under even the most challenging circumstances.

At the same time, a Virgo Moon must feel useful in order to feel at peace. You get pleasure from providing professional and personal services to others. Yet if you are not managing your stress levels well, you can become compulsive and easily slip into workaholic overdrive, desperate to keep busy and unable to relax. Personal shortcomings and the imperfections of others become intolerable, and you can be overly critical of yourself and everyone around you.

Virgo Moon derives a tremendous amount of security by having all the facts. The fast-changing times ahead will be a challenge for you to adapt to, because everything will be so unpredictable and chaotic. But since Virgo is a mutable sign, you can find comfort in adapting to change. In fact, you enjoy refining yourself and the circumstances around you. Virgo Moon needs to understand that there are things you simply cannot understand. Once you find a way to trust that things will ultimately turn out even more perfectly than if you had planned everything personally, you will be vastly happier in your service to the world.

Unlike Gemini, who has trouble on focusing on just one thing at a time, Virgo does best when working on a single task, especially when stressed. If you get overwhelmed, you should stop and remind yourself of your original goal, check in to see if you have lost your way in all the details and small projects. You are dedicated to improving everything that you touch, and your instinctive understanding of healing tools can be of great value. In fact, the Virgo Moon can not help but use this incredible natural wisdom for the benefit of others.

Moon in Libra Under Stress

Moon in Libra craves beauty and peace, seeks to cooperate and please others, and to be in relationships that are mutually supportive. Persons with Moon in Libra are gifted in creating harmony in any environment.

The Libra Moon works to smooth differences between people and conditions by using an innate and highly developed aesthetic sense.

Overall, your gift for helping others restore their inner equilibrium makes the role of therapist and counselor a likely one. Your sensitivity about justice and equality means you are often sought out to help mediate the problems that arise as groups of people come together in the Millennium to accomplish goals in new ways.

Under stress, this need for harmony and peace in the Libra Moon can become a compulsive tendency to avoid controversy at all costs. You may become overly dependent upon a partner, compromising yourself to satisfy the partner's every desire. You also may suppress your own emotions for fear of unleashing discord. Because relationships are so critical to your emotional well-being, you may need to become very selective about your personal contacts. Tend only to those relationships that give you the space to be truly yourself.

With Uranus and Neptune in Aquarius and Pluto in Sagittarius all transiting in supportive aspect to your Libran Moon, change need not be overly stressful for you in the next few years. New contacts will foster your efforts and unexpected opportunities will present new venues to use your mediating abilities.

Moon in Scorpio Under Stress

If your Moon is in Scorpio, you are always at your finest during emergencies and comfortable with deep and powerful emotions. Scorpio has a very transformative energy, so the Scorpio Moon can step into adversity and can meet any emotional challenge.

Moon in Scorpio can be its own worst enemy when not deeply engaged in something constructive or transformative. A sideline passion, fortunately, can defuse or redirect your intensity. During the years Pluto was in Scorpio—from 1984 to 1995—you were tested and tempered by personal hardship and became somewhat comfortable with the fires of adversity and change. This is why you are particularly suited to working as a psychotherapist or healer, having so often had to heal yourself.

The Moon in Scorpio is exhaustingly intense. You thrive on crisis and sometimes can manufacture a problem in order feed this thirst. It is better to focus on the suffering of others rather than your own. When you are in deep pain yourself, you can have a tendency to disappear to lick your wounds. The Scorpio Moon does best when it find ways to let up on its intensity.

Moon in Sagittarius Under Stress

Moon in Sagittarius souls are natural philosophers. Your instinctive style is to uplift others, to enthuse and inspire. You want to make the world a more beautiful place and passionately believe that positive thinking will make a difference in even the most difficult times.

The childlike enthusiasm and persuasiveness of the Sagittarius Moon could come in handy in convincing even the most pessimistic soul to think positively. You are a natural teacher whether you express this through writing or some other media or in a classroom. Any chance to share your perspective is uplifting to you and helpful to others.

Be sure this upbeat philosophy doesn't cause the Sagittarius Moon to lose track of the nature of the hard work and hard times ahead. You may have a tendency to overlook difficulties in your own life and family, focusing instead on the rosy future or the big issues in society.

Sagittarius Moons have strong moral convictions, and this zeal can become fanatical under stress, so do temper your judgments with compassion. Remember, the answers you think you have may not be the right answers for everyone in every situation. Under stress, your liver is your weak organ. A tendency towards excess and an unwillingness to admit to any physical limitations may tax the filtering mechanism of this organ.

Too much of a good thing may not be the politically correct approach as our resources dwindle in the new Millennium. Indeed, if the economy takes a downturn, as it may well do with Saturn in Taurus, we all could benefit from learning to conserve.

Moon in Capricorn Under Stress

The Capricorn Moon needs the security of a structure and consistency, particularly in the home environment, to feel whole. Traditions are very important to Capricorn, and often a Moon in Capricorn indicates that a grandparent was a significant early influence. To your credit, you can be very mature at an early age, happy to take on responsibilities, and soothed by hard work.

Under stress, a Capricorn Moon can become controlling. When feelings and needs emerge in others or yourself, you may not allow them to enter your consciousness. Depression and self-criticism, even self-imposed isolation may be your responses to stress. And while you can be a great workaholic, you can tend to discard all other difficulties in favor of your devotion to work.

Emotionally, Moon in Capricorn needs to be involved with people and projects with similar goals to your own. When you find your niche in a special group that supports your ambition and needs your assistance, you will feel you have come home.

Moon in Capricorn, however, never gives up. You have the tenacity to work through hard times and ultimately capitalize on them. You have the ability to get a job done no matter what is going on in your personal circumstances. Your ambition will pull you through difficult times. You are instinctively drawn to make something out of adversity, and will work harder than any other Moon sign to manifest an opportunity.

Moon in Aquarius Under Stress

People born with their Moon in Aquarius possess a natural inventiveness that reacts well to sudden shifting moods and circumstances. You feel most at home when required to respond quickly and act for yourself. Of all the signs, you are the most willing to take up opportunities to do new things in unconventional ways.

Fortunately, Moon in Aquarius souls may finally come into their own with this Millennium shift. While peer groups or communities forged by common interests or shared causes will have more strength than family ties, your friendly but unattached nature will allow you to interact spontaneously with a wide range of people. After all, it is the Age of Aquarius we are moving into, where your style of being and relating will be the norm.

On the other hand, the Aquarius Moon may have trouble acknowledging emotional needs and tend to be fearful of intimacy. To cope, you can tend to rationalize or intellectualize your feelings, or throw yourself frenetically into a myriad of stimulating distractions. Another common Aquarian stress response is to make sudden abrupt changes which help to extricate you from habits or contacts which have become stifling.

A healthier response may be to use your natural affinity with the collective consciousness and get involved with a project that you really believe in. You can truly feel at home in difficult times if you have the sense of being in it together with others of your ilk.

Moon in Pisces Under Stress

Watery, compassionate, mystical Pisces Moons can relate to feelings of despair or terror under stress and are likely to be of immense help during crisis. Pisces Moons have an emotional need to help others, and their powerful healing presence will be of use in the accelerated times ahead.

Pisces is the most sensitive Moon sign in the Zodiac. You are so sensitive that you often act as a psychic sponge, mopping up and absorbing upset and emotional trauma in any environment you enter. Some Pisces Moons actually pick up the energy of disasters occurring at a great distance or in the depths of the collective unconscious.

The most important thing for Moon in Pisces during times of stress is to protect your sensitivity. If you are called to give out your profound sympathy, make sure that you take the time to get rid of what you have already absorbed. Solitude helps. Being near water, particularly fresh, flowing water, is soothing and cleansing. Learn to immerse yourself or refill yourself with art or music. It is easy for you to neglect your body, or, when the going gets rough, deal with the pain of your hypersensitivity by escaping into addictive or self-abusive habits.

This time of accelerated change will require you to learn to pace yourself. Your rich inner life is a deep source of nourishment— through dreaming, through time alone, through immersion in your passion for art or music. The flowing movement of dance or the peaceful contemplation of time in nature will also be avenues to those rich inner realms of yours. Later you can share your own process of rejuvenation with others, so that they can become their own source of regeneration, and that will be one less soul for you to save.

Yes, these are challenging times. But consider this; in Chinese, the symbol for the word "crisis" is identical to the symbol for the word "opportunity." The Chinese also have a blessing, which is sometimes used as a curse: "May you live in interesting times."

Aquarius and its ruling planet Uranus will dominate the skies as we make this Millennial shift. The best preparation for the explosively unpredictable effect of these two is to know yourself well. And the knowledge of self provided by your astrological map, in particular your Moon sign, is a valuable tool in this struggle.

If you are aware of your weakest points—the problems likely to arise when you are under pressure—you won't be caught unprepared. This should also help you care for yourself during stress, and understand what soothes you and restores your equilibrium. Knowledge is your best preparation for stress, and knowing the position of your Moon provides what you need to know to cope with crisis.

Making Sauerkraut by the Moon

Louise Riotte

It seems high time somebody said a good word for cabbage. Moon-ruled cabbage has a fascinating history, and its numerous good qualities should be more widely appreciated.

A member of the mustard family, Brassica erica, or cabbage, has been cultivated for more than 4,000 years. It is the most ancient of all vegetables, having an aristocratic history that gives sense to the saying "Of cabbages and kings." The pharaohs considered cabbage an aid in drinking, and they ate large quantities of cooked cabbage before their drinking bouts on the premise that this would enable them to guzzle more beer and wine without getting drunk. The Egyptians actually went so far as to worship the vegetable as a god, building altars for it. While the Greeks after them claimed that cabbage heads "sprang from the sweat of Zeus," and recommended cabbage as a cure for baldness and many ailments. Our word for the vegetable originates from the Latin caput, or "head," and the Romans considered cabbage an aphrodisiac. Apictus, apparently, vouched for it in his De Re Cocuinaria, and Roman priests dedicated young cabbage to Priapus, the phallic guardian of fertility. Cabbage that is allowed to grow and blossom becomes an important honey plant for bees.

Cauliflower, kale, Brussels sprouts, and broccoli (all of which are cancer fighters) are all forms of cabbage that also have been esteemed as aphrodisiacs by Europeans. The vegetable's sexual associations can be found in the French slang expression chounette, "the little darling," for the female sex organ and chouserie, which was used by Rabelais and others for "copulation;" both these expressions stem from chou, the French word for cabbage. Similarly, the word cabbage also once also served in English slang as a synonym for the female sex organ. It is said that cabbage should be planted in the garden immediately after a marriage, if a couple wishes to have good luck in their union and their garden.

For hundreds of years the medicinal qualities of the cabbage have been recognized by such people as Charles Cornell, author of Aphrodisiacs In Your Garden. Ancient herbalists have sung the praises of cabbage at great length and often used it as an ingredient in aphrodisiac preparation.

Evolving from the wild cabbage, the garden variety has retained its popularity, and as a year-round pick-me--up must be included in any garden intended to give vigor.

One good quality of cabbage is its ability to transform into that delicious and nutritious concoction, sauerkraut. According to Robert Hendrickson, author of *Lewd Food*, China's Great Wall was built by workers fed on strength-giving cabbage preserved in rice wine. The Tartars were the first to use salt instead of wine as a preservative in the recipe, and they introduced it to Germany, where cabbage itself is still called kraut or kohl. Today, the most famous cabbage dish in the Far East is probably Korean kimchi, a highly spiced dish which is prepared in mid-November and buried in the ground in large earthen jars to preserve it through the cold months. South Korean troops fighting in Vietnam had to be supplied with kimchi to bolster their morale, for a Korean meal without it is unthinkable. There is even a Korean expression: "We can live a whole year without meat, but without kimchi we can hardly live a week." High in vitamins B and C, kimchi is said to be most stimulating (there is a funny episode of *M.A.S.H.* which dramatizes this), but many foreigners find it burns their palates and complain that it has an unpleasant odor.

Patricia Telesco gives the following unusual recipe in her book, *A Kitchen Witch's Cookbook*, which highlights some of the health-giving properties of cabbage.

Cabbage of Temperance (Egypt, Germany)

1   large green apple, diced
1   tablespoon sugar
1   onion, chopped
    Salt
1   tablespoon butter
1   raw potato, shredded

3 pounds red cabbage, shredded
½ cup crumbled bacon
4 tablespoons wine vinegar

Sauté the apple and the onion in the butter until lightly browned. Stir in the cabbage, vinegar, and sugar. Simmer for 10 minutes. Add the salt and 1½ cups butter. Simmer until the cabbage is fully cooked, about 2 hours. Add the potato, simmer until thickened, about 15 minutes. Garnish with the bacon. Yield: 6 servings. This recipe has the magical attributes of moderation, prudence, and self-control, and it is best made before starting a diet, certain winter rituals, and Thanksgiving Day.

How to Keep Cabbage Fresh

An old-time method for keeping cabbage consists of leaving two or three inches of stalk and hollowing out the pith while taking care not to cut or bruise the rind. In this method, you should tie the cabbages up by the stalks and fill the hollow daily with fresh water (cabbage is a plant of the Moon whose element is water). The cabbages will keep in a cool place in this way for several months.

Curative Effects

The curative effects of cabbage and sauerkraut are well-documented. In his book *Medicinal Value of Natural Foods*, Dr. W. H. Graves says of sauerkraut juice: "The lactic acid has a cleansing tonic effect on the bowels. Indicated for diabetes, constipation, colitis, catarrh, dyspepsia, high blood pressure, and poor complexion." Sauerkraut juice is said to be good for teeth and bones and is rich in vitamin C, B1, B2, K, and A. Honey added to diluted sauerkraut juice makes a good drink. Also, a mixture of one-third sauerkraut juice and two-thirds tomato juice makes a fine laxative drink.

Russian women are said to eat a great deal of cabbage because they believe it will give them a beautiful complexion. Cabbage is a good muscle builder and valuable for teeth, gums, hair, nails and bones. It also aids in cases of asthma, tuberculosis, gout, scurvy, constipation, kidney and bladder disorders, obesity, diabetes, toxemia, lumbago, skin eruptions, poor complexion, stomach ulcers, and inflammations.

Sauerkraut

There are so many delightful ways to serve cabbage—raw in salads or cole slaw, as part of a New England traditional boiled dinner (does anyone remember "Maggie and Jiggs" of comic strip fame and their fondness for corned beef and cabbage?), sautéed with Chinese vegetables, and, of course, there is that much-beloved German dish, sauerkraut.

Sauerkraut is considered important enough to have its own rulership—that of Saturn and the Moon. The rulership by the Moon is easily understood, but the connection to Saturn is more questionable. Saturn is much maligned and is called a malefic, but this is not to imply evil. Rather, these terms designate the difference in the effect of Saturn compared with the effects produced by another planet, Jupiter. The vibrations of Saturn are identified with contraction (in making sauerkraut the cabbage shrinks in size), cohesion, and stability (sauerkraut "packs" and has good keeping qualities), while Jupiter is associated with the sensations of relaxation, satisfaction, and contentment. Both planets have an influence that is needful for certain operations.

If you want to endow your sauerkraut-making operations and the final product with good keeping qualities, it is well to know that the best-tasting sauerkraut is made just after the Full Moon, in one of the fruitful signs—Cancer, Scorpio or Pisces. If you want to improve the flavor of sauerkraut even more, place grape leaves over the crock in which it is fermenting (grape leaves are also said to firm the kraut and add bacteria for fermentation).

Making Sauerkraut

Grace Firth, in her book A Natural Year, gives a recipe similiar to the one my husband and I used. First, remove the green leaves from the cabbage and shred each head by quartering it and holding the core while you grate. Then, wash the shredded cabbage and shake it out. Sprinkle with salt and layer into a crock lined with grape leaves. Sprinkle a bit more salt between handfuls of salted cabbage. Cover the cabbage mixture when the crock is full with grape leaves and a wooden lid that fits inside the crock. Top everything with a stone to keep the cabbage immersed in brine, and cover with cheesecloth. As the salt draws juice from the cabbage, liquid will form and cover the lid. This must be skimmed from time to time. If too little brine forms, add salted water—two tablespoons per quart of water. The crock should be kept in a cool, dry place. Choose your location knowing that the sauerkraut will become smelly as it ripens.

In about three or four weeks, or when no more foam appears on the brine, the sauerkraut is ready to eat or be canned. We canned cured kraut when it was white-yellow and free from white-white spots.

To can the sauerkraut, heat the fermented cabbage and pack it hot into hot, sterilized quart jars. If there is not enough juice to cover the sauerkraut, add a holding brine made by dissolving two tablespoons of plain salt in a quart of boiling water. Process quarts in a boiling water bath for thirty minutes after each jar has been lidded and sealed. We always liked to put a fresh grape leaf into the bottom of each quart. Another nice addition is a pinch of caraway seeds in the top of each quart. Grace Firth also suggests strewing juniper berries in the kraut but says they make it taste like a hot martini. To each his own. I suggest you try all three suggestions.

I know you are puzzling over how much salt to add. A quick rule of thumb on the amount of salt needed for sauerkraut is about one to one-and-a-half cups of salt for each forty to fifty pounds of cabbage. Also, it is best to use plain salt rather than iodized.

Sauerkraut Begins With Cabbage

If you are a sauerkraut aficionado (and this is a vast brotherhood and sisterhood all over the world), it is a foregone conclusion that you will want lots of cabbage. Once harvest begins, there are always some heads ready to pick somewhere in the garden. If you grow your own plants, make sure to start them early indoors and set them out three or four weeks before the last spring frost date in the same signs in which they were started. Cabbage plants once up are slow-growing, hardy souls and can take a light freeze.

Cabbage is not always an easy plant to grow so you should sow the seed in the first quarter under Cancer, Scorpio, Pisces, Libra, or Taurus. Often, you can buy started plants, and these should be transplanted in the same quarter end signs. Cabbage now comes in many varieties. I particularly like the crinkled Savoy types which grow better than others and keep their quality well into hot weather. Flowering cabbage is beautiful and delicious with crinkled heads of red, green, and white. This type of cabbage needs cool weather to bring out the color and should be planted in July so it will come to maturity in the autumn.

Spring cabbage grows best in soil that has been deeply dug and liberally manured. Just before planting time, fork the ground over lightly. Sow (or transplant) thinly, and when putting the plants in their permanent positions allow for eighteen inches between them. Cabbage is a heavy feeder and additional side dressings of compost during the growing season will contribute to size and succulence.

It would seem inevitable that cabbage, often pale green in color and resembling the Moon in shape, should be ruled by the Moon and Cancer. As previously mentioned, cabbage will grow best if planted in the first quarter under Cancer, Scorpio, Pisces, Libra, or Taurus. Cabbage needs ample moisture, as evidenced by the first three signs, and watering should be regular and even.

If too much water is taken up at one time, however, as during a heavy rain, the cabbage is likely to show its indignation by splitting. Still, if splitting seems imminent there is something you can do. Cabbage heads, like all vegetable heads, grow from the inside out. If there is a heavy rain, especially after a prolonged dry spell, growth takes place rapidly around the core, but the outer leaves do not grow as quickly and therefore split. If you see a crack in a young cabbage, hold the head and twist the whole plant halfway around as you would turn a faucet. Don't be faint-hearted—take a firm hold and twist the plant. This will break off many of the plant's roots and thus slow the inner top's growth. Give the plant another quarter turn in few days if the cracking continues. This retarding action is influenced by Saturn.

With a little planning and careful cutting you can get extra production from your cabbage plants. Harvest a few of your cabbage heads while they are fairly small (early spring-planted cabbage is best for this), leaving five or six of the outer leaves on the plant. From each leaf a small head will form that is very tasty. Plant the cabbage plants according to variety and far enough apart to have enough space at maturity. While the cabbages are small, you can plant a few onions in the spaces between them and harvest the onions as green scallions for your table, or let the onions grow—they'll help repel the white cabbage butterfly that likes to lay its eggs on cabbage.

Hyssop, a bushy plant with dark blue blossoms, will deter white cabbage butterflies if planted with cabbage. Geraniums will repel cabbage worms, and nasturtiums are beneficial to members of the cabbage family. The parasitic wasp trichogramma is an efficient destroyer of the eggs of many moth, butterflies, and cabbage worms. Hot pepper spray added to water, along with crushed garlic and a little soap to make the spray stick, is also useful in deterring pests. Thyme, wormwood, and southern wood will also repel white cabbage butterflies. Other aromatic companion plants are celery, dill, chamomile, sage, peppermint, rosemary and potatoes. Cabbage dislikes strawberries, tomatoes, and pole beans. Rye flour sprinkled over cabbage plants will dehydrate cabbage worms.

Cabbage and its college-educated cousin, cauliflower, are also subject to clubroot. If this occurs you may try replanting it in a different part of the garden during either the third or the fourth quarter under Cancer, Scorpio, or Pisces. Dig to a depth of twelve inches, and incorporate plenty of well-rotted manure into the soil. Rotate cabbage crops every two years. If cabbage plants do not head up well, it is a sign that lime, phosphorus, or potash is needed in your soil. Boron deficiency may cause the heart of cabbage to die.

Cabbage in America

In her book *A Treasury of American Indian Foods*, Virginia Scully mentions a type of wild cabbage which grew in gypsum soils at altitudes of up to some 7,000 feet. A tall, sturdy plant with delicate yellow-fringed flower clusters and drooping stalked pods, this plant can be eaten as we eat cabbage. However, Native Americans in the country learned by experience and experiment that the cooking water had to be changed several times to avoid emetic results when digesting the cabbage.

Native Americans also used this wild cabbage in caring for their feet. To a nomadic people this was of prime importance, particularly among the Rocky Mountain tribes. In making a remedy, the hard fibers of the cabbage were cut out before being boiled or placed on a hot stone to soften. The leaf was then folded and fastened with a bandage to the hollow of the foot. The same procedure was followed with pounded hot garlic. The remedy was intended to relieve pain.

Cabbage as we know it, meanwhile, was brought to the Americas by the French navigator Jacques Cartier in 1536, and many varieties of cabbage have been cultivated here ever since. Although a recognized rich source of many important nutrients, today this aristocratic vegetable does not enjoy the same prestige it did in days gone by. At the very least, cabbage and its delicious byproduct, sauerkraut, should have better press.

Louise Riotte (1909–1998)

All of us at Llewellyn Publications lament the passing of our long-time contributor and friend, Louise Riotte. Ms. Riotte began gardening many years ago as a child in her mother's parsley patch. She began her career as a writer in the late 1960s, writing articles on gardening-related topics as way to make extra money. Ms. Riotte wrote her first book in 1970, a text on egg-decorating that was published by Drake Publishing in New York. Her most well-known book, *Carrots Love Tomatoes,* has sold more than 500,000 copies since its first publication in 1975.

Louise Riotte's texts, written in an old-fashioned manner on a Remington Rand typewriter with a well-worn ribbon, were filled with wisdom, wit, and the kind of personal recollections that are always entertaining. In looking back, we feel privileged to recount that Ms. Riotte began contributing texts to Llewellyn's *Moon Sign Book* in the early 1980s. She also contributed to Llewellyn's *Organic Gardening Almanac* from 1993 to 1996. Ms. Riotte was a fountain of information on a wide range of topics, and she has written at least twenty books, including twelve for Storey Communications in Vermont. Though she will be sorely missed by fans, Storey Communications plans posthumously to publish a book this coming year by Louise Riotte on herbs and astrology for animals, and Llewellyn Publications plans to continue offering articles written by her in the *Moon Sign Book 2001.*

Mailbox-Scaping

Deborah Duchon and Anna T. Duchon

During our frequent neighborhood walks, the best part is seeing our neighbors' gardens. At the same time, however, we have become intrigued with mailbox landscaping. This relatively new phenomenon seems all the rage, and we are curious why is it that people find this so appealing.

We should mention that we, aunt and niece, are dabblers in anthropology, the science of myths, rituals, and symbols. In our quest to better understand homeowners' penchant for decorating their mailboxes, we decided to apply an anthropological perspective to a systematic observation of mailbox landscaping. That's just a fancy way of saying that we thought we'd look at mailbox landscaping the way an anthropologist would.

The Anthropological Perspective

What is the cultural role of a mailbox? A mailbox, after all, is a utilitarian object whose job is to be a box for mail. It doesn't do a better job whether it is decorated or plain. A pretty mailbox surrounded by flowers does not do a better job of holding mail than a plain black box. In fact, it's generally more important that such a utilitarian object function well rather than look good. For mailboxes, it is vital they meet postal regulations in height and size and keep the mail sheltered from the elements. Decorations should not interfere with the function of the mailbox.

But utilitarian objects have meaning apart from simple utility. Let's look at another everyday object. A fork, for instance is a very useful tool. Like a mailbox, it serves only one practical purpose: for eating, or, more specifically, for carrying food from the plate to the diner's mouth. There is not too much to a fork. It usually has four tines and is made of metal. It is important that it be sturdy enough to spear meat. That's about it.

So, are all forks the same? No. There are salad forks and dinner forks and fish forks. There are forks made of stainless steel, plastic, silver, and even gold. Some are very ornately decorated. Some are cheap and some are expensive. Forks made of expensive materials have more status than forks made of plastic. If you invite company over for a fancy dinner party, you bring out your "best silver," not just plastic ware.

Humans seem to need to decorate everyday things. It's one of the facts that make us human. We decorate everything—homes, offices, cars, food, and bodies to name a few examples. In our culture, we expect women to do most of the decorating, as is evident is the expression, "that needs a woman's touch." But men certainly take part in decoration at times. Men work to ornament in their yards by pruning and landscaping. Another interesting example is the way soldiers, the epitome of "manly men," decorate their weapons and their uniforms. And prisoners, otherwise dysfunctional members of society, often decorate their cells. Yes, to decorate is to be human. A famous philosopher might have said, "I decorate, therefore, I am."

Mailboxes can be seen as items that just beg for decoration. Like an old stump or a rock, there it sits, right next to the driveway, sticking up in the air, a large box on a narrow post ready to be brightened up.

Anthropologists don't really understand why people decorate. All they know is that we do it. Some anthropologists tie it to a search for identity or self-expression. We considered this notion—that maybe mailbox landscaping is a form of self-expression.

A Little History

In older houses, mailboxes were usually built into or next to the front door. The mailman walked up and down the street, lugging a heavy bag and hand delivering mail directly to the door. Over time, this system became too expensive for the Postal Service, and new regulations were put in place which required that mailboxes be "planted"—so to speak—next to the street, where mail delivery could be accommodated by truck.

In rural areas, the system of street-side delivery had existed for years. Interestingly, mailboxes in rural areas were often uniquely decorated, mounted on an old plow, or with flowerpots or old tires attached.

More recently, suburban homeowners have adopted and adapted rural techniques. Gone are the plows and tires. In their place stand mailboxes as stylish sculptures festooned with flowers, from top to bottom. Some are nestled within a bed of shrubbery. Perhaps this trend is another kitschy attempt to recreate our rural roots, like country kitchen fashion or cross-stitch. On the other hand, the mailbox landscaping trend may be a more basic extension of the house to the street.

The Symbolism of Mailboxes

When mailboxes moved from the front door to the street, it was, in our opinion, a little like moving the front door, or at least a part of the front door, to the street. The front door of a home is a very symbolic place. It is the threshold to our private lives. The front of the door is public, readily visible and easily approached. But the inside part of the door is very private, and should only be seen by those who are allowed within.

It's the same with mailboxes. The outside is public and visible, but the inside is between you and your postal carrier. To carry the similarity forward, we often decorate the outside of our front door with such items as holiday wreaths or other floral decorations. We do this as a form of self-expression and as a declaration of our participation in the community. So it goes with mailboxes. As an extension of our front doors, they invite us to express ourselves to others and saying "This is how we want to be seen by our neighbors."

On the other hand, the front door is a barrier, a warning that says, "Private property. Do not enter." Homes which have alarm systems often display warnings to that effect on the front door. Many front doors in fact flaunt mixed messages. We have seen entrances decorated with both "Welcome to our Happy Home" and with security decals!

In the end, this raised a question. Do people decorate their mailboxes as a simple form of self-expression or because they see the mailbox as an extension of the front door? Or is this phenomenon some combination of the two? We set out to discover the answer, and want to share our research findings with you.

The Research Begins

We started out by using a method known as "unobtrusive observation." In other words, we drove around and looked at a lot of mailboxes in many different neighborhoods, searching for patterns. We drove through the suburbs and residential neighborhoods in towns on quiet streets, and on major thoroughfares in prosperous as well as not-so-prosperous city areas.

We found a few interesting things. For instance, we noticed that mailboxes are more often landscaped on small, quiet streets where there is not much traffic. On the other hand, the mailboxes on busy streets are rarely landscaped. Come to think of it, you rarely see people in their yards

We found a few interesting things. For instance, we noticed that mailboxes are more often landscaped on small, quiet streets where there is not much traffic. On the other hand, the mailboxes on busy streets are rarely landscaped. Come to think of it, you rarely see people in their yards when they live on major roads. We concluded that people will tend to landscape in areas of the yard where they spend more time, and obviously people who live on busy streets either feel like they are "on display" or just don't like hanging around with a lot of cars whizzing close by.

We also suspect that it is much like the difference between small town folks and big-city folks. People who live in a small town are friendlier to each other and decorate the outside of their homes and mailboxes in ways that show they are friendly. On the other hand, people who live in big cities, or on busy streets, are more interested in protecting their privacy. They will be less interested in calling attention to themselves by having a "mailbox garden" that other people may notice. In fact, we found that mailboxes on busy streets tended to be in worse condition overall than on small, friendly streets. They were often rusty or simply in need of a good paint job. Some looked as if they were falling apart.

Some interesting mailboxes were decorated in ways that reflected some interest of the owner. This encouraged our view that mailbox decoration is a means of self-expression. For instance, cat and dog lovers fixed up their mailboxes to look like little cats or dogs. Other mailboxes resembled boats or airplanes. Lots of mailboxes look like little houses, which bolsters our opinion that mailboxes serve as an extension of the main house, specifically of the front door. These mailboxes were outfitted with peaked roofs, shingles, front doors, and painted-on windows complete with shutters. Simpler modifications include banners and prepainted mailboxes with pictures of birds or flowers on them.

We also became interested in the so-called "fortress" mailboxes, which are all the rage now. These are mailboxes actually built into a brick, stone, or stucco pillar placed at the entrance to the driveway. They are de rigeur in the fancy cluster homes that are springing up in trendy neighborhoods. Like other decorated mailboxes, they don't function any better than a standard box-on-a-post, but they are imposing structures. It is as if the owner is telling the world that the people who own these mailboxes are solid citizens and pillars of the community, so to speak.

We became concerned because we had noticed that many mailboxes had been hit by cars, often repeatedly. They were dented or nicked or leaned over unsteadily. We agreed that in a collision between a mailbox and a car, it was better for the mailbox to be damaged than the car, because it would be far less costly to repair or replace a mailbox. But a fortress mailbox changes the balance. A collision between a car and a brick pillar is going to damage a car, sometimes seriously. We decided that fortress mailboxes might look fancy, but are ultimately a waste of money and can often raise the asking price of a new home by more than $1,000. Imagine our surprise, then, when an article appeared in the local newspaper reporting that a local community is seeking to halt the building of more fortress mailboxes because they are a traffic hazard. It seems that there have been a number of cases of cars that, veering slightly off the side of the road, have been totaled or started chain reaction accidents because they struck a fortress mailbox.

We learned a lot about decorated and landscaped mailboxes, but we still didn't answer our question: are mailboxes decorated as a form of self-expression, as an extension of the house, or a combination of both? We decided that we needed a little more systematic research method.

We went through the city directory and found people whose names were colorful such as "White," "Brown," "Green," or "Gray." I had once known someone whose last name was Brown who seemed to think it was very important to have a brown mailbox, so we thought we'd see if other colorfully named persons felt the same way. We noted the addresses and set off. After checking twenty-five addresses we found only one, the Gray family, had painted their mailbox to match their name. This downplayed the importance of mailbox decoration as a form of expression. We did, however, notice one thing that we had not noticed before we started looking at color. People often paint their mailboxes to match their house. For example, green mailboxes often go with houses that have green shutters. Same goes for most of the colorful mailboxes that we saw. We saw red mailboxes with matching shutters, white with white, and so on.

Our ultimate conclusion is that, indeed, the mailbox is an extension of one's house. The outside is public, but the inside is private. It holds our hopes and dreams and secrets. Because it is such an important space in our lives, it is important to take good care of it. For those of us who are gardeners, that means landscaping the mailbox. So go ahead, plant a few bulbs, lean a trellis next to the post for a climbing vine and add a good-looking rock for effect. After all, mailboxes are an extension of us.

Physical Predilections of the Moon

Leeda Alleyn Pacotti

Remember when your mother told you to be careful at play and to stay away from strangers? She insisted that you handle sharp objects carefully, avoid poison, and wear good shoes. She even told you not to stand under anything dangling from a rope. Back then, you thought she was a pessimist, but she wasn't. She was actually talking to your Moon.

While radical congenital difficulties are shown by the Sun, your Moon is also an important indicator of your health. Your Moon Sign describes acute illnesses, accidents, and acquired health problems either developed through habit or derived from your environment. Interpretations of the Moon's placement by sign usually explain the relationships, environments, or emotional responses you have with persons and situations met in life. Although not minute-by-minute accounts, they are the background routines that tend to have a strong influence on how you act and react. Actions, reactions, and inaction lead to patterns and habits that can develop into illness by inhibiting your good health and awareness. If you continually ignore you interactions with your surroundings, you could find yourself in a chronic state of discomfort.

Illness, Accident, and Recuperation

At your birth, the Moon was in a specific sign. Throughout your life, that natal position has described how your body responds to passing or repetitive illness and accidents. Though the Moon in your chart indicates these difficulties, it also provides you with certain strengths and personal healing techniques to withstand them. For instance, each Moon sign endows a healing light. During hospital stays or prolonged rest, you can bathe in the glow of colored light to enhance your recuperation and healing.

With such information, you can intervene against illness and use your strengths to remain healthy. An insightful or holistic health-care provider, with an understanding of medical astrology, will give you more information and help you ward off health problems by training you to change destructive habits or minimize them with preventative care.

Moon in Aries

The Aries temperament is known for its impetuosity. You jump straight into any fracas at top speed without thinking and often get in over your head. In the body, Aries rules the head, the muscular system, and the adrenal glands, which pump adrenaline into the bloodstream and account for a body's vigor, speed, and abrupt changes.

Speed characterizes the types of ailments and accidents of the Moon in Aries. Your illnesses tend to be immediate or heated: facial neuralgia, twitches, fevers, gum infections, headaches, hemorrhaging or bleeding, inflammations, measles, muscular spasms, convulsions, postnasal drip, sharp pains, sores, toothaches, tinnitus of the ears, vertigo, or weakness of the eyes. As a matter of habit, you can exhaust yourself by attempting to simulate adrenaline rushes through overconsumption of stimulants in coffee, tobacco, and pills. Accidents, including animal bites, burns, cuts, crashes, and scalded skin, are quick and happen regularly.

To avoid health problems, you benefit from regular health examinations and early attention to aches, pains, and twinges. When recuperating, bathe in a bright red light or wear combinations of scarlet and hot pink to evoke your sense of hope and energy.

Moon in Taurus

Deliberation and deliberative action describe Taurus. Being a person of few words, you prefer to take your time, observing activities and resources before you step into situations. Once committed, however, you are in for the long haul. Taurus rules the throat, neck, and the thyroid gland, which regulates the body's metabolism through rise and fall of internal temperature.

Taurus illnesses tend to be durational. Generally, your throat is the target of soreness, dryness, and tonsillitis. Your propensity to wait and see is not a blessing, as throat problems could cause abscesses or membranous growths. Your ability to wait also reflects in your body, where you usually tend to put on pounds from low activity or an underactive thyroid. At the same time, you generally are not accident-prone since you cannot be rushed into anything and always look ahead for problems.

For the most part, you are usually in good health, although you need regular exercise to release accumulated poisons in the body. Overall, you health depends on and benefits from plain eating and living. During periods of recuperation, use a pale-blue light or wear pale blue, which will increase demonstrativeness, kindness, tenderness, and sympathy.

Moon in Gemini

Butterflies must be Geminis; they're constantly in motion and just as hard to catch. As a Gemini, you are adroit at juggling many activities at once and need constant stimulation. Not much of an eater, you are thought to survive on your nerves alone. Gemini rules the arms and shoulders, the brain, the neural network throughout the body, the lungs, and the autonomic activity of respiration.

Many of Gemini's ills concern the breath and nerves, which are overwrought and gradually weakened by constant activity and poor rest. Your desire to be ever-ready for the new, different, and exciting makes you open to lung problems such as asthma, bronchitis, or pleurisy of the lung lining. Without proper attention, you run the risk of viral pneumonia and tuberculosis. Nervous disorders manifest as weak arms and shoulders, with wandering or shooting pains, sleeplessness, and insomnia. You experience accidents such breaking your arms or collarbones, and you can be prone to automobile fender benders.

Rest is your greatest medicine; however, when recuperation is required, a light or clothing in pale yellow will steady your mind, increase mental continuity, and release you from the bonds of sensation.

Moon in Cancer

Adaptable Cancer has a magnetic attraction to the suffering of others. In an attempt to create comfort, you tend to take on others' worries and illnesses. Cancer rules the breasts, mammary glands, and the preparatory digestive system, which includes the alimentary canal, stomach, pancreas, and gallbladder. The stomach, pancreas, and gallbladder prepare food for absorption in the intestinal tract.

Your ailments arise from irregularity. Overeating, junk foods, long intervals between meals, or repetitive eating leads to chronic illness such as food allergies and digestive malabsorption. Imbalanced nutrition overburdens the stomach and deteriorates the pancreas and gallbladder, causing heartburn, indigestion, and stomach ulcers. Food that is not properly treated in the first digestive process cannot be absorbed in the intestines, which leaves the body's cells starved, creating edema (water retention), or a general malfunctioning and abnormal growth (cancer). Accidental poisonings arise from tainted foods and impure water.

Set a standard for pure living; know where your food comes from and how it is prepared. Breaking the habits of junk food and sweets early will minimize your tendency to heavier weight later in life. Any time you are in recuperation, use a pale-violet light or clothing to expunge magnetic transfers from others and soften your awareness of serious illness.

Moon in Leo

Leo has a very regal presence and is the epitome of commitment, directing projects or developing relationships to their natural end. You play and work hard, sometimes overdoing things to the detriment of your health. Leo rules the back, spinal column, and heart—major supports for other body systems. It also rules the thymus, an endocrine gland which generates the immunological defense system in early childhood.

Your illnesses relate to strain on the back, spine, and heart. Because you have a strong constitution, overexertion and physical stress can result in lumbago of the lower back or heart afflictions, such as palpitations and heart attack. When your body is not replenished with food, water, or rest, you are prone to fainting and, in extreme hot weather, sunstroke and heat exhaustion. Generally, you are not prone to accidents, being more careful about your actions than most signs.

You are endowed with a great vitality and strong constitution, although overdoing it can wear you down. Because you keep a youthful self-image, you need to develop a measured pace for work and play after midlife. During any recuperation period or when you need a rest, use an orange light to deter disease and to strengthen weak tissues.

Moon in Virgo

Virgo is logic and practicality personified. Because everything you do is governed by processes, schedules, and rules, relaxation often turns to overindulgence in eating, dancing, or some other constant activity. Virgo partially rules the nervous system, transmitting messages throughout the body from the brain to the small intestines, absorbing nutrients from foods and supplements for transmission to the bloodstream to other parts of the body.

Because you are prone to worry over deadlines, your illnesses manifest as intestinal difficulties, such as colic, constipation, cramping, dysentery, obstructions, and spasms. Mental worries also arise from your need to overcome restrictions and stay within the rules, resulting in nervous tension and an overworked brain. Accidents happen when you are traveling on foot, during hikes and climbs, or when you are overworked.

Easily influenced by your general surroundings, you need a balanced diet to compensate for physical and mental stress and to keep your digestive pathway working properly. Conquer tendencies to overindulge in foods, which burdens operation of the small intestines. Enhance recuperation by using a light or clothing in pale orange to remind you to choose wise conduct in meeting your health needs. Follow this with pale yellow, which will turn your mind toward caring, differentiating thought.

Moon in Libra

Libra loves exclusive, one-on-one relationship. Your sense of fair play makes you sensitive to others' viewpoints. However, your greatest need is to express your singular opinion at the exclusion of anyone else's attitude. Libra rules the urinary system, including the kidneys, bladder, and lumbar region of the back.

Even the most egalitarian of minds can debilitate into stone. Libra demonstrates the principle: You are what you think. The root of illnesses and health problems begin when you fail to speak up, disagree, or stand apart. Libra experiences kidney stones, a general physical debility or weakness, and muscular paralysis. Accidents, arguments, and upsets generally produce kidney troubles. Problems of urinary release create a sympathetic elimination problem in the colon, which can become ulcerated.

You need to exercise and flex your body as much as you do your mind. Because you respond to other people, your recuperation can be easily disrupted by visitors. Use sky-blue light or clothing to keep environmental influences at bay and repel others' psychic emanations.

Moon in Scorpio

Scorpio is characterized by intensity, passion, and retaliation. You do nothing halfway, and the ever-present urgent voice makes you plumb the depths of individual expression. Only afterwards do you feel justified in throwing yourself into a frenzied celebration of life and love. Scorpio rules the large intestine and the reproductive system, including the sexual organs, ovaries, testes, and prostate gland.

Scorpio has been dubbed the hypochondriac. But your ailments are often hidden, either because they are taboo, such as venereal disease, or because your cellular patterning has been disturbed over time by chemical poisoning, radiation sickness, or toxicity from fumes. Other physical problems include swellings, ruptures, and hemorrhoids. You experience accidents near chemical plants, environmental spills, and nuclear sites.

Lethargy affects the colon, which operates more efficiently with exercise. Although others fear your intensity, do not frustrate your sex drive, or you run the risk of becoming cruel or violent. Your recuperations are usually self-imposed, because you need to release deep energy vortices in your body and aura and these problems are not addressed by most physicians. Use rose crimson-colored light or clothing to strengthen your self-control and open your heart to a more impersonal, universal love.

Moon in Sagittarius

Sagittarius is the court jester of signs. Always ready with a smile or joke, you juggle and back-flip through the demands of the day with boundless energy. Keeping up expectations takes a toll, however, so you often give into overindulgence in food and activities. Sagittarius rules the controlling parts of the body such as the hips and thighs, which permit standing; the pituitary gland, which gives direction to all other glands; and the largest organ in the body, the liver, which filters and purifies the blood.

Many of your ailments arise through indulgence in food, work, or play. Excessive eating, besides adding weight, inhibits proper food absorption, which can introduce unprepared particulate matter into the bloodstream and overburden the liver. Consequently, you may fall ill with blood disorders, such as anemia from poor absorption of minerals, or sluggish blood from poor purification. If bodily systems are hampered, these blood ills could compound into leukemia. Failure to exercise your body can result in lameness, with aches through the hip joints, thighs, and into the knees. Accidents, especially broken legs, occur during vacation, sports, or anytime you try to overcompensate for sedentary habits.

Clearly, you need exercise to keep blood moving and prevent excessive weight. During recovery periods, use an indigo blue light or clothing. This outwardly soothing, yet inwardly stimulating, color will spark your imagination and help you study the causes of your ills, which frequently stem from past life influences on your present conditions.

Moon in Capricorn

Like the ant who prepared for winter, Capricorn understands the necessities of life. While others stick to the moment, you take responsibility in planning for the future while securing the present. Capricorn rules the teeth, knees, skeletal system, and the preparatory and support structures for the body. Its organs are the skin, which possesses finely tuned sensing abilities, and the spleen, which removes dying red blood cells.

When you are oversensitized, you experience skin problems, such as acne, eczema, hives, impetigo, psoriasis, and shingles. These, if ignored, can fester. Structural problems come from inadequate hygiene or movement, resulting in arthritis, dental troubles, bone disorders, obstructed movement (especially the knees), and rheumatism. Dull, heavy aches or pain, colds throughout the body with chills, and all wasting diseases fall within the province of Capricorn. Mishaps and accidents include blows, bruises, contusions, dislocations, falls, and being struck by falling objects.

Born in the midst of wintry cold, which imitates the chill of the grave, you need spring green to uplift your recuperative periods and stimulate a fondness for knowledge. Bathe in or wear this color throughout any illness or when you know aches and pains have their root in too much responsibility. Always remember that spring follows winter.

Moon in Aquarius

Unpredictable and inventive, Aquarius is the most independent sign. Alone but never lonely, your imagination begins where others meet a boundary. Aquarius rules the lower leg, ankles, circulatory system, and pineal gland, which controls the stability of the mind and is the entrance for divine revelation.

Because your mind is electrified with unusual ideas and knowledge, your illnesses manifest from the body-mind relationship. Hysterics, nervous disorders, and psychological disturbances are common, but can be laid to rest when you sort out your thinking. Circulatory problems such as blood diseases, hardening of the arteries, lower leg pain, stroke from clogged arteries, and varicose veins can hamper you. In keeping with the unusual and unidentifiable, uncommon disorders, rare diseases, incurable disorders, and catastrophic epidemics are controlled by Moon in Aquarius. Accidents include electrocution, fractures of the shins and ankles, machine accidents, and all unexpected or strange hurts.

True to Aquarius, your healing light is very different from other signs. Because your nerves and psychic aura are easily strained from persistent receptivity, use multicolored, plaid, paisley, and variegated cloth over a white light, or wear any of these designs, to cultivate tolerance and perception from many viewpoints.

Moon in Pisces

Delicate, kind, and compassionate are qualities that exemplify Pisces. Understanding is your by-word, and your perspective on things lets you see the common threads running through humanity's suffering, pain, and ills. Though deeply immersed in alleviating the pain of others, your recognize the need to protect and nurture your heart and its unbounded charity. Pisces rules the feet and sensory transmissions through the nervous system. It also governs the thalamus, which transmits stimuli between the associative parts of the brain and the sensory organs.

Melancholia, mental complaints, and obsessions are attributed to Pisces. Because the thalamus participates in associative thought processes and viewpoint development, look for difficulties with the sensory organs, including the skin, or pressure on the brain from infected sinus cavities, concussion, or skull fracture. Physical ailments relate to the feet, including bunions, chilling cold, dampness, and a wandering gout throughout the feet or located in one toe. All foot problems for Pisces are severe, potentially impeding walking and standing. Accidents occur during travel or recreation on water, and while inebriated, drugged, or improperly medicated. During these periods, sensory messages become confused, entangled, or garbled, or the feet cannot maintain their stance.

Owing to your bodily hypersensitivity, all medicines must be carefully administered. Your feet affect the entire body, so shoes must be well-fitted and supportive. Being drawn to and sometimes overwhelmed by any animal or person in pain or suffering, you must stay hopeful and kind. Through healing periods or any time you replenish your bodily or psychic energies, use a silvery gray light or clothing that will clarify, refine, and purify your relationships with loved ones and companions.

Sympathetic Triggers of Illness

Every twenty-eight years, the Moon reveals a fascinating journey and signals of developing illness by making a full progression through the zodiac. This movement is regular, producing periods of sympathetic illness related to your natal position. If you are aware of this possibility, you can treat the sympathetic illness with the appropriate remedy and take steps to avert or minimize a future health crisis.

As the Moon makes its zodiacal trek, it comes in waxing square, opposition, and waning square, before returning to its natal place. The approximate age periods of the waxing square are 7, 35, 63, and 91. The age periods for the opposition are 14, 42, 70, and 98. The waning square occurs at ages 21, 49, 77, and 105. The Moon returns to its natal position at ages 28, 56, 84, and 112.

As these aspects occur, you should look at the sign in which the Moon has progressed. Study will help you understand how illnesses or accidents in the progressed sign exacerbate ailments of the natal. For example, when the Moon progresses to opposition in Libra from natal Aries, kidney or urinary difficulties may arise related to overstressed adrenal glands. Unless you take precautions, you may have problems when the Moon returns to Aries. Also, as the Moon progresses to waxing square in Sagittarius from natal Virgo, Sagittarian overeating and sense of sluggishness will disturb the absorption powers of the small intestines. Clearly, Virgo needs a careful diet to withstand the influence of expansive Sagittarius. And during the Moon's passage into waning square from Scorpio to natal Aquarius, long periods of lethargy or sedentary habits will disturb the circulatory system, which can later manifest as varicose veins in the legs. Walking regimens will keep both problems at bay.

In general, the square and opposition fall within the same quadruplicity. In other words, cardinal signs have cardinal aspects; fixed signs, fixed aspects, and mutable signs, mutable aspects. The following list shows the natal sign and the sympathetic signs for waxing square, opposition, and waning square:

Moon in Aries: Cancer, Libra, Capricorn

Moon in Taurus: Leo, Scorpio, Aquarius

Moon in Gemini: Virgo, Sagittarius, Pisces

Moon in Cancer: Libra, Capricorn, Aries

Moon in Leo: Scorpio, Aquarius, Taurus

Moon in Virgo: Sagittarius, Pisces, Gemini

Moon in Libra: Capricorn, Aries, Cancer

Moon in Scorpio: Aquarius, Taurus, Leo

Moon in Sagittarius: Pisces, Gemini, Virgo

Moon in Capricorn: Aries, Cancer, Libra

Moon in Aquarius: Taurus, Leo, Scorpio

Moon in Pisces: Gemini, Virgo, Sagittarius

No magic is involved here. Self-awareness, a good diet, positive health habits, and a good health-care provider will ward off many of these ills. Staying alert to your surroundings will keep you from having accidents. But, foremost, remember Mom and all her words of wisdom. Play carefully, work carefully, and get plenty of good rest. You're going to be just fine!

Choosing Day Care by Your Child's Moon Sign

Leeda Alleyn Pacotti

The baby needs a change, and your two-year-old has decided "no" is not good enough. Your four-year-old has discovered that frog innards make a dandy finger-paint, and to top it all you've decided to get a job and begin the search for day care.

Faced with the modern nightmare of juggling so many responsibilities, you are confronted with two important issues: first, acting as a self-supporting member of society, and second, nurturing the growth of your child or children. Because you recognize the special role you play in your child's life, the search for a sympathetic, beneficial day care can be, at best, confusing.

Still, you should not give up hope. Astrology provides some helpful indicators regarding child care. Aside from the practical questions of costs and proximity, you can choose day care appropriate for your child and avoid unnecessary or costly mistakes that could hamper or disrupt your child's development.

Your child needs to learn important life lessons and develop life-long talents and abilities. His or her Moon sign explains the role of the parent and surrogate nurturers in making your child receptive. The sign of your child's Moon shows what area of life needs to be formed and strengthened through your example and influence, what obstacles need to be overcome, and what activities or environments help your child gain command over his or her abilities. From the powerful directive of the Moon sign, selecting a nurturer and a proper day care environment that will augment your duties as a loving parent becomes a gratifying task.

Moon in Fire Signs: Children with the Moon in Aries, Leo, and Sagittarius

Children with the Moon in Aries, Leo, and Sagittarius see you, their parent figure, as a source of courage, direction, and fearlessness. They believe

you encounter life with gusto, seeking it out rather than waiting for it. Problems, obstacles, and challenges are opportunities for you, a powerful attribute in dealing with life's great blind-sides.

Fire children are curious and adventurous, constantly trying to understand how the world fits together and how they fit into it. Once they attain this understanding, it forms the basis of their spiritual expression for the rest of their lives. These children relate intuitively to all perceived life, benefiting from environments where they can nurture pets, small animals, and plants. Their day care facilities need to be near open space, so they do not feel hemmed in or stifled.

The child with Moon in Aries recognizes that you, the parent, hold some prominence in life. You are probably a supervisor or an executive, usually in a professional field, or you're being groomed to step into these responsibilities. With great amounts of inner poise, self-discipline, and self-assurance, your talents for creative approaches and ice-breaking humor attracts attention. Your child's eyes light with admiration of you.

This child needs to learn consideration of others, especially because he is so focused on your mutual relationship. Taking others into consideration will teach him to think before acting. Through social interaction, he will learn to use his original way of thinking to join the group and not alienate himself. day care programs should include nature walks, field trips to zoos, and care for small animals. His day care nurturer must be energetic, youthful in attitude, and perceptive of each child's individuality. At the end of the day, you may discuss his daily encounters and help him assess his experiences so he will gain self-assurance.

The child with Moon in Leo knows how to take charge and has observed you doing it many times. You are a manager or administrator, involved with project planning. You know how to motivate others, are excellent in finding the basis of problems, and understand the need to see things through to the end. You are not a quitter, and your child wants to be just as dependable as you.

This child easily organizes and directs a group, but she needs to learn to be part of it. Participation in group effort will teach her humility and allow her to limit her responsibilities. Once she understands everyone is important to the outcome, she'll let herself relax by dabbling in art forms. Basket weaving, painting, and band activities let her play while observing how components are organized into an integrated whole. Her best day care nurturer is a well-schooled and poised person over forty. When you receive her full report on the day's events, be sure to ask about her recreational activities and why she liked them.

The child with Moon in Sagittarius sees you as a citizen of the world. You are a project leader, able to see the broad spectrum. You know to look under the surface for the real cause of problems and are willing to go great lengths in time and effort. You bring great knowledge into your problem solving. Your child recognizes that your imagination and vitality spring from an inner well of faith.

This child learns cooperative effort through problem solving, where he can talk about his ideas and solutions and surprising methods. He's always ready to implement them immediately. Contribution and recognition in joint effort will help him gain confidence, direct his quick mind, and create inner faith that his ideas are worthwhile. He benefits from team activities in sports, riding ponies, and quiet moments of reflection. His day care nurturer, who may be a grandparent, must show wisdom from life experience and a philosophical depth in helping this child think through problems. When the two of you are home again, help him keep a memory journal of his accomplishments.

Moon in Earth Signs: Children with the Moon in Capricorn, Taurus, and Virgo

Children with the Moon in Capricorn, Taurus, and Virgo see you, their parent, as a source of responsibility, steadfastness, and care. You meet each challenge with strength and keep sight of life's other chapters. You know how to divide your energies for the needs of the moment, an important personal resource in maintaining a calm, steady pace.

Earth children are reserved and retiring, waiting for the world to disclose itself before exerting themselves. They are preoccupied with time,

measurement and method, and gain an understanding of how their physical limits and strengths influence the outer world. This understanding
lets them select appropriate careers and interests. These children relate
strongly through the five senses, benefiting from environments with tac-
tile experiences, boundaries, and short projects. The day care facility
needs to have structured areas for play, class, and garden. Greenery is essential to foster a feeling of stability in these children.

The child with Moon in Capricorn sees you as a natural leader. You
probably own your own business or have an influential position as an executive or advisor. While you are probably not a direct supervisor, you
have influence on nearly everyone. Up front, you understand competition, customers, and staying power. Beneath the surface, you cultivate the
interrelationships of politics and commerce. Your child sees you as larger
than life itself.

This child benefits from day care when she is the older member. She
seeks to take responsibility for those less experienced or knowledgeable—
not as a parenting experience, but as an opportunity to relate to the needs
of others. Being ambitious, she will take charge; on the other hand, her
responsiveness to others will teach her to weigh her own feelings and
preferences against their needs. This child has hidden fears that she cannot measure up to your example. Responsibility for others will teach her
to confront these fears. Her activities should include clay modeling, one-
on-one games, and role-playing. Her day care nurturer can be a grandparent or an older person who is capable of explaining life's intricacies.
At day's end, draw out her experiences and help her find humor in them.

The child with Moon in Taurus quickly gravitates to your command
of the physical world. You create new physical forms in either the building trades, durable goods, or interior design. No matter your arena, you
have a steady eye on the appreciated and appreciable, recognizing value
and durability. Your lack of pretense makes you a person of few, but
choice, words. Your child knows that what you do and say has meaning.

This child needs to play or work alone. Because he may have mis-
taken your orchestrated actions as stubborn, he needs to learn how to flex
his energies in many ways. Although he is quiet and respectful toward
others, he needs to observe from them how his actions and opinions have
value. His activities will include voice control through poetry recitation
or singing, building toys, and cultivation of potted plants. His day care
nurturer needs to have an outgoing, good nature and show a quiet sense

of humor. After his daily activities, keep him focused on what he did well and how those accomplishments will serve him when he is older.

The child with Moon in Virgo is mystified by your careful methods and concentration; she thinks it's magic. You are the first and last line of defense in any operation. You excel in knowing how each segment in the overall scheme fits together and why. Your child marvels at how much you know and how you keep your knowledge at your fingertips.

This child learns through thinking and talking. Even from her limited perspective, her opinions will be carefully thought out, logical, and meticulously spoken. Because she is sure she is correct, she needs to learn the art of conversing with others. This child benefits from large groups, where she can be exposed to a variety of people and appreciate their differences. Because she has no discipline problems, she can concentrate on learning and applying rules. Her activities include preparing for school by writing her alphabet and numbers, reading, and using phonetics. Her day care nurturer must be a joyful individual, consistent and considerate with rules. After her recounting of the day, give her some quiet time alone for drawing or puzzles.

Moon in Water Signs: Children with the Moon in Cancer, Scorpio, or Pisces

Children with the Moon in Cancer, Scorpio, or Pisces see you, their parent, as a source of flexibility, depth, and quiet courage. During each day, you keep an eye on the universe and demonstrate respect for the consequences of your actions. Your inner vision penetrates the moment, letting you decide how much effort to put into each activity and assess how to maintain your personal energies in the midst of a seeming chaos.

Water children are gentle, peaceful, concerned, and intensely involved in their immediate environment. The consequences of their actions consume them as they observe life's connections. Understanding how emotions cause consequences helps them gain an inner stance that they will maintain through all interpersonal relationships. These children relate through emotional expression, benefiting from environments with singing, poetry, and animal tending. Because of their emotional sensitivity and tendency to absorb the problems of others, these children need solitude with a sitter at a private home rather than group day care.

The child with Moon in Cancer recognizes that you know how to protect yourself in the world. Your career involves nurturing, either in the food, furniture, or hotel trades. Whether you are in management or sales, there is a high demand for your resilience and ability to solve problems without sacrificing the operations of the group. Your child readily senses your inner security and trusts you for safety.

This child is very sensitive to his environment and influenced by his surroundings. His greatest difficulties lie in developing perspective on life's problems, often absorbing the problems of others as his own. Because he is helpful and caring with other children, especially younger ones, he must learn that the events around him are not all part of his life. This child responds easily to admonitions and benefits from uplifting fairy tales and fables; other activities include swimming and finger-painting. Because he may be attached to you, his best sitter might be a relative or a sitter in his own home. At the end of his day, have him describe the problems he encountered. Because his emotional tensions build easily, show him playful, healthy ways to let his emotions out.

The child with Moon in Scorpio is awed by your strength to stand alone. Your work frequently demands that you think privately and deeply, which suits you for medical research, state-of-the-art technologies, or cutting-edge businesses. Whether you are the sole proprietor or a valued employee, you scout the unknown terrain ahead in preparation to lead the pack. Your child understands your intensity and forthright truthfulness will prepare her to meet life's challenges.

This child is very much a loner, with a strong sense of life's limited term. Her greatest challenge is to control her emotional energies, especially anger and coarseness, because what she knows about life's cycles can be shocking to the young minds of her peers. Emotions rage within her. Her early years should be filled with time alone to investigate her world and her mind, while observing the actions of others from a distance. This child needs activities which introduce the element of the unknown while permitting her to find solutions—such as working with puzzles and studying animal habits through observation. Her best sitter is a grandparent or a tutor, who takes her on field trips and stimulates her. When the two of you are together at the end of the day, give her some private time to readjust to life with you and help her talk frankly about her experiences and questions. Your time with her day care provider will entail extra planning to meet this child's needs.

The child with Moon in Pisces appreciates your compassion and charity. Your career is built on service to others in your work in a hospital or other service institution. In any situation, personal, social, and business, you are sought out to listen and soothe. The crises of others are no problem, and you have plenty of understanding regarding the life's variety. Your child knows the advice, concern, and love you bestow on others never empties the well within you; there will always be plenty for him.

This child lives on the seas of cosmic emotion, which pull his heart open. He is more quiet, peaceful, and understanding than most, preferring to step aside while others work. Unfortunately, his desire to get in no one's way can cause him to avoid life's difficulties and fail to become dependable, persevering, or steadfast. His simultaneous residence in the realms of the real and unreal bestow him with high imagination, which he must learn to share with others and observe how they are uplifted by his unseen experiences. This child must not be harshly punished but shown how to rectify hurts. He benefits from mythological and religious stories that connect the timeless to his surroundings. His best sitter is a retired or part-time elementary teacher, skilled in molding minds, who can introduce him early to reading. When you are together, have him reconstruct his day in storytelling and be uplifted by his scope of description, kindness, and deep perception.

Moon in Air Signs: Children with the Moon in Libra, Aquarius, and Gemini

Children with the Moon in Libra, Aquarius, and Gemini see you, their parent, as an example of mental discipline, focus, and curiosity. You use your mind to create imaginative solutions while keeping your perspective on varied tasks and schedules. You understand how the focus of mental power elevates you from the toil and strain of destructive routine.

Air children are inquisitive and curious, allowing everything to capture their mind. They need to develop attention, concentration, and absorption, and gain command over their considerable mental power and knowledge. From these strengths, they will select pursuits and interests that require a disciplined, creative mind. Structured class work, individualized instruction, and group socializing allow these children to go deeply into themselves and learn to deal with others. For them, a school environment with occasional field trips works well. In all air signs, the day care instructor must appreciate these children's needs for a hands-off approach. Their minds need direction, not rote learning.

The child with Moon in Libra respects how you keep life in balance and are never overwhelmed, even under changing circumstances. Your career signals diplomacy and consideration, whether you are in law, upper management, or arts. At work, you are the fulcrum at the center of many seesawing opposites. You are able to see all points of view without losing your own. You understand how a variety of parts integrate into a whole, whether the parts are people or materials. Your child knows fairness is your hallmark and relies on the safety of its boundaries.

This child is calm but influenced by others. Because she prefers one-on-one relationships, she needs to learn to broaden her focus and cultivate many friends in the group. Her desire to avoid conflict must be overcome so she will learn to speak up. To gain an understanding of structure, she will enjoy visual teasers, origami, and abstract exercises which emphasize geometry and architecture. Her day care instructor must exhibit poise, grace, and compassionate intellect, providing this child with an example of how to maintain her bearing, while accounting for the needs of others. When you are together, let her talk when she is ready; quiet moments with you are more refreshing for her than giving daily reports.

The child with Moon in Aquarius enjoys your quirky turns of mind and is amazed at how you turn unconnected events and resources into incredibly helpful advantages. In your career, you can't be pinned down. You are constantly inventive, exhilarating, full of surprises, and prefer self-absorbed research in technological development. Whether you are at work or pitching in to help others, you are the one called when a problem won't resolve itself. Your keen insight, usually unexpressed, yields remarkable, fully logical solutions once others see them unfold. Your child loves your scientific side and feels free to experiment with his own ideas.

This child, though inquisitive and inventive, is sensitive. He will make many friends but cull them carefully so they will not interfere with his solitude. Throughout his life, he will always compartmentalize his interests and be somewhat aloof in personal relationships. He needs to learn how to express his feelings, show warmth to others, and speak of the passion that turns his mind toward invention. Structured nap periods let him conserve and replenish his nervous energy, which can be seriously depleted when he is absorbed in projects or groups. A perpetual tinkerer, he needs toys, safe objects, and projects that can be taken apart and innovatively rebuilt, such as erector sets. He will also enjoy reading about other cultures and people. His day care provider should have interests in mathematics and science. At home, provide a corner or wall to display his projects and encourage him to explain what he has made.

The child with Moon in Gemini is intrigued by your ability to cope with constant change. Your day is filled with short trips, telephones, and talk. Social connections pop out at every turn, demanding attention from your facile mind. Your career in news reporting, celebration planning, or urban development keeps you on your toes. As an employee, you are the one who opens doors with your abounding friendliness. Your child loves the easy openness you exhibit to everyone who comes your way.

This child is physically and mentally hard to pin down. She derives great benefit from structure; without it, she will be constantly talking, moving, and flitting about without focus. She will know everyone, although she may not care to develop a relationship beyond knowing names. Her greatest needs are to listen, concentrate, and persevere. Classroom groups in which children learn to respect others' conversations and report back what they understand will help order her churning mind and restless body. Because she enjoys verbalizing, she needs to pattern her thoughts into coherent conversations or narratives through storytelling, bird song studies, and writing her alphabet. Her day care nurturer needs to be a resilient younger person who can keep up with her physical restlessness. At home, she needs a short report period to tell you about the day's events and a nap to rest her brain.

When you have more than one child in day care, you have a special situation. It is unlikely that all your children will have the same Moon sign. However, they may have Moon signs in the same element (fire, earth, water, or air). In this situation, choose a day care facility described for the element.

When none of the Moon signs are in the same element, choose a day care facility of benefit to the oldest child, who will be the first to venture out in the world alone. Your younger children, who are less socially involved, will still find enjoyment and curiosity if you request specific developmental activities for them. As the next child nears preschool age, consider finding new day care for him or her.

Introducing the intuitive into your search for day care can result in a wonderland for your children. Instead of dread and worry, the best choice will give you a welcome, personal fulfillment.

Baking by the Moon

Heyde Class-Garney

With our demanding work schedules and busy lifestyles, baking has become almost a luxury today. And with the world's increasingly fast pace, it is no surprise that people often overlook the effect of the Moon on such a simple task as baking.

Still, on those occasions you do find time to bake, you will be most satisfied if you plan according to the Moon. This may sound silly, but the Moon can help you produce some delightful, hot goodies from the oven.

For instance, during its first and second quarters, the Moon will tend to foster growth in all things. If you choose to bake during the early, or waxing, quarters of the Moon, the growth and expansion of your baked goods will be encouraged. At the same time, you can do no wrong if you bake during a Cancer Moon. This Moon is well-known for its nurturing qualities and just so happens to rule over the stomach, which explains why some people have a tendency to overeat two days of the month. But more to the point, during a Cancer Moon you will feel very inclined to feed people, and baking will seem like hardly any work at all.

Unfortunately, the Cancer Moon only comes around once a month for only two days, and there is only a fifty percent chance that this Moon will occur during the Moon's waxing phase. If you have to bake some other time, you can settle for the other cardinal signs of Aries, Libra and Capricorn, as they will produce excellent results during the Moon's first and second quarters. On the other hand, you are not strictly confined to any particular time for baking. You simply have to realize that if you want the best possible breads, cakes, and cookies, the above times are optimal.

In fact, there are many factors to consider when baking. Some bakery products, for instance, are more suited for use during particular Moon signs. In all the zodiac signs, foods have certain types of energies and are ruled by their own element and planet. The trick is to chose bakery goods that correspond to the energy of each sign as the Moon passes through it.

In the list below, different factors have been taken into account for you. With a little ingenuity, creative genius, and some good timing you can have some great baking experiences.

Moon in Aries

Don't be surprised if you find yourself wanting to bake up a storm today, as this fire-ruled sign will inspire you to spark up the oven. As the Moon moves through Aries, zesty, robust, and flavorful baked goods will serve. So if the spirit moves you, go ahead and bake up rich poppy seed cakes, zingy gingerbread cookies, and any spicy goodies you can think of.

Moon in Taurus

Pass up any store-bought baked goods; in this Venus-ruled sign you will feel an appreciation for the finer things in life, including a day spent in baking bliss. Taurus is a resourceful and practical sign, with a good dash of determination mixed in, so as the Moon moves through this sign you are primed for some productive baking. Comfort foods such as apple and cherry pie, oatmeal cookies, and banana bread, will be perfect for this placid, domestic Moon sign.

Moon in Gemini

As the Moon moves through Gemini, your day is most often spent in quick conversation and frantic errand-running, so it's more likely that you will talk about baking rather than actually doing it. If you actually do find the time to bake during a Gemini Moon, you will be lucky indeed. Your best bet will be with quick-baking recipes that require little preparation—almond cookies, date muffins, and packaged cake mixes.

Moon in Cancer

The Cancer Moon is tailor-made for the proverbially happy homemaker. As the Moon travels through this sign, you will want to provide for everyone. Comfort foods that make your loved ones smile are best to bake now—blueberry muffins, pumpkin cookies, coconut cream and lemon meringue pies. Your entire family will be very receptive and will thank you wholeheartedly. They may be so happy, in fact, that they clean out the garage or attic.

Moon in Leo

You can show off your baking expertise during the Leo Moon. Your creativity and ability to shine are at their height now, and you will find your efforts most highly appreciated. Showy, complex baked goods are best now; the aromas of pineapple upside-down cake, chocolate chip cookies, and cinnamon-raisin bread will produce many a winning smile.

Moon in Virgo

This practical and health-conscious sign is the best time for using your more detailed and wholesome recipes. Since perfection is Virgo's middle name, you can easily keep up with the highest of standards. A sensible choice for the day would be an earthy whole grain bread. Millet, oat, or barley are good grains to work with. And if you feel the need to spoil yourself, some peanut butter cookies or pecan pie would work nicely.

Moon in Libra

With so many options, making a decision can prove impossible during the vacillating Libra Moon. But if it is any help, remember that Libra governs sweet foods. So you may choose to indulge at this time with any and all types of cookies, cakes, pies, cobblers, sweet breads, or pastries, though your thighs won't enjoy the extra calories. Of course making up your mind which treat to eat first could also prove quite a feat.

Moon in Scorpio

As the Moon moves through Scorpio, it is not at all unusual to feel drawn to the kitchen today. The intense nature of the Scorpio Moon may cause an obsessive and overwhelming need for a sweet fix. So you may well be satisfied by treats such as snappy ginger cookies, savory onion bread, and hearty blackberry pie. These items will fit your mood quite well too.

Moon in Sagittarius

During the Sag Moon, you will feel optimistic and an adventurous, so you may very well find yourself exploring new baking horizons. Plenty of fresh ideas will fill your head, and your ordinary recipes will scarcely seem enough. Look to bake expansive, filling foods, such as macadamia nut cookies or carrot cake, or specialty foods from other countries, such as lebkuchen. But be wary of the tendency to overdo it—during Sagittarius it is difficult to say "enough is enough."

Moon in Capricorn

Hard work and responsibility fuel the Capricorn Moon, so you may feel a little dizzy from your achievements today. In fact, you'll be so efficient you might find yourself planning your baking needs for the next few months. So while you have the bug to organize, you should take advantage of it to conquer your most difficult baking endeavors. Preparing corn or cheese bread will leave you with a feeling of success. You may even chose to enter your rhubarb pie or cranberry muffins in the country fair.

Moon in Aquarius

Don't be shocked if the notion to bake hits you like a bolt from the blue during the Aquarius Moon. This Uranus-ruled sign thrives on doing the most original, unexpected, and unconventional things. So go ahead and experiment with that unusual recipe, and share your efforts with your many friends and neighbors. Remember, the sky is the limit, so tingle your taste buds with some caraway muffins or mulberry pie. Or you can invent your own recipes for pistachio nut or hazel nut cookies.

Moon in Pisces

What a wondrous day to trust your intuition and use your imagination. Ruled by dreamy Neptune, Pisces will help you conjure up new recipes, or will help you go with the flow and surrender to your innermost desires. During a Pisces Moon, you will find yourself baking water-ruled foods such as peach cobbler, apricot coffee cake, or sweet potato pie. So bake on to your heart's content!

Bathing by the Moon

Heyde Class-Garney

There's no time like the present to pamper and spoil yourself, and bathing is a wonderful experience that can be designed to help with relaxation, healing, mediation, and even romance. You can, if you choose, use the Moon in achieving your desired bathing goals.

First off, you should know that the Moon rules over our emotions. The Moon is governed by Cancer, which is well known for its nurturing qualities and sensitivity, and by the element of water, which helps generate our natural healing powers and increase our receptivity. So the Moon is a perfect tool to help intensify or temper our bathing experiences.

Furthermore, the quarters of the Moon are important to take into consideration when bathing. For example, from the beginning of the New Moon until the Full Moon, the Moon's energy is increasing. This, then, is the time to make fresh beginnings or to look for growth. On the other hand, when the Moon reaches its Full Moon stage in the middle of its cycle, energy will be at its height. This climax of energy is intense and could generate "waves," so any bath around the Full Moon will be an intense experience in one way or another. In the third and fourth quarters, the Moon shrinks back to its New stage and energy decreases. This is an excellent time for all matters that need to decline or rest.

There are also many tools available for us to use while bathing if you want to encourage the Moon's assistance. Aromatherapy is the art of using scent to transform or to renew our health and emotions. Certain scents, in the form of essential oils, complement the Moon's role in our bath. You should feel free to experiment until you find ones that have the desired effects, though you should also note that caution is necessary when using essential or fragrance oils. Some oils should not be used in the bath at all, as they can burn and irritate the skin, and pregnant women should avoid them altogether. Below are listed some of the safer essential oils; be sure to check an aromatherapy book if you use different oils or are uncertain.

By taking these precautions, you will have a happy bathing experience. To begin, run your bath water and then use a dropper to add oil. Never add oil into running water, as the stream can evaporate the scent.

Swish the water with your hands to help disperse the oil. Only when you are certain that the oil is safely dispersed, then you can get in the bath.

Herbs serve the same purpose as essential oils but are used in plant form or as an infusion, which is the liquid remaining after boiling and straining the herbs. Herbs sometimes are taken internally or used only externally. You should use some caution and consult an expert before using unknown herbs. You can place herbs directly into the tub, but always try using a cotton sachet to hold the herbs to avoid a huge mess.

Color is another factor that can affect your Moon bathing experience. Each color has specific attributes and produces different reactions. Such items as music, incense, and candles can enhance the mood of your bath. You can place crystals in your bath water to transform the experience. You can experiment with all of these items to develop the perfect bathing atmosphere for you.

A final consideration in Moon bathing is determining the best Moon sign for certain bathing goals. As you must know, a Moon sign has its own energy and unique vibration and creates proper conditions for particular types of events. Planning your bath by the Moon sign can enhance and multiply the effects you want. Keep in mind, however, that in any month the Moon passes though all the astrological signs at a rate of about two-and-a-half days per sign. Each Moon sign creates a mood that has a direct impact on our emotions only for so long. So to get the right influences that you want, you have pick the right day.

The list below explains the dominant features governing the Moon, and by extension your emotions, through the month. It will be simple for you, then, to figure out the optimum times for certain types of baths.

Moon in Aries

Aries is a sign with all the vigor you can handle. The Aries Moon, therefore, is the best time to meet all your energy needs. For even more physical energy while bathing, add a few drops of patchouli essential oil to your bath water. For an added dose of gusto, try a little red food coloring as well. Red will provide you with plenty of energy, enthusiasm, and courage. After twenty minutes or so, you should emerge revitalized.

Moon in Taurus

For a truly relaxing bath, the Taurus Moon is the best time. You can make it simple by adding 2 tablespoons of Epsom salts and a pinch of baking soda to your water. A light amount of gardenia fragrance oil will help you unwind and generate feelings of peace. Also, jasmine oil is excellent for relaxing the body but may make you sleepy. After your Moon in Taurus bath you should be ready to plant your feet back on the ground.

Moon in Gemini

If you are seeking mental alertness, then bathe under the Gemini Moon. During this bath you will jump from one thought to the next and be fully aware of yourself. To enhance this process, add 5 drops of lavender essential oil to your bath water.

Moon in Cancer

If there ever was a time designed to increase your self-love, nurturing Cancer Moon would be it. Of course this Moon sign is excellent for increasing fertility and compassion, and promoting sleep. Add magnolia fragrance oil to increase your ability to give and receive love, and plumeria oil to promote a peaceful and loving attitude. Add five drops of either oil to your bath and a rose quartz crystal for tuning you into loving feelings.

Moon in Leo

No other Moon sign knows as much about pampering as Leo. Born to be a king, or queen for that matter, Leo knows how to live like royalty. For starters, the Leo Moon is a perfect time to take a sunshine bath and flood yourself with radiant light. You may also use six drops of vanilla oil to increase your sex life, love, and energy levels. Rest assured, after this bath you will emerge full of life, joyously erotic, and brimming with love.

Moon in Virgo

Health matters take precedence during a Virgo Moon, so any aliments will benefit from bathing in this Moon sign. To help cure a cold, add eucalyptus or lemon oil to the bath water and breathe deeply. Use the lemon oil sparingly, just a drop or two, as it can irritate the skin. Visualize good health while allowing the healing water and aroma to calm your body and soothe your ills. You will be back to normal in no time.

Moon in Libra

Love and romance take a high priority today. So if you are planning a romantic evening, try indulging in a sensual bath beforehand to get in the mood. Ylang-ylang or jasmine oil added to the bath will further ignite the flames of passion, and six drops of rose oil will help you see yourself in a good light. Rose oil is also effective as an aphrodisiac and helps to channel our thoughts to the healing power of love. You may even throw in a handful of rose petals or light some incense for an added effect.

Moon in Scorpio

The Scorpio Moon is a perfect time for a meditation bath. With the intense and probing power of this Moon sign, you can quiet your mind and sharpen your focus. To help get yourself in the mood, light a white candle while you run your bath water, and add either frankincense or sandalwood essential oils. Both oils are superior for meditative purposes. And if you add some soft music, you can lean back and really enjoy yourself.

Moon in Sagittarius

If you are feeling depressed or overworked, then Sag Moon will come to your rescue. This Moon sign fosters optimism and joy. Clary-sage essential oil will assist you in doing this trick, as it is great for alleviating depression and creating feelings of euphoria and peace. Just add four drops to your bath water, lie back, and revel in the rising good feelings.

Moon in Capricorn

Under the Capricorn Moon, a bath attracts success and prosperity. Bathing now can help you focus your attention on pressing financial matters. So break out the ginger and patchouli oil and sprinkle them into your bath water. You will achieve prosperity in no time.

Moon in Aquarius

Aquarius is known for its objective state of mind, so if you're seeking peace from a chaotic life, the Aquarius Moon will bring matters back into focus. Lavender oil will aid in this too, so add four drops to your bath along with blue food coloring. The color blue promotes tranquility.

Moon in Pisces

To induce sleep and awaken your spirituality, bathe under the Pisces Moon. Chamomile essential oil facilitates meditation and acts as a sedative. A few drops of sandalwood essential oil will draw you into a peaceful bliss. Keep in mind, however, that the dreamy nature of the Pisces Moon sign is suitable for any bathing purpose. So you may pick out a favorite oil and relax in the soothing warmth and intoxicating scent.

Sauerkraut

K. D. Spitzer

September is the month for making sauerkraut in country families. By then, most other fruits and vegetables have been put by and saved for the winter months. In making sauerkraut, the heads of cabbage need to be slightly less than mature—firm and heavy.

Turning cabbage into sauerkraut is a relatively simple task. Traditionally, the preferred cabbage to use in making sauerkraut is one of the white varieties. The canny gardener needs to choose a day when the waning Moon is in a dry sign somewhere along the Pisces to Virgo axis because cabbage gives off a lot of smelly moisture in dry storage. Choosing the proper Moon day to harvest will certainly help reduce this.

The creamy green heads of cabbage are harvested during the sultry days of summer before they swell so with Vitamin C that they burst open and invite the rain or cabbage moths to invade and destroy their hearts. Cabbage keeps well in a root cellar, and by fall families have enjoyed fresh slaw at their cookouts and steamy cabbage soup during the first nippy nights after the kids have gone back to school.

Still, it is in sauerkraut that the country cook sets in motion the perfect culinary marriage. After the crocks of layered cabbage and salt have been set to ferment, the hogs which have been fattening since spring will be dressed, and our busy country cook will use her own secret blend of seasonings to stuff fat sausages and to smoke hams and hocks. When the cabbage has been transformed and the pork cured, the two will be combined in incomparably delectable ways.

The origin of cabbage is lost in the mists of time. We know that ancient Chinese ate it, and it was a beloved staple in Egypt, Greece, and Rome. Cato waxed on enthusiastically for pages about its virtues. Indeed, cabbage juice was the hangover cure of choice in the Mediterranean and the vegetable was a favorite dish for all stomach complaints.

Modern research has proven the ancients were onto something, as cabbage juice has demonstrated its effectiveness in treating stomach ulcers and digestive disorders. Another name for it is colewort, and it has also been used externally for wounds, acne, and even arthritic joints.

The Celts established it as a mainstay in the British Isles before 400 bc and the French gave it its anglicized name from its word for head: caboche. It arrived in the Americas with the earliest explorers, but the middle Europeans raised it to its present status as a culinary art form.

Today, cabbage remains at the heart of an ongoing controversy. In Alsace, a region tossed back and forth between French and German for centuries, sauerkraut is cooked in goose fat with fresh pork and seasoned with juniper berries and wine. And being more French than German, Alsatians are certain their choucroute dishes are superior to the Germans'. The Germans, however, make a comfort food that satisfies even the most critical epicure by combining simple fermented cabbage with wurst, sausage and smoked ham, and potatoes and other vegetables. That is, the Germans are known for their sauerkraut.

Sauerkraut is not just a condiment to drape over a boiled hot dog. If your only exposure to this preserved veggie has been the sour glop sold in stores, then a rewarding treat awaits you once you put out the small required effort.

Good sauerkraut can stand on its own as a vegetable or give marvelous flavor to soups, stews, and casseroles. Sauerkraut can be arranged in stuffed cabbage rolls or peppers before baking. For a zesty and healthy apéritif, mix the drained brine from the sauerkraut with tomato juice in equal parts, and add some lemon juice, horseradish or Worcestershire sauce, salt, and pepper. This drink packs a megapunch of Vitamins C and A, calcium and magnesium, and has only thirty-five calories per cup.

If you dislike sauerkraut, you were probably never counseled to rinse the brine from the cabbage before using it. This is more important when using bulk sauerkraut, but it will certainly refresh the canned version without sacrificing flavor. It is quite common to add caraway seed to dishes with sauerkraut. You can even make sauerkraut with the seeds in it. While many cooks are taken with the flavor marriage of cabbage and caraway, they are unaware of the digestive and antiflatulent qualities of this herb and the power punch it adds in combination with the cabbage. The same is true of fennel seed, which is also often added to cabbage dishes. Digestive aids are always a welcome addition to rich dishes.

It is not necessary to make sauerkraut in forty-gallon crocks. Although the process is still the same, you can make small amounts in quart canning jars and store in the fridge for a couple months or process it in a boiling water bath for long-term storage.

To follow an old housewife's tradition of storing food for medicine as well as nourishment, tuck a few jars of fermenting sauerkraut in your food cupboard. For starters, sterilize as many quart canning jars as you feel you need. Strip the outer green leaves from the cabbage, cut the rest into quarters, and remove the core.

Naturally, wherever cooks gather there is controversy; with cabbage, the controversy is over the thickness of the cabbage shreds. Most cooks believe the thinner the shred, the faster it will ferment. In parts of the country that were settled by Germans, it is still possible to find old cabbage shredders in antique shops. Otherwise, you can use the narrowest blade that came with the food processor to make the thinnest shreds or cabbage for your sauerkraut.

It takes about two pounds of cabbage for each quart of sauerkraut. In this recipe, you must use $1\frac{1}{2}$ tablespoons of kosher or canning salt per jar. Any more or less salt will ruin the batch. Pack the shredded cabbage in the sterilized canning jar, pausing to sprinkle the $1\frac{1}{2}$ tablespoons of salt over the cabbage layers. By the time you fill the jar, the cabbage will have begun releasing its juice because of the salt. Press the cabbage down in the jar and pierce with the long tines of a meat fork to release any trapped air. If the juice does not cover the cabbage, add cold water to top off the jar. When the jar is as full as you can pack it, cover the shreds with one of the outer dark green leaves you removed from the head. For a Martha Stewart effect, cover the top with a washed, fresh grape leaf.

Screw on a new, sterilized lid on each jar, and put the jars in a dark place where the temperature averages 65–70°F. This is a good temperature for the fermentation process. The cooler the temperature of the area you have stored the sauerkraut, the longer it will take to ferment. If you set it to cure in an area that is too warm, it can spoil. Once you have canned the cabbage, put the jars on newspaper as the fermentation process may cause them to bubble over. After eight to ten days, you need to open the jar and check for flavor. Has the cabbage become as tart as you like it? If not, reseal and let it continue to ferment.

When the brine drops back from its previous level, the process is complete, and there will be no more fermentation. If the top of the cabbage or the surface of the brine is pink or discolored, lift off and discard that part. If there is scum at the top, skim it off. Check also for any off-odors, and discard all the contents of the jar in that case.

Once you have determined that the fermentation is complete, you can store the jars in the back of the fridge or in a fruit cellar. To store in the fruit cellar, dump the sauerkraut in a pan and heat just to boiling. Repack in a newly sterilized jar, seal and water-bath process the jar (5 minutes for pints and 20 minutes for quarts.) The jars need a cool dark place for storage; do not allow to freeze.

Now you can experiment and see if your taste buds prefer sauerkraut or choucroute. These are peasant dishes—you won't find chili peppers or cumin or cilantro here to mask their honest flavors. Instead, you will find these dishes to be nourishing and healthy (especially if you trim the fats from the meats).

Sauerkraut meals are even better when served the second day. Make it on your day off and make enough for a couple meals.

Choucroute Garni

2	pounds sauerkraut
6	strips of thickly sliced bacon
2	large onions, chopped
2	Granny Smith apples, peeled and diced
6	peppercorns
12	juniper berries (If you can't find juniper berries at your market, add ½ cup gin)
3	whole cloves
1	bay leaf
2½	cups white wine
6	Italian sweet sausages or large uncooked breakfast links
6	trimmed smoked pork chops

Soak sauerkraut in water for 15 minutes; drain well and squeeze dry with your hands. Meanwhile, using a heavy pot, cook the bacon until nearly crisp; remove and set aside. Sauté the onion in the bacon fat until it's translucent. Add the dry sauerkraut and stir well; let cook about 5 minutes. Toss in the apples and stir. Add the peppercorns, cloves, bay leaf, and juniper berries in a tea ball or muslin bag. Pour in the white wine, cover and let simmer 1 hour. Add a little water if it starts to get too dry. Do not let this dish boil.

In a small frying pan or on the grill, cook the sausages, pricking to let the fats escape. Add to the sauerkraut, along with the bacon and smoked pork chops. Cover and let simmer another 30 minutes. At this point, discard the tea ball, let the dish cool, and then refrigerate. The next day, bring to room temperature and then let simmer 30 minutes until hot.

To serve, place the sauerkraut in the center of the serving platter and arrange the meats around the outside. Serve with new boiled potatoes, garnished with butter and parsley. Make sure you have Dijon mustard for the meats. Serves 6.

Here's another recipe to make a day before serving.

Sauerkraut Stew

1	pound lean spare ribs, separated
	Pinch freshly ground black pepper, or more to taste
1	pound lean cooked ham cut in one-inch cubes
2	pounds kielbasa cut into one-inch chunks
2	bay leaves
10	juniper berries
1/2	teaspoon dried thyme
1/2	teaspoon caraway seeds
1	tablespoon Hungarian paprika
1	large onion, thinly sliced
14	ounces sauerkraut
1	small head of cabbage cut into small chunks
1	Granny Smith apple, peeled and diced
2	cups water
3	cups small stew-sized chunks of potato

Sprinkle the spare ribs with a few grinds of black pepper and brown in a heavy Dutch oven. Be careful not to scorch as it will ruin the flavor of the whole dish. Toss in the ham and kielbasa chunks and stir. Place the bay leaves and juniper berries in a tea ball or muslin bag or tie up in cheesecloth. Add the rest of the seasonings to the meat, along with the onion. On very low flame, sauté the onion until translucent. Do this with the lid on the pot, but be careful not to scorch.

Then add the sauerkraut along with its brine, the cabbage, diced apple, and water. Cover and let simmer about 1½ hour. Toss in the potatoes, cover again and let simmer another 30 minutes or until the potatoes are tender. If you need to add more water, do so with the potatoes. You don't want the stew to be dry, but also not too soupy. Serves 6.

This next soup can be made into a vegetarian version that is as tasty as it is healthy.

Vegetarian Sauerkraut Soup

2   scrubbed diced carrots
1   large onion
2   garlic cloves, minced
½   cup fresh minced parsley
4   cups sauerkraut, drained
2½ quarts beef (or vegetable) stock
4   cups tomato juice
½   teaspoon dried thyme
    Salt and pepper
5   tablespoons uncooked rice
    Sour cream (optional)

Combine all ingredients in a heavy soup pot, mixing well. Cover and let simmer 1 hour. Let sit for several hours and then reheat to serve. Ladle into bowls with a good sized dollop of sour cream. Serve with chunks of dark crusty bread and sweet butter. For vegetarians, substitute a home-made vegetable soup stock. For hearty meat eaters, add kielbasa or other wurst at the first cooking time. Cut into bite sized portions before adding.

Here is a vegetarian dish which can be served as a main dish or as a hot salad.

Sauerkraut and Beans

1   tablespoon olive oil
1   small onion, minced
1   garlic clove, minced
¼   teaspoon fennel seed
2   cups sauerkraut, rinsed and drained
1   cup tomato juice
2   cups white cannellini or red kidney beans, cooked and drained
2   tablespoons minced parsley

Sauté the onion in oil in a large skillet until translucent. Add the garlic and fennel seed, stirring so the garlic does not burn. Then stir in the sauerkraut and tomato juice, cover and let simmer for 15 minutes. Add the beans, recover and simmer 10 minutes. Place in a serving bowl. Serve warm with a tossed salad and warm, dark bread.

Any one of these delicious dishes needs a good bread. If your town doesn't offer an artisan bakery with hearty whole-grained loaves and your bread machine is still in the box, bake this quick bread while the sauerkraut is simmering. It's not a yeast-leavened bread, but for a savory quick bread it's very substantial. Let the bread cool fully after baking before slicing. Wrap it in foil when reheating.

Rye Quick Bread

    3 cups rye flour
    2 cups unbleached white flour
    2 teaspoon baking soda
 1½ teaspoon salt
 1⅓ cups buttermilk
    3 large eggs
  ⅔ cup molasses
    6 tablespoons corn oil
    1 tablespoon caraway seeds

Mix dry ingredients together. Beat milk, eggs, molasses, and oil together. Add to dry ingredients and stir until well moistened. Mix in the caraway. Divide the batter between two greased 9-inch by 5-inch loaf pans. Bake in a 350°F oven 50 to 60 minutes. Check often; if the loaf begins to brown you may have to cover loaf with foil. Let it stand for 10 minutes before removing from pan and let cool completely before slicing.

Wassail

K. D. Spitzer

Good luck to the hoof and horn,
Good luck to the flock and fleece,
Good luck to the growers of corn,
With blessings of plenty and peace.

From the pre-Christian, agrarian orchards of England comes a fertility festival celebration known today as wassail that occurs in the season of the waxing Sun. A corruption of the old Saxon words "waes hael," we understand wassail to mean "be healthy" or "be whole," and today the term survives as a toast that indicates "to your good health."

Back in old England, winter rituals were designed to ensure the fertility of crops and encourage the return of the life-giving Sun. Therefore, on a cold crisp night in January with the stars twinkling overhead, farmers and their friends would move through the fields, shouting in loud voices and making merry with lighted torches, drums, and other noisemakers. Sometimes, the farmers also brought sticks or shotguns and threw or shot them skyward, filling the night with missiles.

Inevitably during the celebration, someone would bring out a hot bowl filled with steaming drink, taking care not to spill a drop. This prized bowl itself had handles on each side and was stored with honor during the rest of the year. Called the wassail bowl, it was brought out from the cupboard on this night for this single purpose.

The merrymakers would wend their way through the orchards to the oldest or largest tree. Using their sticks, they would beat the trunk of the tree or flail at any attainable branches. In time, someone would make a speech reminding the tree of past glories and the quantity and size of its fruit. The other revelers would chant to inspire the tree to greater fruitfulness in the coming season.

This ritual is traditionally celebrated on Twelfth Night (January 5), which, of course, is twelve days after Christmas, although some people used the old Gregorian calendar and celebrated it on January 17.

The timing of this holiday is intended to act as an end to the wild chaos, when the otherworld ruled the human realm. These are dark and dangerous days, when the returning Sun is still unstable and may stay lost to the night forever. It's necessary to encourage the Sun to make a strong stand in the sky, warm the awakening trees to encourage a full and healthy crop, and help return the earth to a midwinter normalcy where no evil demons can walk the night.

Awakening the fruit trees, apple and pear, from their winter sleep is a serious job for the orchardist. It's not the orchards that produce table fruit that merit this attention, but more importantly, it's the trees that provide the fruit for holiday cider and perry!

> Blow, blow, bear well,
> Spring well in April,
> Every sprig and every spray
> Bear a bushel of apples against
> Next New Year's Day.

Participants in wassail will discharge their shotgun into the branches. This will not only serve to awaken the sap in the trees, but will also dispel any evil spirits still lurking there. They will blow whistles and pound on small drums.

Meanwhile, they will pass the wassail bowl among themselves to provide encouragement on this cold crisp night and to keep their energy high. Often the participants will share their wassail bowl with the tree, along with offerings of bread and fruit and cheese. Sometimes a small boy is placed in the branches to receive these offerings for the orchard.

With their breath and the bowl steaming, the wassaillers raise their voices in a chant to the trees.

Old apple tree, old apple tree
We've come to wassail thee.
Stand fast root. Bear well top.
Pray God send us a howling good crop.

Every twig, apples big
Every bough, apples enow.
Hats full, caps full, three bushel bags full
Barn floors full and my pockets full too!
And a little heap under the stairs.

The wassail bowl is filled with a traditional drink called Lamb's Wool. It's made simply of old (or new, if necessary) ale that is spiced with nutmeg, cloves, and ginger; it is then heated, along with sugar and beaten eggs or cream, and served with roasted apples floating on top. The name Lamb's Wool comes from the appearance of these split apples softly bobbing on the surface of the drink.

The oldest known written recipe for Lamb's Wool comes from the English court of Charles I. It has lasted virtually unchanged into this decade (see below). In ancient times the ale was actually a mead ale; that is, the mead was fermented with hops. The bowl itself was as big as a punch bowl and called a mazer. It was handed down from father to son and considered a prized family heirloom.

After exhorting the trees and before leaving the orchard, the participants must bow to the tree and thank it for all its hard work. These wassaillers can be content that they have done their part to restore safe nights, longer days, and a bounteous harvest.

There are other traditions that have evolved from these rituals and have become part of the Twelfth Night celebrations. In some cases, the wassail bowl is carried through the streets by merrymakers who knock on doors and offer a drink to the inhabitants. Often, riddles and word games are part of the festivities and the wassailers are offered candies and other sweet treats as a reward.

Here we come awassailing among the leaves so
green.
Here we come awand'ring, the fairest to be seen.
Love and Joy come to you and to you your wassail
too

And God bless you and send you a happy New
Year,
And God send you a Happy New Year.

Here in the U. S. A., orchardists have joined with Pagans to once again perform a ritual in the cider and perry orchards. Though shotguns are no longer used, these wassailers ride on hay carts into the fields to wake up the trees. They make a Lamb's Wool from nonalcoholic cider.

Lamb's Wool

10	small apples
10	teaspoons brown sugar
2	quarts ale or hard cider
½	teaspoon grated nutmeg
1	teaspoon ground ginger
3	whole cloves
3	allspice berries
1	3-inch cinnamon stick
2	cups superfine sugar
½	cup water
6	eggs separated
1	cup brandy

Wash and core the apples; fill each one with a teaspoon of brown sugar. Place in a baking dish and cover the bottom with ¼-inch of water. Bake in a 350°F oven for 30 minutes or until tender. Combine the ale with the spices, sugar and water in a heavy saucepan. Do not allow to boil, but heat for 30 minutes to allow the spices to perfume the ale. Beat the egg yolks until light. Beat the whites until stiff and fold into the yolks. Strain the ale and gradually add to the eggs, beating constantly. Add the brandy. Pour into a punch bowl and float the apples on top. Serves 8–10.

Wassail, oh wassail all over the town
The cup it is white, the ale it is brown
The cup it is made of the good ashen tree
And so is the beer of the best barley.

Wassail!

LEISURE &
RECREATION

Leisure & Recreation
How to Choose the Best Dates

Everyone is affected by the lunar cycle. Your lunar high occurs when the Moon is in your Sun sign, and your lunar low occurs when the Moon is in the sign opposite your Sun sign. The handy Favorable and Unfavorable Dates Tables on pages 29-51 give the lunar highs and lows for each Sun sign for every day of the year. This lunar cycle influences all your activities: your physical strength, mental alertness, and manual dexterity are all affected.

By combing the Favorable and Unfavorable Dates Tables and the Lunar Aspectarian Tables with the information given in the list of astrological rulerships, you can choose the best time to begin many activites.

The best time to perform an activity is when its ruling planet is in favorable aspect to the Moon—that is, when its ruling planet is trine, sextile, or conjunct the Moon (marked T, X, or C in the Lunar Aspectarian), or when its ruling sign is marked F in the Favorable and Unfavorable Days Tables. Another option is when the Moon is in the activity's ruling sign.

For example, if you wanted to find a good day to train your dog, you would look under animals, and find that the sign corresponding to animal training is Taurus, and that the planet that rules this activity is Venus. Then, you would consult the Favorable and Unfavorable Days Tables to find a day when Venus (the ruling planet) is trine, sextile, or conjunct (T, X, or C) the Moon; or when Taurus (the ruling sign) is marked F in the Favorable and Unfavorable Days Table; or when the Moon is in Taurus.

Animals and Hunting

Animals in general: Mercury, Jupiter, Virgo, Pisces
Animal training: Mercury, Virgo
Cats: Leo, Sun, Virgo, Venus
Dogs: Mercury, Virgo
Fish: Neptune, Pisces, Moon, Cancer
Birds: Mercury, Venus
Game animals: Sagittarius
Horses, trainers, riders: Jupiter, Sagittarius
Hunters: Jupiter, Sagittarius

Arts

Acting, actors: Neptune, Pisces, Sun, Leo
Art in general: Venus, Libra
Ballet: Neptune, Venus
Ceramics: Saturn
Crafts: Mercury, Venus
Dancing: Venus, Taurus, Neptune, Pisces
Drama: Venus, Neptune
Embroidery: Venus
Etching: Mars
Films, filmmaking: Neptune, Leo, Uranus, Aquarius
Literature: Mercury, Gemini
Music: Venus, Libra, Taurus, Neptune
Painting: Venus, Libra
Photography: Neptune, Pisces, Uranus, Aquarius
Printing: Mercury, Gemini
Theaters: Sun, Leo, Venus

Fishing

During the summer months the best time of the day to fish is from sunrise to three hours after, and from two hours before sunset till one hour after. In cooler months, fish do not bite until the air is warm, from noon to 3 pm. Warm, cloudy days are good. The most favorable winds are from the south and southwest. Easterly winds are unfavorable. The best days of the month for fishing are when the Moon changes quarters, especially if the change occurs on a day when the Moon is in a watery sign (Cancer, Scorpio, Pisces). The best period in any month is the day after the Full Moon.

Friends

The need for friendship is greater when Uranus aspects the Moon, or the Moon is in Aquarius. Friendship prospers when Venus or Uranus is trine, sextile, or conjunct the Moon. The chance meeting of acquaintances and friends is facilitated by the Moon in Gemini.

Parties (Hosting or Attending)

The best time for parties is when the Moon is in Gemini, Leo, Libra, or Sagittarius with good aspects to Venus and Jupiter. There should be no aspects to Mars or Saturn.

Barbecues: Moon, Mars
Casinos: Venus, Sun, Jupiter
Festivals: Venus
Parades: Jupiter, Venus

Sports

Acrobatics: Mars, Aries
Archery: Jupiter, Sagittarius
Ball games in general: Venus
Baseball: Mars
Bicycling: Uranus, Mercury, Gemini
Boxing: Mars
Calisthenics: Mars, Neptune
Chess: Mercury, Mars
Competitive sports: Mars
Coordination: Mars
Deep-sea diving: Neptune, Pisces
Exercising: Sun
Football: Mars
Horse racing: Jupiter, Sagittarius
Jogging: Mercury, Gemini
Physical vitality: Sun
Polo: Uranus, Jupiter, Venus, Saturn
Racing (other than horse): Sun, Uranus
Ice skating: Neptune
Roller skating: Mercury
Sporting equipment: Jupiter, Sagittarius
Sports in general: Sun, Leo
Strategy: Saturn
Swimming: Neptune, Pisces, Moon, Cancer
Tennis: Mercury, Venus, Uranus, Mars
Wrestling: Mars

Travel

Long trips which threaten to exhaust the traveler are best begun when the
Sun is well-aspected to the Moon and the date is favorable for the traveler.
If traveling with others, good aspects from Venus are desirable. For enjoy-
ment, aspects to Jupiter are preferable; for visiting, aspects to Mercury.

To prevent accidents, avoid squares or oppositions to Mars, Saturn, Uranus, or Pluto.

For air travel, choose a day when the Moon is in Gemini or Libra, and well-aspected by Mercury and/or Jupiter. Avoid adverse aspects of Mars, Saturn, or Uranus.

Air travel: Mercury, Sagittarius, Uranus
Automobile travel: Mercury, Gemini
Boating: Moon, Cancer, Neptune
Camping: Leo
Helicopters: Uranus
Hotels: Cancer, Venus
Journeys in general: Sun
Long journeys: Jupiter, Sagittarius
Motorcycle travel: Uranus, Aquarius
Parks: Sun, Leo
Picnics: Venus, Leo
Rail travel: Uranus, Mercury, Gemini
Restaurants: Moon, Cancer, Virgo, Jupiter
Short journeys: Mercury, Gemini
Vacations, holidays: Venus, Neptune

Writing

Write for pleasure or publication when the Moon is in Gemini. Mercury should be direct. Favorable aspects to Mercury, Uranus, and Neptune promote ingenuity.

Hunting & Fishing Dates

From/To	Quarter	Sign
January 10, 4:59 am—January 12, 1:48 pm	1st	Pisces
January 18, 11:01 pm—January 20, 10:58 pm	2nd	Cancer
January 27, 12:01 pm—January 27, 11:57 pm	3rd	Scorpio
January 27, 11:57 pm—January 29, 11:17 pm	4th	Scorpio
February 6, 11:02 am—February 8, 7:17 pm	1st	Pisces
February 15, 7:45 am—February 17, 9:11 am	2nd	Cancer
February 23, 8:58 pm—February 26, 7:10 am	3rd	Scorpio
March 4, 6:30 pm—March 6, 12:17 am	4th	Pisces
March 6, 12:17 am—March 7, 1:54 am	1st	Pisces
March 13, 1:51 pm—March 15, 4:43 pm	2nd	Cancer
March 22, 6:17 am—March 24, 3:43 pm	3rd	Scorpio
April 1, 3:12 am—April 3, 10:22 am	4th	Pisces
April 9, 7:16 pm—April 11, 8:30 am	1st	Cancer
April 11, 8:30 am—April 11, 10:16 pm	2nd	Cancer
April 18, 2:35 pm—April 20, 11:58 pm	3rd	Scorpio
April 28, 12:06 pm—April 30, 7:54 pm	4th	Pisces
May 7, 2:14 am—May 9, 4:01 am	1st	Cancer
May 15, 9:16 pm—May 18, 7:09 am	2nd	Scorpio
May 25, 8:07 pm—May 26, 6:55 am	3rd	Pisces
May 26, 6:55 am—May 28, 5:08 am	4th	Pisces
June 3, 11:30 am—June 5, 11:45 am	1st	Cancer
June 12, 2:55 am—June 14, 1:18 pm	2nd	Scorpio
June 22, 2:52 am—June 24, 12:55 pm	3rd	Pisces
June 30, 10:09 pm—July 1, 2:20 pm	4th	Cancer
July 1, 2:20 pm—July 2, 9:38 pm	1st	Cancer
July 9, 8:48 am—July 11, 7:06 pm	2nd	Scorpio
July 19, 8:44 am—July 21, 7:09 pm	3rd	Pisces
July 28, 8:30 am—July 30, 8:23 am	4th	Cancer
August 5, 4:04 pm—August 6, 8:02 pm	1st	Scorpio
August 6, 8:02 pm—August 8, 1:30 am	2nd	Scorpio

Hunting & Fishing Dates

August 15, 2:41 pm—August 18, 12:44 am	3rd	Pisces
August 24, 4:59 pm—August 26, 6:17 pm	4th	Cancer
September 2, 12:55 am—September 4, 9:08 am	1st	Scorpio
September 11, 9:34 pm—September 13, 2:33 pm	2nd	Pisces
September 13, 2:33 pm—September 14, 7:00 am	3rd	Pisces
September 20, 11:16 pm—September 23, 2:00 am	4th	Cancer
September 29, 10:30 am—October 1, 5:50 pm	1st	Scorpio
October 9, 5:36 am—October 11, 2:51 pm	2nd	Pisces
October 18, 4:37 am—October 20, 2:59 am	3rd	Cancer
October 20, 2:59 am—October 20, 7:42 am	4th	Cancer
October 26, 7:23 pm—October 27, 2:58 am	4th	Scorpio
October 27, 2:58 am—October 29, 2:40 am	4th	Scorpio
November 5, 2:13 pm—November 8, 12:02 am	2nd	Pisces
November 14, 11:21 am—November 16, 1:19 pm	3rd	Cancer
November 23, 2:33 am—November 25, 10:33 am	4th	Scorpio
December 2, 10:23 pm—December 3, 10:55 pm	1st	Pisces
December 3, 10:55 pm—December 5, 9:17 am	2nd	Pisces
December 11, 8:48 pm—December 13, 9:09 pm	3rd	Cancer
December 20, 8:12 am—December 22, 4:57 pm	4th	Scorpio

Editor's note: This chart lists the best hunting and fishing dates for the year 2000, but not the only possible dates. To accommodate your own schedule, you may wish to try dates other than those listed above. To learn more about choosing good fishing dates, see the fishing information on page 152. To learn more about hunting dates, see the animal and hunting information on page 151.

Transits of the Outer Planets to Your Moon

Kim Rogers-Gallagher

Of all the planets in the heavens, the Moon is undeniably the most protective. She is the Earth's personal guardian, a lovely lady whose orbit is predictable and reassuring—yet constantly different. The phases of the Moon show us her connection with the ups and downs in our "moods"—a great Moon word. She represents our relationship with our Mom and our family, tells us where and when we're most comfortable expressing our emotions, and describes our attitude about children and childhood—not to mention how we cope with what life tosses at us. In short, this beautiful silver orb is in charge of the inner person, the silent side of us that feels, wishes, sighs, and remembers.

Each and every transit to the Moon in our charts, then, will affect our mood. When chatty Mercury comes for a visit, we're more apt to talk about our feelings. When Venus arrives, there are plenty of hugs on the way. But these two are familiar inner planets. What about the outer planets—Uranus, Neptune and Pluto? What is unique about the way each of these three touch the Moon?

A Look At "I Will Never" Syndrome

First off, all three outers tend to stay put for a long time. For instance, the quickest of the three, Uranus, spends seven years in a sign. This means that, given a five-degree applying and separating "orb of influence," Uranus can be in the neighborhood of one of your natal planets for two-and-a-half years, and considering retrograde motion he will make three to five passes across the exact degree, shaking things up all that time.

Neptune and Pluto, meanwhile, hang around a particular degree for even longer—at least three years, in fact—during which time your planet either gets religion (Neptune) or gets intense (Pluto). Regardless of which of the three outers comes to visit, your natal planet forgets what life was like before they arrived and begins acting out a lot of the qualities of the "visitor," just as we often pick up expressions or habits, or even accents, when we live with someone.

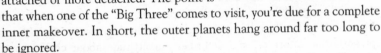

In the case of the Moon, your feelings, reactions, and responses will undergo gradual but dramatic change when this orb falls under the influence of an outer planet. For instance, your relationship with family members or your Mom may shift somehow. You become more dependent, more independent, more attached or more detached. The point is that when one of the "Big Three" comes to visit, you're due for a complete inner makeover. In short, the outer planets hang around far too long to be ignored.

But there is another thing that happens under outer planet Transits that is rather specific to Uranus, Neptune, and Pluto. This is called the "I Will Never" Syndrome. It happens when the Universe provides opportunities for us to eat our words. You know exactly what I mean here, because we've all done it. We've all taken oaths we swore we'd never break, made resolutions we were sure we'd keep, and promised ourselves never to do certain things—and then done them. This is part of being human. In the spirit of adding a bit of humor to this discussion, let's take a look at this common affliction through a fictional reenactment. Let's have some fun with the outers—since they often seem to be having so much fun with us.

Picture this scene: three rather striking figures are perched casually around the edge of a circular cloud, staring into the center, sipping some wonderful potion and bemusedly watching the activities of the mortals below. Occasionally, they elbow one another and point. You've heard of "surfing the net"? Well, this is the celestial version, aptly called "surfing the populace"—though you must be an Immortal to get online. Our cast today is made up of, you guessed it, Uranus, Neptune, and Pluto.

Uranus is dressed in a long white robe that is cold and frosty, crackling when he walks like a sheet that's been left out on the clothesline too long in February. He, of course, isn't really watching the "show;" he's pacing, tapping, and staring off straight ahead into space, his icy blue eyes fixed on the future and automatically understanding exactly what the end result of all actions will be. To his left is Neptune, dressed in her most beautiful pink gown, trusty pink-smoke machine on her right and a bucket of pink dust on her left. She is ready to snap the machine on and toss a handful of dust down on whoever or whatever she sees acting a little too realistic for her tastes. Occasionally, she sniffles and dabs a tear away when one of the mortals does something dreadfully romantic or wonderfully compassionate. Finally, there's Pluto, dressed all in black and glaring over the edge of the cloud, his arms crossed and head cocked forward.

Sound far-fetched? Well, maybe not. Maybe the gods and goddesses really do check in on us, listen to our conversations pretty darned carefully, trying to catch any of us making this proclamation:

"I will never…"

Doesn't matter what you finish that sentence up with—if you start a sentence with those three words, somebody in the cloud-circle hears it. Then, whoever's on duty at the moment, whether it's Uranus on a swingshift, Pluto on nightshift, or Neptune just snapped out of a daydream, will hear you utter those fateful words and, leaning over to tap Mercury on the arm, will say, "Jot that down."

While Mercury understands exactly what's about to go down, he's still operating from a very keen memory of what happened to Prometheus the last time Mercury tried to help a mortal. That is, Mercury angered the other gods by helping Prometheus steal fire, causing them to chain Prometheus to a cliff for eternity with a vulture pecking at his breast. So he snaps to attention and dutifully records what he's supposed to. Whatever we just decided we'd never do is now public record, up on Mount Olympus. That done, the outers return to sipping nectar, and the matter is over for now.

It's over, that is, until one of the big three happens to be in the neighborhood of the planet ruling whatever we were righteously proclaiming would never do. When this happens, the outer stops, pulls out a galactic cellphone, and dials up Mercury, saying something like "Hey Merc, I'm in Seattle. Pull Whatzername's 'Never' file and fax me the Moon section, wouldja?"

Sure, it sounds crazy. But what other explanation could there possibly be? How else could the outers know just how to turn our lives completely upside down? In the end this mystery has only one certain ending: change. Since the outer planets all have a "Make Things Change" clause built right into their contracts, they must show up when it's time to shake us up and force a bit of evolution on our souls. Sometimes we're conscious of how badly we want out of our present circumstances, and sometimes we're not. If we are conscious of needing a break in our routine, then we're more able to gracefully surrender what seems to be leaving our lives and set to work creating new stuff. On the other hand, if we're not conscious of how badly we want out we may view these changes as "bad." Either way, however, change happens, and our best bet is to cooperate. In a nutshell, then, the outers represent times when we make evolutionary leaps. We need to decide where to leap and how quickly.

The Outer Planets and "Buffers"

Often, "temporary people" arrive under outer planet transits to provide the jolt necessary to get us moving. I like to think of them as "buffers." They walk into your office or your local hangout one day and say something that captures your attention and holds it. Within weeks, you find yourself inspired to make radical changes in your life—changes you would have thought were totally out of the question not too long ago. Often, these buffers do their job by providing just exactly what you is sorely missing from your life. It might be freedom, romance, even financial independence—it all depends on the house or planet that an outer is visiting.

In the case of the Moon, someone often arrives who perfectly epitomizes your heart's desire. This Buffer will support you in your quest for fulfillment and tell you why you're right to want it, and stand by, nodding, with their arms folded while you, A) quit the job you've had and hated for five years, B) end your marriage because you want to try true independence, C) move to Sri Lanka, or D) who knows. Then they leave just like that. And you shouldn't doubt for a minute that they're permanently gone no matter what they've told you. Sure, you're crushed and scared, having just trashed what you thought you'd never be brave enough to even change a little bit. And now you're alone with all these holes you've punched in your life.

What you'll discover later on is that these folks aren't meant to be permanent fixtures in your photo album. You'll eventually realize that you really wouldn't have wanted them to be, anyway. buffers function as exciting, pretty "lures" that bring you messages you're in the mood to hear, even if the message seems "crazy" at the time. In the end, if it feels right to you, if it feels like you're breaking free of long-held restrictions and now doing what you truly want to do, you will listen.

Buffers usually bring about a serious tumult in your life, but they're also pretty darned exciting. When they arrive at the Moon's door, then, it's best to learn to recognize them, enjoy them, and think about what they're providing for you, emotionally speaking. Just don't try to hold on when they get up to leave.

All that said, we should take time to look at what happens to your Moon when each of the outer planets is in the neighborhood. We'll start with Uranus—Mr. Unpredictable himself.

Uranus/Moon—Get Yourself Free

How about moving to Zimbabwe, Milwaukee, or Sri Lanka? Or would you rather move into a tree house with seven other people? Although these ideas might sound out of the question to many people, if Uranus is any-where near your Moon right now you're probably nodding your head and wondering what the weather is like in Zimbabwe this time of year.

Your natal Moon is the place where you have the need to root your-self to the physical world and find security through a home. Our Moons typically want to stay put, causing us to limit our emotional "growth" at times. These grounded Moons will find themselves living with the most uncertain of circumstances under the influence of Uranus, the cosmic loon and wildman.

Under this conjunction, as with the conjunction of Uranus to all our planets, we often feel for some time that a change is coming, and we in-ternally "plan" for it. We don't actually "do" anything that those around us can see, however, until the very last minute—which can make our be-havior seem totally out of the blue. But we know better—that we've been growing tired of this place, and this living arrangement, and have been waiting for the right time to change our circumstances. When Uranus' lightning bolt strikes, then, we're ready. Why wait? We fix dinner on Tuesday night and pack everything up and move out Wednesday.

The tougher Uranus transits can make you feel as if you're living with a constant earthquake beneath you. You're never quite sure of what you're going to do until the transit happens: You want to move—or do you? You want to have a child—but do you really? You'll feel these jitters as you're considering the action and afterwards once you realize exactly what you've done. In general, Uranus grabs our Moon by the collar and turns her around. We feel nervous, and excitable—and we may even find it dif-ficult to sleep.

Under easier transits to the Moon (which depend on our natal Moon's opinion of Uranus to start with), we're less nervous about what we're going to do and more stimulated by how drastically our lives are going to change afterwards. We're looking forward to Zimbabwe or happily making plans to move into a commune. Group living situations like communes and ashrams, by the way, are a distinct possibility when Uranus touches the Moon, by any aspect. The influence of Uranus on the Moon makes her crave freedom. A rather Uranian urge to adopt—a stray critter, perhaps, or a computer, or a child from Bosnia—may also come along now. Any transit to the Moon ordinarily brings out our nurturing instincts, and Uranus makes us want to nurture someone or something unusual in an unusual way.

Our relationships with our Moms (and the rest of our families, and with women in general) are also due for a quick turnaround under any Uranus visit. If we're very close to Mom, we might find that we feel suddenly "distanced" from her. On the other hand, if we're not ordinarily close we may find ourselves spending more time with her and seeing a "new" side of her personality. Or she may do something that knocks our socks off—like announcing that she's leaving Dad, or joining the Peace Corps—and allows us to see her as a real individual instead of just Mom.

No matter where you move to, what you adopt, or what happens with your Mom, remember that this time is all about you uncovering a previously hidden side of your emotional nature. Don't be afraid of the changes that happen now. Enjoy the scenery and the newfound freedom of discovering who you are when you're no longer on familiar territory.

Neptune/Moon Transits: Dream A Little Dream Of Me....

Our Moon is our soft underbelly and the planet that we use when it's time to react to whatever is out there and not to take action. When Neptune visits the Moon, she is prone to punching holes in an already delicate shell. Needless to say, this is a time of extreme sensitivity to the environment—for better or worse—when our susceptibility to the outside world heightens to the point that harsh sounds can actually make us wince. In fact, I've often found that many folks just can't listen to music any harsher than Mozart during this period.

Since the Moon is the instinctive side of our nature, and Neptune is the queen of intuition, under the easier transits we can often have truly "psychic" moments. We tend to just "know" things without knowing why. We may think of someone we haven't seen in awhile, and then bump into them a day later, and sometimes our dreams can be quite prophetic.

On the other hand, this combination in a tougher aspect can conjure unexplainable discontent. We sigh, feel nostalgic, and are unable to see things clearly at all. Like Dorothy, when Neptune touches our Moon all we want to do is close our eyes and say, "There's no place like home." The best thing to do when your Moon is under the wistful spell of this pink goddess is remember that you're not supposed to see or understand anything intellectually under Neptune, much less do anything. You're supposed to feel—everything—romantically, vaguely, and dreamily. You should expect to sigh a lot and wish for the way things were—or the way they could be—and to forget the arguments that drove you from your significant other. Under Neptune, you'll remember your former partner's smile, how nice it was to walk down that sunny hallway at work in the morning, and how great the view was from the old front porch. In short, our memories become extremely selective and very tender when Neptune waves her magic wand over us. Since the Moon is the compassionate side in all of us, our sympathy towards other creatures deepens, too.

So what's the point? Well, at the end of this transit, we often wake up and realize that everything about us is different—that we've been sensitized from the inside out. The reality we were living at the beginning of the transit has been gradually eroded right out from under our feet. Often this pertains to our homes and to the loved ones in our lives. Of course, after the smoke clears, which isn't until well after Neptune has moved on, we're able to see things a lot more clearly and begin to understand. In the meantime, enjoy your new (or improved) "psychic" abilities—and be tender to critters, kids, and your elders.

Pluto Transits: Just Let Go

You remember Pluto, right? The Head of the Department of Extremes and Inevitables? The guy, dressed like Darth Vader, who is in charge of such delectable topics as Death, Destruction, and Endings? Well, although you might want to avoid Pluto, you simply cannot. That is what the "inevitable" part is all about. Nor should you try to avoid him—because what you'd really be ignoring is your opportunity to evolve. Ultimately, Pluto's visits are really all about evolution and enlightenment.

Certainly, enlightenment can't happen if you don't shed some weight—and Pluto wants you as light as possible when he arrives since he's trying to help you climb up the evolutionary staircase. Over the course of a transit, he'll come by to visit several times—to check your progress and measure your weight. He wants to make it easier for you to rise to all occasions. And if he has to, he'll do much of the work himself.

It's like this: Pluto doesn't like it when nothing happens. He considers this stagnation and knows it is not going to help you on your quest to change. As a matter of fact, Pluto is a very big fan of changing everything. You might even call him ravenous about it; he doesn't want just a little bite of anything, he wants it all at once. He wants you to give up everything that is superfluous.

Needless to say, when you feel Pluto's breath on your neck, it's time to evaluate what's really necessary in your life—and to ditch what isn't. Remember, the harder you try to hold on, the more ungently he'll shake you off.

That's what Pluto transits are all about: they're a three-stage process of life, death, and rebirth. But you shouldn't let this scare you, because you'll get to keep the real things—everything that is really you. It's the useless things that pass away under Pluto transits—so any existing condition in your life that is going to die is already outworn and useless. Yes, this sounds harsh, but deep down inside, you always know what's got to go under a Pluto transit even if it's tough to accept that now's the time. It's also true that if you try to hold on to someone or something useless it will probably still leave your life, or you'll learn down the road why it would have been better to let go then.

So tap into that inner-knowing and let change happen. It's always best to cooperate with the rising tide. Make plans for Stage Three—Rejuvenation Time—and gracefully release what's got to go.

Pluto/Moon

Your Moon represents your home, your family, your relationship with your Mom, your security, and your children. It's the urge or need we all have to emote, and one of the most tender parts of us—a place where we prefer to react rather than act. When Mr. Intensity arrives, then, it's a given that you're going to be feeling everything—and probably expressing your feelings—in far more urgent a fashion than you're used to. Pluto turns up the volume on your feelings when he touches the Moon, to the point where you might find yourself feeling obsessed with someone or something and emotionally tied up in knots. Most often, the focus will be on family issues having to do with your Mom, your kids, or your home. But Pluto also empowers—so if your usual reaction to what the world tosses at you is to sit back and take it, you won't be doing that anymore. With Pluto here, you'll change from the inside out—much the way a caterpillar transforms into a butterfly after its evolution inside the cocoon is over.

When Pluto conjoins the Moon, you're going to be emotionally supercharged, and everything about your outer life will reflect this inner change. You'll take a new, stronger stance within your family, put your foot down and make final decisions at home, and be amazingly in charge of your feelings. You may become a mother (or a father) now, too, and this one will powerfully change your life whether or not you've ever had children before. You may move now, too, or make drastic changes to your home's foundation, or tear one down and build a new one in its place. You may also begin to realize the profoundness of the emotional connection you share with your mother. She was your original shelter from every storm and your life-source. You felt everything she felt while you lived inside her womb—so with Pluto here, you'll subconsciously begin feeling with her again.

Under the tougher transits, you may find your relationship with your Mom undergoing drastic change or a complete turnaround. You may separate from her in some way, too. At times, you may also feel as if you're a prisoner in your own home—but you won't take that for very long. There is nothing like a tough Pluto/Moon transit to open your eyes to any kind of emotional manipulation and inspire you to deal with it. On the other hand, if you've ever been prone to using manipulation on others—via guilt, for example—now may be when you see what you've done and take steps to change it. You may alienate family members with this new attitude, or bring them closer to you—but something about the way you approach family situations will be altered dramatically. Since Pluto also coincides with times when we feel as life is forcing us to let go, a child of some kind may also leave our lives when Pluto touches the Moon. In fact, if ever there were a catalyst for empty nest syndrome, this is it.

When Pluto touches the Moon by easier aspect, you begin to see just how powerful a tool changing your feelings can be. You'll still be transforming from the inside out, but the changes will come easier. Your whole attitude about children in general and your Mom in particular will be drastically different, and although your home life and your relationship with your family will still be altered in some inevitable way, you'll be a bit more accepting of the situation from the beginning. You may plan a move across the country, contact your Mom after several years of stony silence, or begin an adoption process.

Most importantly, keep in mind that everything that happens to you now, whether you see yourself or the outside world as the choreographer of these events, is going to be a symbolic instant replay of the birth process you went through when you separated from your Mom's body. As any physical birth is a tough process, full of immense joy and great fear, be aware that both you and your Mom were terrified. She'd created a brand-new person and had to deal with a tremendous responsibility. And you were about to leave home for the first time, to separate from the safest place you'd ever lived to begin a brave new journey towards fulfilling the purpose you chose when you decided to hop inside a human body. When you feel a push towards change in your present life coming along, then, think of it as a contraction of sorts—the end of a period of comfort and protection but the beginning of a wonderful journey. And as the contractions come at closer intervals—that is, as events start to pile up towards a new beginning—take deep breaths and get ready to see a whole new world with entirely new eyes.

Auntie Estelle's Sure-Fire Predictions for the Millennium

Estelle Daniels

(Cue fanfare and flashing lights)

With the millennium approaching, every psychic worth their salt is supposedly focusing on what the next thousand years will bring for us unsuspecting humans. After all, there is great lucrative potential in making predictions about the millennium. And so a season of silliness has arrived and the predictions seem to be right in tune with it.

Well, I'm a big important psychic—just ask my mother, she'll vouch for me—and I want my share of this pie. So in the best tradition of supermarket tabloids and sleazy get-rich-quick entrepreneurs everywhere, I present my predictions.

(Warning: Do not try these predictive systems at home. Check with a professional or get yourself a deck of Tarot cards, they're easier. And you might not be able to duplicate my results. These predictions are for entertainment only and should not be used to buy stocks, book passage on an alien spaceship, or build a bomb shelter in your back yard.)

Here We Go

While researching this article, I embarked upon a quest not only to get the best most radically up-to-date predictions available, but also to try and use lost means of divination. My research led me far and wide, and I bring to you, gentle reader, the results of my exhaustive searches—both the fruitful events, and those that didn't do so well.

First off, I wanted to get the lowdown on natural disasters. After wondering awhile about what would be the best way to predict those, I got myself a little book and found several possible methods. For instance, there is geomancy—a divinatory system involving stakes driven in the earth at random points and connected with string. A geomancer uses the resulting shapes in making predictions.

Since my yard is too small for such a practice, I went off to the nearest park with an armful of stakes and a big ball of twine. Hoping to minimize my own brain functions and psychic interference, I went early—well before my morning cup of coffee. I hammered stakes until I ran out, looked for the twine I had misplaced, then began tying the stakes together. Unfortunately, before I could finish the police came and arrested me for causing a public nuisance. I guess some of those random placements crossed the jogging paths and joggers didn't see the twine and, well …

I tried to convince the police that I was conducting important research which could alter the course of mankind. They just said "yes, yes" and "of course" as they gently led me into their police car. My attempt at geomancy was incomplete, so I was worried how I would get my predictions? All seemed lost at the start.

But all was not lost. As you may know, every jail cell is equipped with a toilet. And since toilets hold water, I cheerfully passed the time until my hearing practicing hydromancy—divination by gazing into a water surface. Here are the results of those divinations as I can best remember them. They wouldn't let me have pencil and paper in jail, something about being crazy and a risk for suicide….

Natural Disasters Worldwide

There will be floods in China accompanied by the loss of life. Commerce and farming will be disrupted. World leaders will express concern for China's growing population and economy. Government officials will promise to look into the matter and formulate a plan to prevent it from happening again in the future.

An earthquake will strike without warning, causing massive damage and loss of life. This will be somewhere in Asia or South America. Scientists will warn everyone that this is a sign that the big one is coming, though they will not be able to say when it will come. It will take time for relief workers to get to the afflicted area, and the world will cry out to help those people. There will be TV reports showing the rescue efforts, damaged homes and buildings, and people living in the streets. The water supply will be disrupted, and people will get sick because of it. The government will call for more earthquake-proof buildings, but the costs will be prohibitive. There will be at least one miraculous rescue of a person trapped in the rubble and rescued days later.

A volcano will erupt unexpectedly. There will be smoke and ash, and possibly a lava flow. Some people will die, and a community will be devastated. People living nearby will be surprised but will vow to stay and rebuild the home they don't want to leave. Scientists will claim that this eruption is small and just a precursor to the big one, though they will not be able to say when that will hit.

After I was released, I saw my court-appointed psychiatrist. He was really nice and insisted that I tell him my dreams. This was perfect because oneiromancy, or divination through the interpretation of dreams, tied in well with my ongoing research. And since I'd dreamt about some interesting people, I was able to make some key predictions.

People

Scandal will rock the Royal Family. There will be calls to abolish the monarchy and some will demand the Queen start paying her own way. There will be criticism of Prince Charles. Paparazzi will continue to photograph the young Princes, but at a distance. A memorial to Princess Diana will be constructed. There will be criticism of the management of the Princess Diana Trusts. Elton John will refuse to perform "Candle in the Wind."

Tragedy will hit the Kennedy family. There will be a wedding and a divorce in the Kennedy family, and one family member will be elected to office. There will be new speculations about the assassination of JFK. There will be a book published about either JFK's assassination or Jackie O.

There will be at least one sighting of Elvis and people will claim he is still alive. Elvis impersonators will appear on TV. People will gather at Graceland on the anniversary of Elvis' death.

A famous celebrity couple will spilt and possibly divorce. They will say their careers caused them to grow apart. There will be claims of unfaithfulness on the part of one or both. The tabloids will pick it up, and there will be threats of a lawsuit.

A child celebrity will threaten to sue his or her parents to get control of earnings. This child will claim the parents spent the money on themselves. A former child celebrity will be in trouble with the law.

Another Hollywood celebrity will be in trouble with the law over drugs. People will say they knew about it for years but that nobody could do anything about it. The celebrity will go through treatment and will say, "I am a new person."

A famous person will be in the headlines because of a deadly disease. This person will choose to go public to help raise awareness for the ailment. People will send cards and gifts in an outpouring of sympathy.

There will be a Hollywood blockbuster movie which will turn out to be a box-office disappointment. There will be a movie which will be an unexpected hit. There will be some surprises at the Oscars this year.

After this round, my doctor said I needed to get some exercise and should try sports. Keeping with my research I discovered this predictive system: Belomancy—divination by the use of arrows. A number of answers are attached to a target and the priest shoots an arrow at the target. The answer pierced by the arrow is the correct one. Armed with this knowledge, I went out and bought myself a box and lots of arrows.

Making a target was easy. I didn't have a priest, however, so I wrote out a lot of predictions myself and put them on the target. I took my bow and arrow, and shot an arrow. It went straight up and I lost it in the Sun, landing nowhere near the target. In fact, I looked and looked, but never did find it. I tried it a few more times, but lost the arrows each time. What good is a system if you can't find your arrows? It wasn't until much later, after some practice, that I eventually began to hit the target with some regularity. These are the results of my divinations by arrows.

Health

Doctors will announce a new breakthrough in the cure for cancer. There will be an announcement about a common food which can cause harm. There will also be an announcement about some other food which can prevent disease. These results will be called inconclusive by some and further study will be called for. A miracle diet will be announced, thousands will try it and few will get lasting results.

There will be a study showing new findings about the links between diet, cholesterol, blood pressure, and heart disease. One doctor will say that the study is flawed. New drugs and techniques for controlling cholesterol will be recommended, but some will say it's too soon to tell if these recommendations will work. Further research will be called for.

An epidemic will break out in Africa or Asia which will cause a possible panic. World Health officials will promise to look into the matter and formulate a plan to prevent it from happening again in the future.

There will be a salmonella outbreak in the U.S. which will potentially affect millions. Food products will be recalled and the government will step up efforts to keep the food supply safe. The outbreak will be traced to a manufacturer who will say they were doing all they could, it was just a freak occurrence, and all safeguards will be strictly enforced from now on. Government officials will promise to look into the matter and formulate a plan to prevent it from happening again in the future. Many of the people who were affected will either accept a settlement or will sue.

After all this exercise I was really hungry, so I thought I'd get some food. After a wonderful Chinese all-you-can-eat buffet, I did get some predictions from the fortune cookies, which is a form of aleuromancy—divination with flour. In traditional aleuromancy, messages are placed in dough and baked, and the message you get is a prediction of the future.

Ecology

A corporation will attempt to log a forested area and an environmental group will protest. There will be at least one demonstration and people may be arrested. Some people will chain themselves to trees or logging equipment. A government official will say that this logging is ultimately good for the ecological health of the area. Another government official will say it is an exploitation of the country's natural resources. Paper prices will go up.

A new toxic waste site will be discovered in an area near where people live. People near the site will claim they have had mysterious illnesses for years, but nobody took them seriously. It will be called an outrage that homes were allowed to be built so close to a dangerous area. Government officials will promise to look into the matter and formulate a plan to prevent it from happening again in the future. There will be a call by environmentalists to clean it up, but the local government will say this will cost too much and will petition the federal government for aid.

There will be a controversy out West about an area set aside for the preservation of an endangered species. Some will say it is blocking progress and the ecologists have gone too far.

There will be heated rhetoric and possibly shots will be fired in protest.

I returned home, fired with zeal and a determination to be more efficient. I would combine aspidomancy and gyromancy for a new round of predictions. Aspidomancy is divination by sitting upon a shield in a magical circle, wherein the diviner pronounces certain spells, falls into a trance and then prophecies. Aspidomancy is divination by spinning, wherein a circle is drawn on the floor and divided into various segments which have drawn in them Hebrew letters, Kabbalistic symbols, astrological symbols, and the like. A person spins around and falls into a particular segment of the circle, and where they fall is then interpreted for messages.

I got a shield from a friend and drew a circle on the floor with all sorts of neat symbols and sayings. Then I spun around and around while sitting in the shield. After a false start—the book said nothing about where you throw up being part of the process—I got the following results.

Crime

There will be a string of bank robberies in a major metropolis. The police will be baffled. Eventually the robbers will make a mistake, and they will be caught.

Violent crime will increase in a major city. There will be an outcry to stop the criminals, reform the judicial system, and make the streets safe again. Governmental officials will promise to look into the matter, new laws will be proposed, and some will actually be put into effect. More prisons will be built.

There will be a ghastly murder which will be traced to a man who was described by his neighbors as quiet and nice, if somewhat distant, and somebody you'd never suspect of being a murderer. His relatives will be shocked but will stand by him. There will be an outcry to prevent such a tragedy in the future.

Police will engage in a high-speed chase and somebody will be injured or killed. This will result from a crime where drugs were involved. Innocent bystanders will be endangered, injured, or killed. There will be an outcry and a demand that something be done. Government officials will promise to look into the matter and formulate a plan to prevent it from happening again in the future.

After such activity, I wanted to try something more restful. I decided to practice ceramancy—divination by interpretation of melted wax. I got a pan and started melting wax on the stove, and then wondered how you got the messages. Did you gaze into the liquid pool of wax and thereby also use elaeomancy—divination by means of a liquid surface. Molten wax is a liquid, so I tried it, but my hair kept getting singed, and when I took the wax off the heat it would start solidifying. I had to throw all the wax out. In the garbage pail, however, the wax formed into some interesting shapes and gave me some ideas. Here are the predictions.

Everyday Life

A new fad will sweep the country which will cause long lines at stores and disappointment among those who miss out. The hot new toy for Christmas will be in short supply, and people will be upset at the thought of disappointing children. One person will be injured in a mad last-minute scramble.

At least one community will experience an outage in their 911 service. There will be an outcry and demand for upgrading of services. Government officials will promise to look into the matter and formulate a plan to prevent it from happening again in the future.

There will be an internet scandal involving a pyramid scheme. There will be a call to clean up the Internet. New software will be marketed which allows parents to screen out objectionable material from their children. There will be an arrest of a pedophile who uses the internet to lure children to have sex with him. There will be an outcry against pornographic sites. People will complain about Bill Gates and Microsoft having a monopoly.

There will be a new live happening on the internet and
many people will try to log on and circuits will be jammed.
AOL may have trouble with too many people trying to log
on during the same time and may experience a possible system shutdown. At the very least, there will be times when
people will have to wait to be able to log on.

Then I was so fired up I thought I'd get further predictions using
molybdomancy—divination by means of melted lead. I took the pot and
a bunch of a friend's lead soldiers, which took a long time to melt. I by-
passed gazing into the molten lead, and just threw the lead into the
garbage. Before the fire got too large I was able to get some interesting
predictions. But these predictions are suspect because my friend said that
there is no lead in lead soldiers anymore.

Natural Disasters, North America

There will be storms with torrential rains and floods that will
all but wipe out a town. The TV will show scenes of people
in water, stranded pets, and people carrying other people.
Some will refuse to leave their homes and will have to be
rescued. The flooded area will be declared a disaster area, but
many will choose to stay and rebuild their homes. Government officials will promise to look into the matter and
formulate a plan to prevent it from happening again in
the future.

There will be a massive fire, which will result in a loss of life.
An investigation will uncover some defect or weakness
which contributed to the fire. There will be an outcry to
regulate things so this type of disaster won't happen again.
Government officials will promise to look into the matter
and formulate a plan to prevent it from happening again in
the future.

There will be a drought which will cause crop losses and
some livestock will die. There will be an earthquake which
will cause damage, possibly even loss of life, and scientists

will warn everyone that this is a sign that the big one is coming, though they can't say exactly when it will come. Residents will vow to rebuild their homes. A new fault may be discovered that poses new dangers to people living in widely populated areas. There will be a call for earthquake-proof buildings but the costs will be prohibitive, and it will be stated that it is impossible to retrofit all buildings to meet current earthquake codes, which may or may not be adequate anyway. Government officials will promise to look into the matter and formulate a plan.

After the toy soldiers, I settled at a friend's house determine to continue my research. I discovered many means of divination through the use of animals. For example, there's armomancy, or divination by observation of the shoulders of a sacrificial animal; hepatoscopy, or divination by means of a sheep's liver; hippomancy, or divination by observation of a horse's pace; alectryomancy, or divination by means of a cock which pecks at grain placed on letters of the alphabet; and augury, or divination by reading the movements and sounds of birds. Then there is the poison oracle, or divination performed by feeding an animal poison to determine a true answer. If the animal dies, the answer is no and the test is administered again to another animal to corroborate the answer.

I suspected using poison and animal sacrifice might get me in trouble with the ASPCA. I thought for a moment that I could go out and watch the birds, but then I remembered that I probably was not welcome at the park anymore. None of my friends owned horses, and I think horse stealing is still a hanging offense. Keeping a chicken probably would run afoul (ha, ha) of the zoning laws. So I was nearly at my wit's end when I remembered icthyomancy, or divination using the entrails of fish. I got my fishing pole and bait bucket and set out. While waiting I also practiced a little aeromancy, or divination by reading atmospheric conditions, clouds, storms, and winds. With my stringer and notebook both full I went home to clean my fish and get additional answers from their guts. It was yukky, but enlightening.

Politics, United States

The elections of 2000 will be seen as indicative of the next millennium. There will be some surprise candidates and a few surprising upsets. Some people will take this as a trend. Others will say it is just a fluke.

Some political campaigns will be dirty and negative. There will be an outcry to clean up politics. There will also be a movement to reform political spending and the way campaign money is collected. Government officials will promise to look into the matter and formulate a plan to fix it once they are reelected.

The presidential election will be seen as crucial for the direction of the country. There will be Republican and Democratic candidates but also an unprecedented number of minor-party ones. Some people will ask Ross Perot to run, but he won't. There will be a movement to get some new blood in politics and to change the status quo, but not much new will happen.

There will be scandal in the U. S. Government. A government official will admit to sexual misconduct. There will be calls for a member of Congress to step down. One member of Congress will be suspected of mishandling campaign funds. In California a statewide proposition will be on the ballot which will be controversial, and some will call it fair while others will say it will just legislate prejudice.

I had to do something with the fish parts left over after I had used the guts for divination. So I cooked them up, wondering if this is what ancient priests did. When everything was done, I sat down to eat and accidentally spilled the salt. After throwing a pinch over my shoulder—no sense alienating the fairies—I remembered alomancy, or divination with salt. So while I ate I made use of my accident, thinking that there are no real accidents—only fortuitous happenings for people to capitalize on.

World Politics

There will be a coup in South America, Africa, Asia, or Eastern Europe. A disaffected minority will rise up and put their leader in power. A bloodbath may result as one group tries to eliminate their rivals. There will be an outcry, but nothing much will be done. The U. S. government will warn travelers to leave the country, and the embassy staff will be reduced to a skeleton crew. There will be TV pictures of people trying to flee the country by plane.

A head of a foreign country will face a massive scandal at home. There will be talk of overturning the government, but most of the world will not understand the issues involved. This may trigger a vote of confidence which will be close. No matter the outcome, the leaders will call for the country to work together for the sake of the nation as a whole and try to put the scandal behind them.

After supper, I was able to get even more information using my-omancy—divination using mice and their movements and actions. I think my friend needs a cat or two in his place.

Finance

There will be fluctuations in the stock market. One stock will plummet after doing well for years. Some investors may find their holdings are worthless and will call for an investigation. Government officials will promise to look into the matter. Some analysts will predict a big crash as in 1929 or 1987. Some analysts will predict a new high for the market. Some people will sell all their holdings and invest in gold.

There will be financial scandal as a bank which was previously considered solid suddenly needs a bailout. Customers will be worried about their savings. Government officials will promise to look into the matter and formulate a plan to prevent it from happening again in the future.

There will be a big strike which will attract the attention of the American people. Management will claim union demands are excessive, and workers will call management greedy and unscrupulous. There will be heated rhetoric and claims of unfair labor practices will be filed. Workers who

cross the picket line will be heckled and police will be on
hand to prevent violence. The issues will be benefits, pay,
job security, and pensions. The company will plead they
cannot afford to give in to union demands and still remain
competitive in their industry.

Well, there it is. Maybe next year when I have a chance again I'll try
a couple more divination systems. There are several I was not able to get
to, but rest assured that I will not fail in my eternal quest to find the truth
before it happens.

(I should explain, just in case, that this was intended as a satirical and
parodic piece of humor. It was written for all you who are sick of the "M"
word by now and just want life to get back to normal. Hopefully it can
give perspective to all those dime-a-dozen gloom-and-doomsayers who
definitely need to get a life.)

Traveling by Moonlight

Kenneth Johnson

The Moon moves through the night sky more quickly than any other celestial object—with the exception of a momentary meteor. Compared to the other astrological planets, in fact, it fairly races through the heavens, circling the entire zodiac within the brief span of a month.

Furthermore, the Moon is forever changing. It begins as a tiny sliver, grows until it is round and full, then wanes again into nothingness. The Moon's wandering, ever-changing nature was evident to Shakespeare, since he refered to the orb as "the inconstant Moon." (But then again, Shakespeare was a Taurus, and he was probably suspicious of change in general.) And it wasn't just the "inconstancy" of Luna which reminded the ancients of restlessness and wandering. In the old days, long distance travel took place by sea, and the Moon is associated with water because of its influence over the tides. So it's no wonder that ancient and medieval astrologers associated it with travel.

In contemporary astrology, however, we are more likely to look to Mercury and Jupiter as indicators of our propensity to wander. All the same, the Moon still plays a powerful role when it comes to the subject of travel. In the first place, your Moon sign is a very personal element in your horoscope—if you want to know what kind of food a person likes to eat, you can look at her or his Moon sign. By the same token, your natal Moon sign has a great deal to tell us about the kind of vacation or journey that suits you best.

Aries

If your Moon is in Aries, you need lots of activity, and some of it might be a bit extreme. Don't overdo it, though. After all the parasailing, rock climbing, hang-gliding, and downhill skiing, you should try to take a break for lunch. Although it's a difficult concept for Moon in Aries to grasp, there's always tomorrow. Furthermore, you might be more comfortable traveling alone, since no one will be able to keep up with you anyway. Don't feel bad about it. Though you probably don't often feel bad about anything you do, be assured the world still needs explorers.

Taurus

If your destination is Europe, you're likely to develop a taste for five-star restaurants and vintage wines as you cruise down a lazy river with nothing much on your mind but a great deal in your stomach. If your destination lies elsewhere, you won't change your style for any reason. You still enjoy the finer things in life, and you like to enjoy them slowly, savoring each moment. With the possible exception of Moon in Pisces, you have the greatest talent for sitting in a deck chair doing absolutely nothing. A vacation which includes great sex doesn't hurt much either.

Gemini

If your Moon is in Gemini, you will probably want to travel to a workshop or a seminar where you can learn a new skill. Travel destinations with tons of museums, clubs, events, and activities are perfect as well, because you hate boredom worse than just about anything. Also, rather than going stir-crazy while sitting alone on a beach sipping a piña colada, you prefer to travel someplace where there will be lots of other people in attendance. In other words, your idea of a vacation may resemble someone else's idea of a mental endurance course. But who cares? There's nothing like keeping busy, is there?

Cancer

A vacation? You mean, travel away from home? Why would anyone want to leave home—where you have your own bed, your own kitchen, your own garden, and everything else that makes life worthwhile? With your Moon in Cancer, you are likely to agree with that well-known hobbit Bilbo Baggins, who believed that adventures were nasty, uncomfortable things. But if the other members of your family insist on leaving the comfort of your roost, you wouldn't want to feel left out. Therefore, Moon in Cancer is the pack-rat traveler of the zodiac. And if you must travel, you will probably want to include some shopping too.

Leo

Leo Moon does best to plan a trip early. After all, you are always sure to get front-row center tickets for the hottest show in town, not to mention an invitation to an exclusive art opening or coveted tickets to the opera. These kinds of activities stoke the Leonine fire, which is a sign ruled by the Sun. Moon in Leo folk literally like the Sun, so a winter expedition to Scandinavia probably won't appeal to you. Besides, everybody else is bound for a sunny destination, and if you're not there with them, who will take notice of you? It wouldn't be much of vacation if people weren't noticing you.

Virgo

Virgo Moon is well-organized for every vacation. You pack your suitcases way ahead of time and include all the important items—vitamins and supplements, medicines and remedies, and of course your disinfectant spray. Of course you're not headed for the wilds of some third-world country, where places can be unsanitary and rife with unmentionable diseases. Instead, you've packed your bags for a visit to the health spa, where no diseases can reach you.

Libra

The Moon in Libra is nothing if not romantic. A Caribbean cruise with gorgeous sunsets, wine glasses clinking in a toast to love, and the glimmer of moonlight on a beautiful beach—these are the travel experiences that turn you on. For you're in love with love itself, and a journey without romance is not worth the trouble. Moon in Libra always carries extra suitcases to accommodate the tuxedo, the evening gown, and all the other accoutrements which make you desirable. And you never travel alone. For a Libran Moon, there's nothing worse than sitting on a deck chair watching the other couples stroll hand in hand.

Scorpio

Scorpio Moon doesn't care where you go as long as the journey is intense. If you're on your way to a psychology seminar, you will expect to be transformed to the very core of your being. If you travel to a sacred site, you will expect a genuinely mystical experience. And if you're on a jungle cruise, you won't be entirely happy until you've sailed uncharted waters. As with the Moon in Aries, other people may find it hard to match your intensity and drive. Often, they simply get in your way. For you, that won't be a problem; you can easily ignore them.

Sagittarius

The really wonderful thing about Moon in Sagittarius folk is that you like to travel just for the sake of traveling. For you, it's fun to travel, because you love to expand your horizons mentally as well as geographically. You're up for anything, from the ridiculously mellow to the ultimately adventurous. That's why you make the best of all possible traveling companions—you're never afraid to cast aside your worries along with your cultural prejudices and simply enjoy the ride. Be wary, though, your curiosity about people and places can never be satisfied, and you may find yourself forever searching for whatever is new over the rainbow.

Capricorn

As with your opposite number, the Moon in Cancer, you may have trouble getting into the whole idea of a vacation. If the Moon in Cancer is reluctant to leave home, the Moon in Capricorn is reluctant to leave work. After all, the purpose of life is to achieve things, to reach new heights, and make more money. Everyone knows that, right? Who wants to sit on some stupid beach somewhere staring at a lot of people who are too lazy to work? The solution for Capricorn Moon? Turn your vacation into a business trip and accomplish something. You'll feel better just knowing that you haven't wasted your precious time.

Aquarius

Like most of the Fire and Air signs, Aquarius Moons regard every journey as an adventure. Everything has to be new and different. You like to explore, investigate, and talk to people you wouldn't ordinarily talk to. Your best traveling companions are likely to have the Moon in Gemini or Sagittarius, because like yourself they are impatient with mere relaxation. To some Aquarian Moons, a scientific holiday may appeal: perhaps a visit the world's largest telescope or to a space launching. Others may be happy simply to go where no one else would even think of going; have you been to Antarctica yet?

Pisces

Moon in Pisces folk are already on a perpetual vacation, so why go anywhere new? On the other hand, the world is just one big beautiful dream, so why not travel around and see what everyone else is dreaming about? One way or the other, Pisces Moon is likely to be happy sitting for hours, relaxing, and merging with the universal flow. Needless to say, you oceanic souls are at your best near the water: other lunar types may need the stimulus of a concert or a club, but the sound of waves crashing on the shore is enough to satisfy you. And at your best, you feel at home anywhere.

Where to Go?

Once you've determined what kind of a trip you want to take the next question is, when shall you go? The thumbnail sketches given above can be used to answer this question as well. If you're going adventuring in the wilds of Utah, begin your journey during an Aries Moon. If you're going on a sea cruise with a new—or old—flame, start when the Moon is in Libra.

There are other guidelines about timing a journey according to the Moon, and if you're fairly skilled at reading your astrological calendar, you may want to take them into account.

First of all, try to leave home and begin your journey during a waxing Moon. The waxing half of the Moon's cycle symbolizes growth and beginnings, and most ventures should be started while the Moon is in its growing stage, between New and Full. Whether you're planning a trip across the state or a trip across the world, the idea is the same; begin during the time when beginnings are favored. According to ancient tradition, one should try to avoid the actual days of the New or Full Moon. The absolute best days for undertaking a journey are the 2nd, 3rd, 5th, 7th, 10th, 11th, and 13th days of the waxing cycle. If you must travel during a waning Moon, count from the day of the Full Moon and try to find the 2nd, 3rd, and 5th days of the waning cycle, and so on as listed above.

There are, of course, certain signs of the zodiac which are more favorable to travel than others, so, if possible, try to travel while the Moon is in Aries, Taurus, Cancer, Leo, Libra, or Sagittarius. Yes, I know that I've spotlighted Cancer Moons as being somewhat reluctant to travel. Nevertheless, ancient astrological tradition affirms that Cancer is a favorable lunar position during which to begin a journey.

If we choose, we can become even more precise by borrowing a page from the vast astrological tradition of India, which insists that travel may be either positively enhanced or negatively influenced while the Moon is in certain constellations. These are not the constellations of the zodiac as we usually understand them, but the ancient "Mansions of the Moon." In this way of looking at things, it's important that the actual constellation or grouping of stars should be used, which means converting to a sidereal zodiac. Our own zodiac, used here in the Western world, is "tropical" rather than "sidereal." The whole issue is too complex and technical to deal with here, so you'll just have to take my word for it when I tell you that I've done all the work for you, and converted the sidereal Mansions of the Moon into lunar positions that you can find in an ordinary astrological calendar. For advanced students, these observations may add some detail to the general considerations mentioned above.

Travel is extremely favored when the Moon is:

- In the last 10 degrees of Aries or the first 6 degrees of Taurus
- In the last 5 degrees of Cancer or the first 10 degrees of Leo
- In the first half of Libra

According to this tradition, there is a section of sky which is especially unfavorable for travel. It occurs when the Moon is between 6 degrees Taurus and 3 degrees Gemini. So we may want to take some care with our travel plans, especially since the traditional notions above, which list Taurus as a favorable travel sign, creates some contradictions. It's all a matter of how detailed you want to be. Those who have a knowledge of astronomy may notice that the part of late Taurus and early Gemini which the Hindus distrusted is right smack dab on the constellation of the Pleiades or "seven sisters." So another way to say it is: Don't start a journey when the Pleiades are obscured by the Moon.

Then there's a tricky number which astrologers call the Void-Of-Course Moon. The Moon is considered to be "void-of-course" when it has made its last aspect before leaving one sign and entering another. For example: during January of the year 2000, Jupiter will be at 27 degrees of Aries. So as the Moon moves through, let's say, Sagittarius, it will make a trine to Jupiter when it reaches the 27th degree of Sagittarius. After that, it won't make any other aspects to any other planets until it reaches Capricorn. During that period of time—which is only about five hours—the Moon is said to be "void-of-course." A really good astrological calendar will usually indicate when this is happening. Try to avoid actually getting into the car and starting your engines for the trip until the Moon has entered the new sign and is no longer void-of-course.

Firewood

Louise Riotte

You are a King by your own Fireside as much as any
Monarch in his throne.

—Miguel de Cervantes, *Don Quixote de la Mancha*

Nothing is lovelier or more comforting than a wood fire on a cold
winter night. Our subliminal minds reach far back into a time
when fire represented not only warmth, but safety as well. Perhaps, too,
our custom of telling ghost stories around a fire originated long ago when
fire was believed to keep evil spirits away.

An open wood fire indoors or out can be made even more pleasurable
if the right wood is cut at the right time. In my earlier book, *Sleeping with
a Sunflower* (Storey Communications, 1987), I explained that since wood
was the only means for most frontier people to stay warm in the winter or
cook food, stocking a woodpile was a very serious matter. There are two
schools of thought regarding the best time to cut timber. Most people
agree that wood should be cut in the time of the waning Moon, but there
is some disagreement as to the correct month. One old book, for instance,
says August in the best month because the weather is generally hot and
dry. Meanwhile, another book claims that for more durable hardwood
timber, October, November, and December are the best months.

The sap of most trees is down in the old of the Moon in December
and January. All kinds of timber that shed leaves in the fall should be cut
in the third quarter of the Moon—in winter if possible. Wood will tend
to be more durable if cut in December. Evergreen timber must be cut in
the second quarter of the Moon in fall and winter.

There may be some people who still recall that in France woodcut-
ters once advertised that "wood was cut in the waning of the Moon."
Today, however, people who take on this demanding profession are
becoming scarce.

Two thousand years ago, Plutarch, a Greek moralist, had this to say
regarding cutting wood:

"The moone showeth her power more evidently in those bodies which have neither sense nor lively breath; for carpenters reject the timber of trees fallen in the ful-moone, as being soft and tender, subject also to wormes and putrefaction, and that quickly by means of excessive moisture."

Cynthia Chung, staff writer at the *Daily Oklahoman*, tells us: "Not all woods are equal. For fireplaces and stoves, some varieties are better than others." According to experts, hardwoods—such as oak (Jupiter-ruled), ash (Venus, Sun, Saturn), hickory (Jupiter), hackberry (Jupiter), black locust (Saturn, Scorpio), maple (Jupiter, Moon), and birch (Jupiter, Venus)—make the best fuel. Hardwoods are denser, weigh more, give off more heat, and burn longer than other woods. Fruit trees such as apple and cherry (both ruled by Venus), and nut trees like chestnut (Jupiter, Venus) infuse the room with a delightful aroma when burned. Oak is a perennial favorite in Southwestern fireplaces, especially the native blackjack oak now that the white and red varieties are becoming more rare and expensive.

Softwoods, meanwhile, such as pine (ruled by Saturn, Capricorn, Mars) and cedar (Pluto), are not recommended for burning since they leave more ash or creosote in chimneys and have a higher moisture content that causes them to pop while they are burning. Think about hardwoods as having leaves while softwoods have needles.

Piñon wood from New Mexico, meanwhile, is becoming more popular in the Southwest, and is often used in clay chimneys and outdoor fireplaces. This wood gives off a strong odor that is also a natural insect-repellent, and it leaves hardly any ash. Mesquite is another wood about which relatively little is known outside of the Southwest. Mesquite burns exceedingly hot, is very fragrant, and is highly prized. It is a leguminous tree or shrub that forms extensive thickets and bears sugar-rich pods that are used as livestock feed. Two kinds of gum from the mesquite tree are used to make candles and dyes. As firewood, mesquite is often sold in chips and used with other wood.

In many places it is becoming difficult to find seasoned firewood. If you are buying, it is best to get your order in early in the season and to check your wood carefully for signs of "greenness."

If you are selling wood, advertising it early in the season may be an advantage in getting the word out about your product. You should write your advertisement on a day when the lunar cycle is favorable, when Mercury is sextile, trine, or conjunct the Moon, and you should not write it when Mars or Saturn are in square or opposition to the Moon. You should also plan for your ad to reach the public on a high lunar cycle day if you want to deal with pleasant people and have a satisfactory sale.

When buying firewood, you should always ask how long ago the wood was cut, or how long it has been left to season. Wood takes six to twelve months to season properly, and seasoned wood has less moisture and weight, and has fewer cracks and splits in its grain. Consumers should also know how much wood they are buying. A full cord of wood is four feet tall, eight feet long, and four feet wide. Wood can also be sold in fractions of a cord. That is, a three-quarter cord will measure four feet in height, eight feet in length, and three feet wide. A half cord, sometimes known as a rick, is four feet tall, eight feet long, and two feet wide, and so on.

Charles Carter, the program administrator of the Oklahoma Department of Agricultural Forestry Services, says to take care when buying firewood. "Consumers should expect to get a receipt, a delivery ticket, or a sales invoice. It should have the name and address of the vendor, the name and address of the buyer, the date of delivery, the quantity delivered, the price quoted, and the weight on which the price is based." Consumers should also check their wood for signs of rot, wormholes, open spaces, knots, or termite damage. If any of these problems occur, a consumer can file complaints with the Better Business Bureau, though using a reputable wood vendor will help avoid trouble in the first place.

Many fireplace-related house fires that occur every year can be avoided by taking sensible precautions. Chief John Long of the Oklahoma City Fire Department offers these rules in building a home fire:

Do not use charcoal or lighter fluids in a fireplace.

Cover the fireplace with a metal screen or glass door.

Clear the chimney regularly of creosote and ash. Where fireplaces are not used often, the chimneys should be swept every two years.

Do not build too large a fire. Most fireplaces aren't meant to heat a house. Four logs is plenty to fill most prefabricated fireplaces.

Do not leave a fire burning when you go to bed. You don't want a fire to be wide awake when you're fast asleep.

Do not cook hot dogs, hamburgers, steaks, and the like in a fireplace. Smoke from meat can coat the inside of a chimney and make the fireplace more combustible.

Do not burn trash, newspaper, or Christmas wrapping paper in a fireplace. Burning ash from the paper will float out the chimney and onto the roof.

Do not allow children to play around the fireplace.

Born of Moonlight

Louise Riotte

The Moon like a flower,
In heaven's high bower,
With silent delight,
Sits and smiles on the night.

William Blake (1757-1827)

Knowing the significance of our own name and tracing it back to its origin is a fascinating pursuit. For instance, the name Margaret comes to us directly from the Greek for "pearl." In turn, this usage derived from a Persian word which can be translated as "born of moonlight." The idea was that pearls were created from dewdrops which pearl-oysters, rising from the sea, received on moonlit nights.

Furthermore, Marguerite, the French form of Margaret, means both "pearl" and "daisy" in French—the flower was probably so-called because of its Moon-like whiteness—which in turn accounts for Daisy as a nickname for Margaret.

How Were Names Invented?

Family names are of fairly recent origin. Given names, on the other hand—according to Winthrop Ames in his book *What Shall We Name The Baby?*—date from the very beginnings of time. Early tribes had to invent titles to distinguish its members from one another. Apparently, this happened universally—from prehistoric Troy to far-off Timbuktu. Families often give names based on some characteristic or quality they hope a child might possess, such as courage, or they might borrow a name from a tribal god, a strong or swift animal, a precious stone, a weapon, or a flower. Favorite names were then handed down from generation to generation (their original meaning often forgotten in the process) while others dropped out of use, so that each tribe developed its own original collection. Then as trade and war brought nations in contact, they lent and borrowed names from each other.

Of the given names used in America today, ninety-nine out of every hundred originated in four languages: Hebrew, Greek, Latin (via the Italian, French, Spanish), and Teutonic (including languages such as Scandinavian, Old German, and Norse). But this now is changing as many African American mothers are giving their children names which have ancient African origins. For instance, one of the contestants in the Miss African American Collegiate Pageant had a beautiful African name which meant "Everything belongs to God." There are, of course, fashions in names, but these are not as transitory as the fashions in other human habits as most names are sanctioned by tradition and cemented by sentiment. Fashions only affect the fringe-names, as it were. In the United States it is especially difficult to follow any clear trends because successive waves of immigrants have brought their unique names with them.

I was curious to know how the Moon and stars might have influenced the naming of children, and what I learned surprised me. The heavenly bodies have played an important part in naming throughout history. For instance, many early peoples named their children for the Moon and the stars or chose names based on some birth circumstance. A Native American mother might name her daughter "New Moon" because she was born at the time of the new Moon, or "Pale Moon" because the Moon was not bright. We can see more of this in a poem by Longfellow:

> By the shores of Gitche Gumee,
> By the shining Big-Sea Water,
> Stood the wigwam of Nokomis
> Daughter of the Moon, Nokomis...

Furthermore, a wealth of names based on the Moon and stars have come down to us today, especially names for girls. Diana, for example, is an ancient name that means "pure goddess of the Moon." The name Diana dates back to the time of the Roman Empire, and continues to be popular in its variations and diminutives: Dina, Deanna, Diane, and Di, made famous by the late Princess of Wales.

Other celestially inspired names you might wish to consider for a daughter include:

Luna—Latin, "of the moonlight"

Estelle, Stella—Latin, "a star;" variations: Estella, Estrelita

Selena–Greek, "the Moon;" variations: Lenn, Selene

Starr, Star–Anglo-Saxon, "star"

Hester–Persian, "a star;" variations: Hetti, Hetty

Hesper–Greek, "night star"

Astra–Greek, "like a star;" variations: Astrea, Astrid, Astred

Esther–Hebrew, "a star"

Celeste–Latin, "heavenly;" variations: Celesta, Celestine

Rita–Greek, "a pearl," or "jewel of the Moon and Venus"

Pearl–Latin, "precious gem;" variations: Pearle, Perle, Perl, Perlie

Pegeen–Celtic, "pearl"

Chandra–Sanskrit, "she outlines the stars"

Electra–Greek, "shining star"

Dione–Greek, "the daughter of heaven and earth"

Urania–Greek, "heavenly"

Danica–Slavonic, "the morning star"

Cynthia–Greek, "Moon goddess;" diminuitive: Cindy, Cynth

Isadora, Isidora–Greek, "gift of isis." Isis was the Egyptian goddess of the Moon, so these names may also mean "gift of the Moon"

Neoma–Greek, "new Moon;" a contraction of Neomenia, which was an ancient festival held at the new Moon; girls born under a new Moon were often so named

Philomela, Philomel–Greek, "the nightingale," which means "lover of the Moon;" in Greek mythology, Philomela was a maiden changed into a nightingale

EXPLORE NEW WORLDS OF MIND & SPIRIT

Just drop this card in the mail to get your FREE copy of *New Worlds*—it has 40 full-color pages packed with books and other resources to help you develop your spiritual potential to the fullest.

New Worlds is your key for opening new doors to personal transformation. Get fresh insight into herbalism, spiritual gardening, health, astrology, and many other subjects, with:

- Articles and "how-tos" by Llewellyn's expert authors
- Tasty previews of Llewellyn's new books and features on classic Llewellyn titles
- Upbeat monthly horoscopes by Gloria Star
- Plus special offers available only to *New Worlds* readers

❑ **Please rush me my free issue of *New Worlds of Mind & Spirit*!**

Name _____

Address _____

City _____

State _____ Zip/Postal Code _____

MSB00

LLEWELLYN PUBLICATIONS
P.O. BOX 64383
ST. PAUL, MN 55164-0383

Phoebe, Phebe–Greek, "the shining or brilliant;" a special
title given to the Greek goddess Artemis as goddess of
the Moon

Selene–Greek, "the Moon;" in Greek mythology, Selene
was the goddess of the Moon

Sidra–Latin, "starlike"

Trella–Spanish for Esther, meaning "a star"

Does the meaning of a name have an impact on the child's consciousness? I believe it does, especially if the child knows the meaning and
is inspired to live up to it. Even if a child does not consciously know the
meaning, there seems to be a subtle effect upon the personality.

Surnames

Matching up the child's name with the family surname is something that
should be given considerable thought. Surnames were first taken by the
nobility, and indeed, according to Ames, were originally spelled "sir
names." Surnames at first were mostly derived from the family place of
residence or stronghold. This custom still survives somewhat today; when
a new English peer is created, he is given an extra title taken usually from
his place of residence.

Commoners eventually followed the example of the nobility, but
even as late as 1405 surnames were not universal when Edward V passed
a law to compel certain outlaws, who apparently thought it easier to escape identification without a surname, to take them. The wording of this
law well illustrates how the new names were devised:

They shall take unto them a Surname, either of some
Towne, or some Colour as Blacke or Brown, or some Art or
Science, as Smyth or Carpenter, or some Office, as Cooke
or Butler.

To this day in some remote parts of Ireland surnames are still rarely
used, and neighbors still call each other by such nicknames as "Michael,
son of Big Peg." Many family names plainly grew out of nicknames based
on some personal characteristic of an ancestor. Peter the strong became
Peter Strong; black-haired John or blond William became John Black or

William White. Other surnames were manufactured by adding "son" to the father's name, thus, Tom's sons became the Thompsons, William's sons the Wilsons, and so on.

According to Professors Albert Mehrabian of UCLA and Herbert Barry III of the University of Pittsburg, two psychologists who have studied how names affect the way we think about people, and coauthors of *The Name Game,* the ideal name should be traditional, popular, and not subject the child to ridicule. You should listen carefully to the sound and rhythm of any names you are considering for your child. Say the full name aloud and let your ears, good judgment, and taste guide you. Most people find that a surname of just one syllable—Jones, Smith, White, Brown— usually combine best with a given name of two or more syllables. For example, Bethany Kent has a softer, more lyrical sound than Bess Kent. With surnames of two syllables—Carter, Hudson, Turner, and so on—try given names of three or more syllables, such as Belinda Pauling, Patricia Meeker, Jessica Irving, Isadora Duncan. With surnames of three syllables, experiment with given names of one or two syllables—Rita Cameron, Astra Luellen, or Pegeen McCrary. Try various names for your child, always remembering this general rule of rhythm and sound: a full name usually sounds most pleasant when the syllables of the given and surnames are unequal in number.

If your surname is simple, says Dr. Mehrebian or Dr. Barry, it's usually wise to keep the child's given name simple. You may be drawn to alphabetical names such as Locelyn or Stephanie, but if your last name is Tead or Stubbs, the combination may sound humorous. It's also well to keep in mind the national origin of your surname and of the given names you consider. For example, Yvonne and Jacques, or Estelle and Marcel, sound fine when combined with a surname such as DuBarry, but awkward when matched with names like Kelly, Cohen, or Cappola.

I found few astrological names for boys—only Isidore or Isadore, derived from the feminine Isadora (Greek, "gift of Isis") had any relation to the Moon or stars. Names for boys seem to favor the Sun. Cyrus, Persian for "the Sun," is a name made famous by Cyrus the Great, founder of the Persian Empire. Sol, Latin for "the Sun," is also a short form of Solomon and dates back to Biblical times. There are also a few boy's names based on the planets and astrological signs:

Mark–Latin, "belonging to Mars," or "a warrior;" variations:
Marc, Marcel, Maven, March, Marco, Marcus, Marek,
Mars, Martin, Martyn, Marty

Leo–Latin, "a lion", or "proud as a lion;" variations: Leon,
Leonard, Leonardo, Lion, Lionel, Lyon, Len, Lennie,
Lenny, Leopold, Lee, and Lapp

The most popular and influential names for boys have meanings relating to power, leadership, and responsibility. For example, take the name Michael, which in Hebrew means "Like unto the Lord." Michael was an archangel, the "Defender of Heaven" and chief of the angelic host; when people hear the name they often think of someone who is successful, honest, cheerful, warm, and very masculine. Can this be because we unconsciously associate this name with the Hebrew meaning? Perhaps, but there is another factor playing a part here as well.

According to Dr. Barry, "block letters with vertical lines are seen as stronger than (letters with) curves." The name Michael has the "k" sound that, even though the letter itself does not appear in the name, is often associated with success. Other "k sound" masculine names such as Mark, Luke, Victor, and Rock continue to be popular as well. "K sound" feminine names include Victoria, Katherine, and Jacqueline.

James, Charles, and Alexander are also ranked high in the success category for men. In general, Dr. Mehrabian notes, the longer a man's name, the more likely he will be considered honest and accomplished. "In society," he theorizes, "we attach importance to height, to substance. In the same way, we may attach importance to a longer name."

It would seem that if you want to create a feeling of respectability and trust, you can best achieve this by using your full name. On the other hand. if you want to be popular and considered "one of the gang," you would do well to shorten it—maybe by using a nickname, such as Bill for William, or Bob for Robert.

Legalizing Names

Until recent years, according to Winthrop Ames, there was no provision for registering births in many parts of the United States, and our parents often had trouble getting the equivalent of a birth certificate. Nowadays, the attending doctor or midwife is legally obliged to record the date and place of every birth, with the infant's full name if determined. If no first name has yet been chosen the infant is entered as "Girl" Smith or "Boy" Jones, and may so remain until some proof of identity, such as a birth certificate or marriage license, is required. Then the complete name must be attested and put on record. It is therefore quite possible for especially indulgent parents to allow a child to grow up and then choose his or her own legal first name.

An adult may change both their given names and surnames legally, but only by permission of a court. This is granted usually on the provision that there is no intent to evade responsibility or take advantage of another's reputation, and if the reason seems proper.

The Magical Hare in the Moon

Chandra Beal

You have probably heard of the hare in the Moon who can be seen, according to Chinese legend, grinding the herb of immortality in a mortar to make the elixir of life. There are many other legends that tell us how the hare got into the Moon in the first place to ponder while you are gazing at the lunar body .

Many eastern cultures tell variations of a story of a rabbit who leapt into a fire in an act of self-sacrifice. The fire transformed him and his essence became the last moonlight of dawn, and so the rabbit was immortalized on the Moon forever.

In a Chinese legend, Buddha takes the form of a hare during a period of spiritual testing and encounters Brahma, a sky god. To please Brahma, Buddha leaps into a fire and becomes food for him. Brahma is so pleased with this selfless act that he paints the image of the hare on the Moon as a remembrance of the event. The hare escapes the flames unscathed because his spiritual transformation has immortalized him. In another version of this story, Buddha is walking in the woods and he meets a hare. The hare offers himself to Buddha as sustenance, which pleases Buddha. In this version, Buddha honors the hare by placing him on the Moon for eternal safekeeping.

The symbol of the hare in these cases seems to reference the basic belief of many religions that in dying you will rise again to life. The Moon also dies and rises again each night in the sky, so the hare directly correlates to the lunar disc.

Another story from Africa records that the Sun and Moon were sisters who once sat side by side. One day they caught a rabbit for dinner, skinned it, and set it in a pot of boiling water to cook. In the meantime, the sisters engaged in an argument. The Sun took the rabbit skin out of the pot and flung it at her sister, the Moon. The Moon retaliated by throwing the boiling water in the sun's face. Ever since this incident, the hare's mottled skin can be seen on the surface of the Moon, while the sun remains boiling hot. Furthermore, the two sisters no longer sit side by side.

There is a widespread view that the hare is the maker and purveyor of the elixir of life, sometimes called *soma*. It is not specified which plant is the "herb of immortality," but it is understood as representative of the divine spark in all of us, or the element of the eternal.

Lunar mythology and rabbit folklore are inextricably intertwined. Often we will see a correlation between hares and the Moon and moonlight, Moon deities, and qualities which we attribute to lunar energy, such as intuition, renewal, and immortality. According to some folktales, rabbits conceive simply by looking at the Moon, or by licking each other's fur. Both hares and the Moon are perceived as fickle since they never appear in the same place or shape for two nights in a row. They are both surprising and varied characters. The hare pops his head up in the forest and then quickly fleets away. Hares are never seen walking but are always leaping about. This is one reason why hares are associated with witches, who often are well versed in Moon magic, are flighty and intuitive, and are able to shapeshift.

Moon festivals are still celebrated today by Chinese, English, and German people around the time of the vernal equinox when people exchange small cakes imprinted with the image of a rabbit.

So next time you find yourself gazing at the Moon, look for the magical hare. You just might find him looking right back at you.

Moonlore

Verna Gates

In ancient Sanskrit, the word for Moon also means measurement. In olden times, calendars evolved around the lunar cycle. In fact, many calendars still rely on the Moon. The Moon has been used through the centuries to measure seasons, planting and harvesting times, and good or bad fortune.

To live by the ancient lunar rules, you must divide your activities according to the phases of the Moon. Starting with the New Moon, your activities should wax until the Full Moon and then wane again.

There are some further rules you should follow. For instance, never let the New Moon catch you with your pockets empty or you will have little all month. Instead, turn the coin and bow to the Moon asking it to leave you as well as it found you. Meat lovers should beware—stock slaughtered at times other than the New Moon will shrivel in the pan. The New Moon is also an excellent time to start taking medicines for a quicker cure. Any New Moon appearing on Saturday or Sunday is a sure sign of rain and ill-fortune. And by the way, it is lucky to look at the New Moon only over your right shoulder, never the left.

In fact, the first ten days after the New Moon are all-important for long and happy life. According to advice taken from an ancient English document, the first day after a New Moon is the time to start anything new—a business, an exercise regimen, a diet, or a fledgling love affair. Be careful of cold winds, however, for illness can easily begin on this day, and it may take more time than usual for you to recover. A child born on the first day after the New Moon is meanwhile destined for health, wealth, and happiness.

The second day after the New Moon is a time for trading and traveling. You will fare well in anything bought or sold on this day, and you will reap great rewards. Sea voyages, risky at other times, are granted smooth sailing on this day. For farmers, plowing and sowing also bring rich results on the second day of the New Moon.

The third day is a time of trouble. You should stay home and begin nothing new, and if you try to lie or deceive anyone on this day, you are likely to be caught. Criminals especially should refrain from making their livelihood on the third day of the New Moon.

The fourth day is a day of building. Houses and buildings started on this day will stand solid. If you have political aspirations for your child, this is a good day for his or her birth.

The fifth day casts the weather. Whatever happens today will tell you if you are in for rain or Sun the rest of the month. If you want a big, healthy baby, it is an excellent day to conceive. The sixth day, meanwhile, brings the fish to the bait and the prey to the hunter. And the seventh day is an excellent time to meet your mate; true love comes from a meeting on this day.

The eighth day after the New Moon should see you in the picture of health. To be sick on this day rings an ill omen of death. On the ninth day you should never turn your face to the Moon, especially when sleeping— you will surely become mad or wake up with your beauty stolen by the Moonbeams, or in a severe case you could transform into the dreaded werewolf. If pregnant, avoid the Moon altogether on this day—there is a risk of birthing a child with a hare lip.

Finally, people born on the tenth day of the New Moon are destined to be restless spirits who wander the earth in dissatisfaction. At twenty days after the New Moon, time will be ripe for getting rid of warts. To do this, you should stare at the Moon without blinking and rub yourself with anything within reach.

The Full Moon, on the other hand, is a ripe time for activity of all sorts. It is an excellent time to marry, as the round orb blesses the circle of love in a wedding. The Full Moon is also an excellent time to harvest medicinal herbs; all living things are stronger during the Full Moon, and there is greater potency in plants. It is the optimum time to harvest the powerful herb mandrake, and it is also a great time to cast spells, get rid of ants, and to rub horse nettle on your horse to strengthen him.

At the same time, the Full Moon is a bad time to be controversial. Assassinations are known to be common at this time—Christ, Caesar, and Abraham Lincoln were all killed under a Full Moon. Also, there's a danger for fat people, who are, according to some beliefs, more likely to die during a Full Moon.

The waning Moon is a time when things cut will not grow. Timber, however, has its lowest sap at this time, and if cut it will dry perfectly. In Europe, in fact, it was once illegal to cut wood during a waxing Moon.

Runner plants that grow counter-clockwise up poles, such as beans and peas, should be planted during a waning Moon. If you have a restless urge, meanwhile, a waning Moon smiles on a move to a new house. According to legend, you will never be hungry if you move into a house during a waning Moon. Any marriage, on the other hand, will have ill-luck if it occurs during a waning Moon.

The waxing Moon is generally the time to plant seeds in order to encourage maximum growth. It is also the best time to deliver a child; there will be less pain for the mother, and the baby will be its most healhty. Your hair will be at its richest and most Full now, so the waxing Moon is the right time to get a haircut. Or as the expression says: "Crop your hair in the Moon's wax, never cut it in her wane, and then of a bald head you'll never complain."

Tricks of the Night

According to the Huichol people of Mexico, the Moon is a trickster that people should be wary of. Kauyumari, the Wolf, has learned this lesson many times at the hands of the trickster Moon. One time, Moon convinced Kauyumari to cut down a honey tree to get to the sweet treat inside. For his effort, Kauyumari exposed water to reflect the Moon's laughing face and the angry countenance of a disturbed bear. Another time, Moon persuaded Wolf to open a basket of food by saying it was the most delicious taste to be found on earth. When Wolf opened the basket, grasshoppers landed squarely on his canine nose. Still another time, Moon sold Kauyumari a donkey that delivered silver coins everyday through his digestive system, so to speak. When Kauyumari lifted the animal's tail he was greeted not with his fortune, but with … well, you can imagine the rest.

Coyoclxauhqui Defeated by the Fifth Sun

The Aztecs' Moon, Coyoclxauhqui, plotted with her four hundred brothers to prevent the birth of the fifth Sun. The four previous Suns had been destroyed. A fifth had been magically implanted into Coatlicue, Snake Skirt, by a ball of down floating earthward from heaven.

Huitzilopochtli, the fifth Sun, was warned in the womb about the plan to kill his mother. Just as the assassins sprang upon her, he leaped from the womb fully grown. Painted blue, he was already a mighty warrior. His weapon was a flaming fire-serpent. He first defeated his sister, the Moon, then turned to his four-hundred brothers, the stars.

Huitzilopochtli's victory ensured the fifth Sun's rising and the current era of Aztec time. He rules our current time, awaiting the earthquakes that will in time destroy his reign as it destroyed the other four Suns.

The Rabbit on the Moon

According to the Oaxaca people of Mexico, Sun and Moon were brothers and lived with an old woman who had adopted them as infants. One day, the old woman asked the boys to cut grass for her husband, who was a deer. The boys put so much force into their work that they scared up a rabbit. The rabbit sprang into Moon's face and left a permanent imprint that you can see today.

The Inca Explain the Eclipse

Mama Kilya, the Moon goddess who mothered all the earth and gave birth to the Inca race, lights the night sky from her perch in the heavens. This great mother served her children by marking time on their ritual calendar. During a lunar eclipse, Mama Kilya was vulnerable to the attacks of an enemy, a giant serpent, who wanted to eat her round flesh. To scare away the beast, her earthly offspring would make as much noise as possible to frighten away the attacker. The image of the Moon was preserved in silver, her nightly color, in the Coricancha, the great temple. She was flanked by the mummies of Inca queens.

The Moon and the Great Flood

Once, according to the Pima of the Southwest, there was no war and little death. People grew numerous and the earth was getting crowded. A council was held in the skies. Sun and Moon were there, and their two sons, Coyote and his twin brother Fox. It was decided to send a great flood on the people to reduce their numbers. The flood lasted four long years. Coyote and Fox, living in the heavens, ran up and down the Milky Way, fixing stars with their fishing nets. Coyote fell off and we can see him when his mother, the Moon, takes him into her lap when the Moon is Full.

The Moon of Death

Australian Aborigines of the Bathurst Islands recite the tale of the Moon Man Tjapara who is responsible for the first death. Tjapara lived with a married couple, Purukupali and his wife. Intent on seducing Purukupali's wife, the Moon man urged her to take a walk into the forest with him. She left her infant son behind sleeping beneath a tree, and the Sun saw fit to punish the illicit lovers by killing the infant with his rays.

In anger over both the betrayal and the death of his child, Purukupali cursed humanity with death. The shamefaced Moon man pleaded with the wronged husband and father, asking him to let the gentle Moon undo the damage of the fierce Sun. If the Moon man could only have the boy for three days, he could restore his life. Purukupali stubbornly refused. They fought over the child. Tjapara, the Moon man, snatched the child and headed for the sea. As he entered it, a twisting whirlpool pulled them from sight. Tjapara turned himself into the Moon but still bears the scars of his fight with Purukupali.

The Maori Reveal the Secret of Phases

Rona, a male deity, could not find his wife. It had been several days since he had seen her working under the feminine Sun. However, Rona noticed that the masculine Moon, which usually guided the Maori men during the hunt, now had unusual spots on it.

Rona left his home in New Guinea to search for his missing mate. His travels ended at the Moon, where he engaged in a fight. Whether the Moon hides his wife or not, we will never know as the battle continues to this day. The two men hack and slash each other, eating each other's body and causing the Moon to shrink. When the Moon begins to grow and to wax bigger, it means the two warriors are nursing their wounds and waiting for the battle to begin anew.

Selene, Bright Light

Sister of Helios, the Sun, and Eos, the dawn, Selene was the goddess of the Moon. Her serene face shone gently down upon humanity, waxing and waning from her virgin crescent to her full maternal glory. The beautiful Selene wore gossamer wings and rode across the sky in a chariot drawn by snow-white steeds.

Selene enjoyed the company of young men. However, unlike Eos, the dawn, who was given to abducting vigorous young men for their vital energy, Selene preferred a calmer state of affairs and fell in love with the simple shepherd Endymion as he slept in a cave on Mount Latmus. She visited him nightly, denying the children of earth their heavenly lamp.

Zeus eventually forced a compromise and gave Endymion eternal youth and life in exchange for Selene's renewed commitment to work as the Moon. Apparently the somnolent state did not affect Endymion's masculine gifts and the couple produced fifty daughters. Some say a single son, Naxus, was also born to the couple.

Raven Finds the Light

Raven, the Trickster Bird, was tired of flying around in the dark; he was already black from bumping into things. But there was no light in the great Northwest, only darkness and freezing cold. No one had ever felt the warmth of the Sun's face.

A rumor spread among the Inuit that a man lived far away with his beautiful daughter. Their hut was always well lit by two bright, shiny balls. Raven decided to find this home and to steal the shiny balls.

He traveled far and asked the forest creatures if they knew how to find the home of the shiny balls. The owl hooted him on, the bear gave him a ride, and the rabbit hopped along beside him. Finally, Raven saw a bright shiny light ahead of him that was rapidly disappearing. He hurried towards it to find a man shoveling snow. The mountain of snow was quickly covering up the two shiny balls.

Raven asked why the man shoveled snow over the shiny balls. He answered that he was just clearing the path to the doorway so his daughter could come in from her walk. He invited Raven inside the warm shelter. He set aside his shovel and opened the bearskin doorway and entered, expecting Raven to be behind him. Raven grabbed the shovel and pitched the dimmer light high into the sky, it spins still with the force of his toss. That's why we sometimes see the edge of the Moon, and sometimes it spins to show us its full face.

Raven snatched the brighter ball and began flying home. The man soon realized what was happening and ran after him. But it was too late. Raven was already pitching out pieces of light into the sky. That's why some days are longer than others, because Raven would toss out a piece of light whenever he thought about it, sometimes sooner, sometimes later.

Thoth, Beautiful of Night

The Egyptian god of wisdom and learning, Thoth, ruled the night sky as the Moon. A celestial scholar, Thoth invented hieroglyphics and is often pictured with his scribal tools of ink palette and pens. On earth, he was depicted as either a baboon or a human with the head of an ibis. Thoth was also king of the scribes. The Egyptian state employed many of these early bureaucrats in order to record important events. It was said that a spiritual baboon supervised the work of the scribes to prevent laziness.

The most famous of Thoth's academic writings was the Book of Thoth; many ancient scholars wrote about this book, but no copy has ever been discovered—though some people believe a copy of it burned in the Great Library of Alexandria.

Written by the Thoth himself, the book supposedly contained all the collected magic and wisdom of ancient Egypt. It was discovered for the first time by a prince, Neferkaptah, who was also a powerful magician and an ardent student of the temple libraries. One day while casually reading and transcribing his temple walls he was tempted by the cackling voice of an ancient priest. This evil one scoffed at Neferkaptah for seeking small wisdom on the walls when he could find the ultimate secrets of the world in the Book of Thoth.

Neferkaptah used his magic to cipher the location of the book and traveled far to find it. He located the book in a golden box surrounded by serpents. One enormous and dangerous snake coiled around the book itself. Neferkaptah summoned his strongest spells to overcome all of the serpents and managed to kill Thoth's guardian snake.

Neferkaptah opened the Book of Thoth and immediately felt its power with the reading of the first spell, through which he could understand the language of every living creature. He became so greedy for the spells, he began gobbling them down—writing them on papyrus, soaking the papyrus in beer, and swallowing them. Drunk with power, he was prepared to sail to the land of rulers, Memphis.

Displeased with Neferkaptah's disrespect, however, Thoth rained down a terrible punishment that one-by-one destroyed his loved ones. Realizing his selfish error, Neferkaptah killed himself. His father buried the family and the cursed book in a sealed tomb. Those who find the burial chamber and discover this book of lunar magic must win it by playing a game of draughts with the spirit of Neferkaptah. But it is believed that destruction will fall on anyone who finds the Book of Thoth.

The Full Moon Gazing Festival

Lynne Sturtevant

S ilk rustles in the evening breeze. A fortune teller whispers secrets to a eunuch as two of the Emperor's concubines giggle behind paper fans. The aromas of roasting duck and baking Mooncake blend with the fragrant smoke from thousands of smoldering incense sticks. Members of the Chinese Imperial court and the Son of Heaven's invited guests glance anxiously at the Hall of Peaceful Contemplation's pointed red roof, as no one wants to miss the first glimpse of the Mid-Autumn Full Moon.

The Chinese Full Moon Gazing Festival falls on the fifteenth day of the eighth Moon in the traditional lunar calendar, corresponding to the Western September Harvest Moon. According to an ancient Chinese belief, the Moon is closer to the earth during the eighth lunar month than at any other time; therefore, it is the largest and brightest Full Moon of the year. As people all over the world gather for September Moon watching parties, they preserve and continue traditions whose origins are lost in China's distant past.

The Lunar Calendar

The ancient Chinese calendar was based on the phases of the Moon. Though the government officially adopted the western dating system in 1912, traditional festivals are still set according to lunar calculations. The Chinese New Year, for instance, occurs on the second New Moon after the winter solstice. By western astrological reckoning, the New Year begins on the first New Moon after the Sun enters Aquarius, and the date always falls between January 21 and February 19.

Months are calculated from New Moon to New Moon. New Year's Day is also called the first day of the first Moon. All other festivals follow accordingly. Festival dates are described in terms of the day and month in which they fall. For example, the Dragon Boat Festival is held on the fifth day of the fifth Moon. The Hungry Ghost Festival is on the fifteenth day of the seventh Moon and so forth.

The Jade Rabbit and the Lady in the Moon

When the ancient Chinese gazed at the sky, they did not see the Man in the Moon, they saw the Jade Rabbit who sits in a grove of cassia trees pounding a mortar and pestle creating the elixir of life. Even though Chinese literary and historical records reach far into the past, there is no explanation how the rabbit got to the Moon or why he went there in the first place. The original story is one of the lost legends of old China.

The Jade Rabbit is not alone. His companion, a woman named Chang'e, was married to Houyi, the celestial archer. Houyi performed several amazing feats, including killing a nine-headed beast, and wanted to become immortal and continue ridding the world of evil threats. So he traveled to the sacred K'un Lun Mountains to obtain the elixir of immortality from Hsi Wang, the West Queen Mother.

Houyi found the West Queen Mother and the members of her court in a palace of pure gold. He met the queen's magician, Tu-o Chen-jen, who lived in The Eight Clouds Sparkling Cavern, as well as the five Jade Fairy Maids, who tended the queen's enchanted garden where the peaches of immortality grew. It took three thousand years for a single peach to form and three thousand more years for it to ripen, but they were the main ingredient in the elixir of immortality. The West Queen Mother gave Houyi a vial of the elixir, but she warned him not to drink it until he had fasted for a year.

Houyi returned to his wife, Chang'e, but he did not tell her where he had been. He hid the elixir on the roof of the house and followed the West Queen Mother's instructions, taking sustenance only from the perfume of flowers. The vial caused a strange glow to radiate from the ceiling, and the mysterious light and Houyi's odd behavior aroused Chang'e's curiosity. She discovered the elixir of immortality and drank the entire vial. As soon as she swallowed the magical liquid, she began floating in the air. Houyi tried to catch her and pull her back, but it was too late. Chang'e flew through the starry night sky until she reached the Moon, where she became a three-legged toad and sits beside the Jade Rabbit under the cassia trees to this day.

Moon Gazing in Old China

The Full Moon Gazing Festival has its roots in ancient agricultural ceremonies. It began as a festival of thanksgiving and was held outdoors after
the back-breaking wheat and rice harvests were done. Because the Moon
symbolizes yin, the female principle, women performed the Full Moon
Festival rites.

In old China, the celebration began on Moon Festival Eve as story
tellers entertained children with the story of Chang'e. The next morning,
an outdoor altar was set up and decorated with rabbit statues, lanterns,
and large paper Moon banners depicting the Jade Rabbit pounding away
at his mortar. Late in the afternoon, family members and guests gathered
for Moon-watching parties, exchanging gifts and performing ritual dances
as they waited for the Moon to rise.

At sunset, the Moon banners were burned. There is an old folk belief
that children who wet the bed should be tapped on the bottom with the
pole of a burned Moon banner. The women placed food offerings on the
altar and burned incense to "heaven and earth."

They served the guests round fruits such as pomegranates, melons and
grape that symbolized family harmony and unity by mimicking the perfect
roundness of the Full Moon. At Moon rise, the mistress of the household
presented her guests with heaping platters of Mooncakes. The cakes were
either round or shaped like the Jade Rabbit. Everyone then enjoyed their
Mooncakes and gazed at the bright Full Moon, waiting for the magical
moment when its reflection appeared in the center of their cups of tea.

Mooncakes

Mooncakes are an essential part of the Full Moon Gazing Festival. In
Imperial China, the Emperor's Mooncakes were elaborate creations several
feet across decorated with images of Chang'e, the rabbit, and the cassia
trees. Today's Mooncakes are round and usually two to three inches in diameter. There are many regional variations. They may be either savory or
sweet, filled with such items as sweetened red bean paste, walnuts, pork
and vegetables, lotus seed paste, and salty duck-egg yolks.

According to legend, Mooncakes helped overthrow a dynasty. In
1353, revolutionary conspirators concealed messages in Mooncakes which
described a planned midnight massacre and included the time and place
the uprising would begin. Word spread quickly as the Mooncakes circu-
lated through the population during the Full Moon Gazing Festival. The
revolt succeeded in toppling the hated Mongol Yuan dynasty.

Have Your Own Moon Gazing Party

Mid-Autumn Moon gazing is a wonderful excuse for a party. In most areas, September evenings are perfect for outdoor entertaining. Your Full Moon celebration can be as simple as burning a stick of incense and drinking tea with a few friends on your balcony, or as elaborate as a full-costume affair. There are many ways to incorporate the themes of old China and the Full Moon. Use the following suggestions as a springboard and let your imagination soar like Chang'e as she flew to the Moon.

Make your party invitations out of red paper. In China, red symbolizes good fortune. If you don't like red, you can use another color. But don't use white. White is the traditional color of mourning and funerals.

Decorate your invitations with Chinese characters. Bookstores and libraries have books on Chinese calligraphy. With a little practice, you'll be able to copy a few simple characters. You can also honor the Jade Rabbit by drawing him on your invitations. If you feel calligraphy and rabbits are too complicated, draw circles for the Full Moon.

A few well-placed decorations will create a magical atmosphere for Moon viewing. Paper lanterns are inexpensive and fun. If you're artistically inclined, create some large paper Moon banners. The exotic fragrance of sandalwood has always been associated with China. So, by all means, burn lots of sandalwood incense to "heaven and earth."

The purpose of your celebration is to enjoy viewing the Moon, so make sure there are plenty of chairs and places for people to sit. It's perfectly appropriate to sit on the ground for Moon gazing. Spread a cloth or blanket on the grass. Red is best, but any color other than white will do.

Serve your guests a variety of round fruits. Arrange apples, oranges, grapes, melon slices, and cherries on a platter. Then sprinkle the fruit with rosewater for an authentic Chinese touch. Rosewater is available in health food stores and at specialty food markets. Or prepare a vegetable tray with cherry tomatoes, radishes, sliced zucchini rounds, boiled new potatoes, mushroom caps, and round crackers. If you want to offer your guests a full meal, serve a formal Chinese banquet or order from a Chinese take-out restaurant. Don't forget the fortune cookies.

Appropriate beverages include black tea, Chinese beer, and plum wine. Keep an eye on your guests and make sure everyone's cup or glass is full when the Moon rises. You must gaze into your drinks and wait for the Moon's reflection to appear.

Eating Mooncakes and special sweets is part of the fun of Full Moon gazing. But traditional Chinese Mooncakes do not generally appeal to western tastes. Lard is one of the main ingredients, which causes authentic Mooncakes to be extremely dense. Also, not everyone appreciates the unusual flavors of the traditional fillings. So the following recipes have been specially adapted to satisfy western taste buds. The Mooncakes are a cross between biscuits and old-fashioned shortcake and the Moon cookies are delicate and sweet. All three recipes are easy and call for common ingredients. That is to say, no salty duck egg yolks are required.

Recipes
Peaches of Immortality Mooncakes

Dough
- 2   cups baking mix
- 2   tablespoons sugar
- 3   tablespoons melted butter or shortening
- $1/2$ cup milk

Filling
Peach preserves or jam

There are a number of baking mixes, such as Bisquik and Jiffy, that are readily available and perfect for this recipe. Combine the dough ingredients with a fork in a medium-sized bowl. Pull off pieces of dough the size of small lemons and pat them into balls. Make wells in the center of the balls and fill them with peach preserves or jam, about a teaspoon of preserves in each Mooncake. Pinch the dough over the peach filling, and roll the dough balls in your hands until they are round and smooth. Place them on a greased cookie sheet. Press the tops to flatten a little. Bake in a 450°F oven for eight to ten minutes. Do not overcook; remove them from the oven before the tops turn brown. Take care to keep the bottom of the cakes from burning. This recipe yields eight Mooncakes each about two inches in diameter.

Three-Treasure Mooncakes

Dough

See the recipe for Peaches of Immortality Mooncakes above.

Filling

¼ cup crushed walnuts
2 tablespoons sugar
⅓ teaspoon cinnamon

Combine the sugar and cinnamon in a small bowl and set some aside. Toss the remaining sugar mixture with the crushed walnuts. Follow the directions in the Peaches of Immortality Mooncakes recipe for forming the Mooncakes, but substitute a teaspoon of the walnut, cinnamon, and sugar mixture for the peach preserves. After you place the Mooncakes on the greased cookie sheet and flatten them down, sprinkle the tops with the reserved cinnamon sugar. Bake according to the directions for the Peaches of Immortality Mooncakes.

Increasing Brilliance Moon Cookies

1 cup softened butter
1 cup sugar
1 egg
2 tablespoons white vinegar
2½ cups flour
½ teaspoon baking soda
1 teaspoon salt

Cream the butter and sugar, then add the egg and the vinegar. Add the sifted dry ingredients and mix until well blended. Drop teaspoons of dough onto a greased cookie sheet. Sprinkle the tops with sugar and bake in a 400°F oven for ten to twelve minutes. This recipe yields between five and six dozen one-inch cookies.

Moon Surfing

Rowena Wall

The Moon is one of the most important elements in a chart. It tells us how we need comforting and how we react to emotional situations. It reflects our daily habits and our relationships with our families. The Moon also tells us all about the women in our lives and describes our deep and most hidden feelings. It describes our desires and has a great deal to do with our personalities.

The Moon is so powerful in a chart that many astrologers believe that the vast majority of us live in our Moons. By this, they mean that we actually reflect more Moon qualities than Sun qualities, implying that as people we tend to live in a more emotional state than a logical one. In fact, I often recognize a person's Moon sign long before their Sun sign.

Is it surprising, then, that many astrologers use the Moon as the focus of their websites? As a matter of fact, the Web provides a heady place for lunar research and exploration and affords us the opportunity to study and learn from some of the finest astrologers in the world. Furthermore, the Web, once you learn to use it, is a pleasant and fun vehicle.

With these facts in mind, my mission with this article is to take you Moon surfing. So hop on board my webmobile and let's get on with our tour. If learning more about the Moon is one of your goals, our destination couldn't be any more instructive.

To Begin

First off, let me suggest that you learn to make use of the "Bookmark" function in your browser. That is, whenever you find a website you really like, you can save a lot of time if you regularly "bookmark" the site by using this function button. It's like a map to your favorite parking spaces. It can be difficult sometimes to find favorite websites later if you do not bookmark them.

To make this search easier for you, I would like to walk you through a typical website. There can be many different parts to a website, though many of them are very simple. My own site, for instance, is rather large and has several features. It includes a number of articles of different lengths on various subjects. The articles rotate every three or four weeks, so you can visit often without becoming bored.

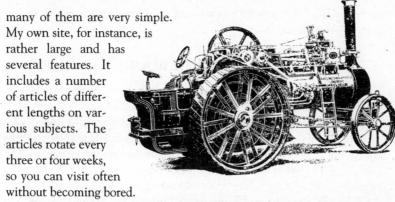

To view this site, simply enter http://www.starflash.com/zodiac on your brouser's address line. On the right side of the screen, you'll find a navigation bar. Select the "Articles" link and you will be taken to an index of the various articles on the site.

The Zodiac Gazette is an "ezine", which is Web talk for online magazine. Ezines provide varied and vast materials for you to read. Some websites, including mine, may also have a bulletin board, scheduled Web chats, and much more. So while you're looking for Moon articles you may want to look around at other things.

For instance, you can link from my site to the "Mercury Hour" site. In case you haven't heard of it, Mercury Hour is one of the oldest and best known astrological journals around. Some very well-known astrologers have written articles for the magazine over the years, some of which are published on the website. The Moon is the subject of several of these articles, so this is a good place to start your lunar explorations. Just direct your browser to http://www.starflash.com/mercury-hour and scroll down to the Articles Link and click on the little hourglass. This will take you to an index of the many articles on the site.

If you want to learn a lot of basic information about the Moon and its role in your astrological life, another good website to check out is at: http://www.thenewage.com/na/gap.htm. This site includes an in-depth article written by Michael Erlewine entitled "Taking Advantage Of the Lunar Cycle." An interesting discussion of lunar cycles and Moonlore in Western Astrology, the article also includes information about eclipses. There is also a discussion of the lunar cycles in Eastern Astrology on this website, and a discussion of the lunation cycle and science which is very technical but informative if you want to make a good start towards understanding the Moon.

A Lot of Information about the Moon

There are a lot of websites that give good information about the Moon in general and about its affect in our lives. The "We'Moon Astrology Web Site," for instance, is located at http://www.teleport.com/~weMoon/astro.html. Here, you'll find information about navigating the month by the Moon and a general overview of lunar astrology for the current and future years. The site is updated at each new Moon, and it offers a great deal of personal insight regarding yours and others' relationship to the Moon. At the same time, Cheryl Robertson's "Whole Moon Page," located at http://www.arrow-web.com/M1/theMoon/, has all sorts of good information about the Moon—including articles on the Moon in Religion, Ancient History, Modern History, the Arts, Legend, and Physical Properties. And "The Inconstant Moon" site, located at http://www.minervatech.u-net.com/Moon/inconstant.htm, has all kinds of information about the Moon.

Barb Novak's "Crescent Moon Astrology" site at http://www.geocities.com/Athens/Acropolis/2242/ has information on the void-Of-course Moon, including an ephemeris of the upcoming v/c Moons and an explanation of how the v/c Moon works. If you're not using the void-of-course Moon in your astrological life, you should look at this site. The use of the term, void-of-course, goes back many centuries in astrology. Maurice McCann has written a book on this subject and a portion of the book has been published on the web at http://astrology-world.com/debate.html. The information on this website includes a debate on the use of the void-of-course Moon in astrology.

A very instructive article by Deborah Houlding, entitled "The Moon," is included on her website at http://ourworld.compuserve.com/homepages/Deborah_Houlding/ Moon.htm. Meanwhile, there are a number of Moon articles on Jane Arnell's terrific astrology site for the Mining Company located at http://astrology.miningco.com. One article, on lunation mapping, includes excellent material to help you use the Moon as a tool. Other outstanding articles on this site discuss "Lunar Burnout," natal Moon phases and other lunar information, and the Moon through the signs. Incidentally, you'll find a number of other excellent astrological resources at this site.

Of course, you may be more interested in scientific information regarding the Moon, and you can get it at the "Space Zone" at http://www.spacezone.com:80/Moon.htm. On this site, you can read how NASA recently discovered water on the Moon's surface, and other interesting tidbits. Another good source of scientific information regarding the Moon is the official NASA site for the Lunar Prospector. Located at http://lunar.arc.nasa.gov, it includes audio and video clips of past, present, and upcoming missions to the Moon. And still another site, located at http://www.bibliography.com/Moon/, includes a very comprehensive list of references on "The Moon in Science Form."

A few sites deal with lunar eclipses. The "Lunar Eclipse Observer" is of particular interest to astrologers because eclipses are utilized in forecasting so often. This site gives the paths of occultation and visibility of upcoming and past eclipses and is located at http://www-clients.spirit.net.au/~minnah/LEOx.html. Bette Denlinger's website at http://www.web-span.com/solsticepoint/astralecl.html also includes detailed and extensive information on eclipses. The "Hitchhiker's Guide to the Moon," meanwhile, located at http://www.shallowsky.com/Moon/hitchhiker.html, includes a lot of general astrological information on the Moon, such as a number of articles and active links to other Moon-related sites. There is a lot worthwhile information for astrologers here as well.

Moon Lore Sites

The Moon is such an important influence in people's lives that there is endless lore surrounding it. We should, of course, expect to encounter a lot of this on the Web. If interested in Moon lore, you might check out the "MoonXcape" website located at http://www.geocities.com/Athens/Acropolis/8756/. This site is all about the Moon—with sections on Moon Lore, Moon Phases, Moon Signs, Moon Rhythms, Moon Ware, Moon Seeresses, Moon Dances, and Moon Links. If you can't get your Moon lore fix here, you'll never get it!

Remember when you were a child and heard stories about "The Man In the Moon?" Well, you can get the lowdown on these tales at http://www.geocities.com/Athens/Forum/2739/Man_in_Moon.html. On the other hand, at http://www.tufts.edu/as/wright_center/fellows/georgepage.html you'll get all the information about the Woman in the Moon. While at http://www.geocities.com/TimesSquare/Lair/9869/Moon.htm, you'll find an extensive list of lunar goddesses from various cultures.

Native Americans had special and descriptive names for the various Moon phases which can be found at http://mars.superlink.net/user/dakota/Moon.htm. You can also read some native legends about the Moon, such as "The Moon Spirit and Coyote Woman," at http://www.things.org/~tanais/storyteller/coyotew.html.

Traditionally, the Chinese put much more store in the Moon than we do, using the Moon in many aspects of their astrology and daily lore. You'll find "Chinese Moon Lore" at http://peacock.tnjc.edu.tw/MOON/Moon.html, including information about Chinese Moon festivals, Moon gazing, and much more.

The well-known author Isaac Asimov wrote in *The Triple Triumph Of The Moon* about just what the Moon has meant to man from the beginning of time. Some of the work from that book has been turned into a website located at http://magna.com.au/~prfbrown/i_asimov.html. Finally, at http://www.lunaranomalies.com, you can join in the debate about whether or not life has ever existed on the Moon. This site, sponsored by a group called VGL, also includes some discussions about other Moon anomalies and phenomena.

Your Personal Moon

The Moon of course plays quite a role in each of our daily lives. "The Moon's Illusion" website tells us about the phases of the Moon and about what the Full Moon can bring to your life. Located at http://darkstar.swsc.k12.ar.us/~kwhite/misc/illusion.html, you can use this site to examine your own chart and see what phase the Moon was in when you were born. This website also allows you to do a little research about what the phases mean to you personally.

If you are in the mood for a little prediction in your life, go to http://www.awarenet.com/in-lunar.html, where you'll find the "Daily Lunar Transit Report." Here, you can enter personal data and get a report on the Moon transits to your chart for any date. This report also includes an interpretation of your lunar transits, and you can receive a report of planetary transits for that day, though without an interpretation.

A very scholarly website about the Moon and her phases is located at http://www.Trinity.Edu/~mkearl/time-1.html. Though you're probably aware of the fact that the Moon controls a great deal in our lives, this site examines how Moon phases affect us both physically and mentally and discusses what it calls "The Natural Rhythms of Our Lives." Furthermore, Donna Cunningham's book, *The Moon in Your Life*, is published in sections on the web at http://www.astrology.net/guest/9003.html. Here, you can read how the author urges us to utilize our Moon cycles for easier and happier living.

If you are a bit Moonstruck, you can get the lowdown on how you got that way at http://members.tripod.com/~timdavis/Moon.html. This excellent website also includes some historical information on the Moon, along with other interesting information. And if you are a Moon Child, or a Cancer, you are not alone. A very long and interesting list of these sensitive and creative people is located at http://www.arrowweb.com/-M1/theMoon/cancers.html. This list of "Famous Lunarians" may give you some insight on your personality or the personality of someone you know.

Another website, called "A View Of The Moon" and found at http://saatel.it/users/lore/Moon.html, gives you a picture of how the Moon looked on the day you were born. By entering your data carefully, you'll not only see how the Moon looked on that fateful day, but you'll also be given a lot of technical information about the orb.

Finally, at the risk of asking a personal question, have you ever considered the effect of the Moon on your or your partner's fertility? If so, you'll want to take some time to visit the "Your Lunar Fertile Cycle" website at http://www.yoni.com/wombMoon/wmlunarfertile.shtml. This site explains how a woman's cycles are linked with the cycles of the Moon, and how one's fertility can be controlled by following the Moon rhythms. Okay, I'll go back to minding my own business now.

Fun Moon Sites

Some Moon-related websites are intended to let you have fun as you pick up some good lunar tidbits. For instance, the New Moon Dungeon interactive game website is located at http://eclipse.cs.pdx.edu/index.html. Full of fun graphics and interesting scenarios, you'll definitely want to visit this site and stay awhile. Meanwhile, you'll find out all the intricate details for creating a Moon Dial on Ian's Lunar page at http://www.geocities.com/Athens/Forum/2739/Make_your_own_Moondial.html. Moon dials, believe it or not, will help you tell time in the middle of the night when there are no clocks around, which I'm sure is something you've always wanted to learn how to do.

The "Transient Lunar Phenomena and Other Weirdness" website, located at http://www.geocities.com/Athens/Forum/2739/tlp.html, has lots of fun information about the Moon and links to other sites that examine the Moon in literature and legend. The "Full Moon Girl" website at http://www.geocities.com/~fullMoongirl/ offers Moon lore and links to sites about werewolves and other fantastic creatures.

Another really great website is called "MetaMaze," located at http://www.metamaze.com/. This site has a long listing of celebrities with their Sun and Moon signs noted, and it has a feature wherein you can enter you own Sun and Moon signs and see who shares these signs. Be sure to check out the section which gives quotations from celebrities too—some of them are quite surprising.

As Often as a Blue Moon

Many astrologers have written extensively about "Blue Moons." Richard Nolle is one of the most well-known among them. His list of Blue Moons and Super Moons for the twentieth Century is located at http://www.astropro.com. There are a number of links to Blue Moon information on this site too.

Another in-depth article about Blue Moons can be found at http://156.26.14.23/lapo/bluMoon.htm. This website has an interesting article about the history and folklore of Blue Moons written by Philip Hiscock, the Archivist at the Memorial University of Newfoundland Folklore and Language Archive. You might want to also check out the "Blue Moon Calculator" website at http://www.obliquity.com/astro/blueMoon.html. This site indicates when Blue Moons are upcoming in the next year and lists previous Blue Moons as well.

For the more adventurous among you, a website at http://pages.prodigy.com/fullMoon/page6.htm has been been devoted entirely to the subject of Mooning. We all know someone who has either Mooned someone or been Mooned, don't we? This site practically obsesses about it. In other words, there's a moral here. You need to be careful who you expose yourself to now days!

Tracking Our Lives by the Moon

A great number of websites offer information to help us keep track of the Moon in its phases and to plan our lives accordingly. The "Daily Moon" website, located at http://www.artcharts.com/no_frame_pages/adailyu-pload/dailyMoon.html, presents the Moon's weekly activities, including phases, rising and setting times, Moon signs, and the void-of-course. There is also a lot of information on this site about other imporant Moon-related issues.

Many of us remember grandfathers or grandmothers who did their farming or gardening according to the changes in the Moon. In fact, there are many modern farmers who use the Moon to guide them in their plant-ing today. You can find out all the interesting details at http://www.geoc-ities.com/Athens/Forum/2739/Lunar_gardening.html. After all, if you're going to put out all that energy planting and harvesting your garden, you might as well learn how to ensure its success. Furthermore, you can use lunar cycles for planning events in your life. If interested, you may enjoy visiting the "Lunar Planning" website at http://www.gbtech.net/~jwelch/-lunar.htm. This site includes discussions of the Full Moon and New Moon cycles, and lunar action periods and voids, and it helps us to know what we can expect to happen on particular dates.

Meanwhile, the progressed Moon is an important factor for forecast astrology. Understanding how this works will go a long way in helping us to understand how to see the trends ahead. Some excellent information about this can be found at http://www.panplanet.com/articles/pluna-tion.html. Several articles are included on this site that will explain the progressed Moon and its affect on our lives. And the International Occultation Timing Association website, located at http://www.sky.net/-~robinson/iotandx.htm, has extensive information about occultations and other Moon data.

If you would like to know more about the Moon when it is in apogee or perigee, then you'll want to visit "The Inconstant Moon" website at http://www.fourmilab.ch/earthview/Moon_ap_per.html.

At the "Moon Calculator" website, at http://www.starlight.de-mon.co.uk/Mooncalc/, you can download free software that will tell you all about the Moon in her different phases. Meanwhile at the "Locutus Codeware" website at http://www.locutuscodeware.com/download.htm, you can download a program called MoonPhas.exe. With this program you can find out how old the Moon is and when the next New Moon or Full Moon is expected.

Finally, the "Moon Wobble/s" website, located at http://www.the-ul-timate.com/space/Moonwobl.htm, explains the theories of distinguished astrologer and mathematician Carl Payne Tobey. His theories, which are increasingly attracting attention, explain that an unstable Moon comes around about every 86.5 days. This concept is being used to help explain many manmade disasters and strange phenomenon that cannot be ascribed to any other astrological event. As such, this website is unique among all the others I've listed.

Well, it's about time to park up the old webmobile. By now, you have plenty to keep you busy for the time-being. I hope that you've enjoyed your excursion, and that your future visits to all these fun and interesting sites will keep you out of trouble! It was great having you along on the expedition. If I can answer any questions or if you need any help, please feel free to write me at starstuff@starflash.com.

BUSINESS
& LEGAL
VENTURES

Business & Legal Ventures

How to Choose the Best Dates

When starting a new business or any type of new venture, check to make sure that the Moon is in the first or second quarter. This will help it get off to a better start. If there is a deadlock, it will often be broken during the Full Moon. You should also check the aspects of the Moon to the planet that rules the type of venture with which you are becoming involved. Look for positive aspects to the planet that rules the activity in the Lunar Aspectarian (pages 28-51), and avoid any dates marked Q or O.

Planetary Business Rulerships

Listed below are the planets and the business activities that they rule. If you follow the guidelines given above and apply them to the occupations or activities listed for each planet, you should have excellent results in your new business ventures. Even if it is not a new venture, check the aspects to the ruler of the activity before making moves in your business.

Sun: Advertising, executive positions, acting, finance, government, jewelry, law, and public relations

Mercury: Accounting, brokerage, clerical, disc jockey, doctor, editor, inspector, librarian, linguist, medical technician, scientist, teacher, writer, publishing, communication, and mass media

Venus: Architect, art and artist, beautician, dancer, designer, fashion and marketing, musician, poet, and chiropractor

Mars: Barber, butcher, carpenter, chemist, construction, dentist, metal worker, surgeon, and soldier

Jupiter: Counseling, horse training, judge, lawyer, legislator, minister, pharmacist, psychologist, public analyst, social clubs, research, and self-improvement

Saturn: Agronomy, math, mining, plumbing, real estate, repairperson, printer, papermaking, and working with older people

Uranus: Aeronautics, broadcasting, electrician, inventing, lecturing, radiology, and computers

Neptune: Photography, investigator, institutions, shipping, pets, movies, wine merchant, health foods, resorts, travel by water, and welfare

Pluto: Acrobatics, athletic manager, atomic energy, research, speculation, sports, stockbroker, and any purely personal endeavors

Business Activities

Advertising, General

Write ads on a favorable Sun sign day while Mercury or Jupiter is conjunct, sextile, or trine the Moon. Mars and Saturn should not be aspecting the Moon by square, opposition, or conjunction. Ad campaigns are best when the Moon is well-aspected in Taurus, Cancer, Sagittarius, or Aquarius.

Adverstising, Newspaper

The Moon should be conjunct, sextile, or trine Mercury or Jupiter.

Advertising, Television, Radio, or Internet

The Moon should be in the first or second quarter in the signs of Gemini, Sagittarius, or Aquarius. The Moon should be conjunct, sextile or trine (C, X, or T) Uranus, and Uranus should be sextile or trine (X or T) Jupiter.

Business, Education

When you begin training, see that your lunar cycle is favorable that day and that the planet ruling your occupation is marked C or T.

Business, Opening

The Moon should be in Taurus, Virgo, or Capricorn and in the first or second quarter. It should also be sextile or trine (X or T) Jupiter or Saturn.

Business, Remodeling

The Moon should be trine or sextile Jupiter, Saturn, or Pluto, or conjunct Jupiter.

Business, Starting

In starting a business of your own, see that the Moon is free of afflictions and that the planet ruling the business is marked C or T.

Buying

Buy during the third quarter, when the Moon is in Taurus for quality, or in a mutable sign (Gemini, Virgo, Sagittarius, or Pisces) for savings. Good aspects from Venus or the Sun are desirable. If you are buying for yourself, it is good if the day is favorable to your Sun sign.

Buying Clothing

See that the Moon is sextile or trine to the Sun during the first or second quarters. During Moon in Taurus, buying clothes will bring satisfaction. Do not buy clothing or jewelry when the Moon is in Scorpio or Aries. Buying clothes is best on a favorable day for your Sun sign and when Venus or Mercury are well aspected, but avoid aspects to Mars and Saturn.

Buying Furniture

Follow the rules for machinery and appliances but buy when the Moon is in Libra too. Buy antiques when the Moon is in Cancer, Scorpio, or Capricorn.

Buying Machinery, Appliances, or Tools

Tools, machinery, and other implements should be bought on days when your lunar cycle is favorable and when Mars and Uranus are trine (T), sextile (X), or conjunct (C) the Moon. Any quarter of the Moon is suitable. When buying gas or electrical appliances, the Moon should be in Aquarius.

Buying Stock

The Moon should be in Taurus or Capricorn, and should be sextile or trine (X or T) Jupiter and Saturn.

Collections

Try to make collections on days when your Sun is well aspected. Avoid days when Mars or Saturn are aspected. If possible, the Moon should be in a cardinal sign: Aries, Cancer, Libra, or Capricorn. It is more difficult to collect when the Moon is in Taurus or Scorpio.

Consultants, Work With

The Moon should be conjunct, sextile, or trine Mercury or Jupiter.

Contracts

The Moon should be in a fixed sign and sextile, trine, or conjunct Mercury.

Contracts, Bid on

The Moon should be in the sign of Libra, and either the Moon or Mercury should be conjunct, sextile, or trine (C, X, or T) Jupiter.

Copyrights/Patents, Apply for

The Moon should be conjunct, trine, or sextile Mercury or Jupiter.

Electronics, Buying

When buying electronics, choose a day when the Moon is in an air sign (Gemini, Libra, or Aquarius) and well aspected by Mercury and/or Uranus.

Electronics, Repair

The Moon should be sextile or trine Mars or Uranus in one of the following signs: Taurus, Leo, Scorpio, or Aquarius.

Legal Matters

A good aspect between the Moon and Jupiter is best for a favorable legal decision. To gain damages in a lawsuit, begin during the increase of the Moon. In seeking to avoid payment, set a court date when the Moon is decreasing. A good Moon-Sun aspect strengthens your chance of success. In divorce cases, a favorable Moon-Venus aspect is best. Moon in Cancer or Leo and well aspected by the Sun brings the best results in custody cases.

Loans

Moon in the first and second quarters favors the lender, in the third and fourth favors the borrower. Good aspects of Jupiter and Venus to the Moon are favorable to both, as is the Moon in Leo, Sagittarius, Aquarius, or Pisces.

Mailing

For best results, send mail on favorable days for your Sun sign. The Moon in Gemini is good, as are Virgo, Sagittarius, and Pisces.

Mining

Saturn rules mining. Begin work when Saturn is marked C, T, or X. Mine for gold when the Sun is marked C, T, or X. Mercury rules quicksilver, Venus rules copper, Jupiter rules tin, Saturn rules lead and coal, Uranus rules radioactive elements, Neptune rules oil, the Moon rules water—mine for these items when the ruling planet is marked C, T, or X.

New Job, Beginning

Jupiter and Venus should be sextile, trine, or conjunct the Moon.

News

The handling of news is related to Uranus, Mercury, and the air signs. An increase in spectacular news occurs when Uranus is aspected.

Photography, Radio, TV, Film, and Video

Neptune, Venus, and Mercury should be well-aspected. The act of photographing does not depend on particular Moon phase, but Neptune rules photography, and Venus rules beauty in line, form, and color.

Promotions

Choose a day when your Sun sign is favorable. Mercury should be marked C, T, or X. Avoid days when Mars or Saturn is aspected.

Selling or Canvassing

Begin these activities during a favorable Sun sign day. Otherwise, sell on days when Jupiter, Mercury, or Mars is trine, sextile, or conjunct the Moon. Avoid days when Saturn is square or opposite the Moon.

Signing Papers

Sign contracts or agreements when the Moon is increasing in a fruitful sign, and on a day when Moon-Mercury aspects are operating. Avoid days when Mars, Saturn, or Neptune are square or opposite the Moon.

Staff, Fire

The Moon should be in the third or fourth quarter, but not Full. There should be no squares (Q) to the Moon.

Staff, Hire

The Moon should be in the first or second quarter and should be conjunct, trine, or sextile (C, T, or X) Mercury or Jupiter.

Travel

See the travel listing in the Leisure & Recreation section.

Writing

Writing for pleasure or publication is best done when the Moon is in Gemini. Mercury should be direct. Favorable aspects to Mercury, Uranus, and Neptune promote ingenuity.

Economic Forecasts for the Year 2000

Kaye Shinker

S ince ancient times astrologers have looked at a select number of charts set for the capital of their country to determine the future economic picture. Financial astrologers look at the same charts, studying the position of the Sun, Jupiter, and Saturn, specifically. First they look at the position of these planets in the solstice and equinox charts when the Sun reaches zero degrees of the cardinal signs Capricorn, Aries, Cancer, and Libra. Next they look at the solar eclipses—there will be three in the year 2000. Finally they look for any special aspects between the business planets Jupiter and Saturn—these will be conjunct in the year 2000, which is called the Great Mutation and happens every twenty years. Financial astrologers study all of these charts to determine market trends in the year ahead.

Jupiter in Aries, Taurus, Gemini

Jupiter returns to Aries on October 24, 1999, and remains there until February 16, 2000, when it reenters Taurus. It remains in Taurus until July 1, 2000, when it enters Gemini and remains there until the end of the year. Things that are in abundant supply are marked down in price; items in short supply carry a premium price. Read the entries below and make your shopping list. For example, you might want to buy a car in August because there will be an abundant supply of them, since Jupiter indicates abundance and Gemini rules cars.

In the year 2000, when Jupiter is still in Taurus, it will seem like there is an abundant supply of everything, which is almost true. Jupiter is expanding in three directions. There will still be plenty of steel, building materials, machinery, and inventions. There will be an abundant supply of banks, currency, wood, beans, costume jewelry, and leather, as well as vehicles with wheels, newspapers, communications devices that plug into the wall, and tires. Do you want to make a million dollars? Then figure out what to do with bald tires!

Industries still suffering from oversupply during the early part of the year will be scrap metals, steel, hardware, durable goods, the military, hair care supplies, and diamonds. Merchants selling these items will be able to offer deep discount prices, in case you want to start a shopping list.

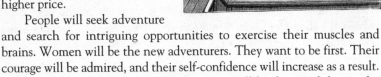

Industries experiencing shortages will be art, apparel, candy, fruit, and copper. These suppliers will be able to fetch a higher price.

People will seek adventure and search for intriguing opportunities to exercise their muscles and brains. Women will be the new adventurers. They want to be first. Their courage will be admired, and their self-confidence will increase as a result.

Issues of personal freedom and privacy will be discussed during the 1999-2000 holiday season. Impatience with authorities and excessive paperwork will be the subjects of many complaints. Oppression by governments, including our own, will make the headlines. People will reject any restrictions to their personal freedom. Laws meant to protect individuals from themselves will be ignored or repealed. Personal honesty will continue to be a topic of discussion, and public figures seeking office will reveal the dust bunnies in their attic.

In February, industries in abundant supply will be banks, financial services, lumber, agriculture, plant nurseries, real estate, and arts and crafts materials suppliers. Banks will compete vigorously for accounts. Folks with stable incomes or government checks will be their first targets. Real estate profits will be high now. If you need to sell property, this is the time to put it on the market. Builders will find material and supplies in abundance and take advantage of the opportunity to fill up vacant lots.

Industries such as container and packaging manufacturers, mortuaries, pipelines, insurance, surgical equipment, mining, and security systems will have shortages. Recycling is expensive and labor intensive. Packaging companies will have difficulty obtaining all the materials required by Green laws.

Taurus is conservative and stubborn and insists on playing by the rules. This will be difficult for entrepreneurs seeking markets for their exciting ideas. Green parties and ecology groups will expand as flowers and trees get the undivided attention of the public. Presidential candidate Al Gore will promote their cause.

On July 1, Jupiter enters Gemini. The roads will fill up with vehicles of all sorts. Billboards will line the roads. Every truck will have "www" painted on its door. Bumper stickers will become cutesy e-mail addresses. Your friends will live by their e-mail.

In short supply will be reservations out of town, passports, sporting goods, and university placements. Everyone suddenly will want to expand their learning options. Everyone will try new sports, and the sports equipment suppliers won't keep up with demand. Shipping to foreign countries will be difficult. There will be a shortage of boats, maritime labor, and airplanes. A sudden demand for books by popular authors will catch the publishing industry with empty shelves.

Gemini loves a party, and any excuse to throw one will be valid. By the time you're ready to usher in the twenty-first century, your party clothes will be threadbare and your dancing shoes full of holes.

Folks with stock in communications companies can expect good dividends. Travel-related industries will also do well. Good advertising will be in short supply. This is political campaign time, and once again the printing industry and the media will make a profit. Politicians will decide to run for office at the last possible minute, and the media will be surprised by the number of incumbents who are defeated. Jupiter in Gemini likes to move the furniture around, and that applies to the Congress and the White House, too.

Saturn in Taurus

Saturn made a preview transit of Taurus from June 10 to October 26 in 1998, and then returned to Aries. Saturn returned to transit Taurus on March 2, 1999, and will remain there until March 2001.

The flower children of the late sixties will return to become flower children in their sixties. Economically secure—with a pension and an IRA or Keogh account to draw on—they will live on the edge. The

Volkswagen bus will be a little bigger, and they will be on the road to peace, love, and adventure.

These folks, however, will not retire from their favorite sport—shopping. Older and wiser now, their key word will be value. Efficiency will be their mantra. Saturn in Taurus makes profit margins difficult to maintain. Luxury items will be new tires and a phone card.

The New York Stock Exchange has a Taurus Sun, and it will be migraine headache time. The stock exchange's problems will be structures, procedures, and interaction with foreign markets, and its employees will be stretched to the limit. Too much money will change hands. Expect transaction delays throughout the year with Y2K problems or satellites taking most of the blame for system crashes. Congress will make changes in the tax laws in order to gain reelection.

October will be difficult, as usual. The year's end will find the Dow Jones Industrial Average past 14,000 points and Standard and Poor's Index will be vying for 2,000. During Saturn's transit of Taurus, shrewd buyers look at price earnings ratios and search for bargains in broader markets. Incidentally, 300-point swings in the Dow will become meaningless; therefore, reporters will begin to use percentages as a better measurement.

The technology-heavy NASDAQ's problems will begin in May. They will be forced to rethink rules and procedures and tighten requirements for new issues. They will certainly have to tighten their requirements for Initial Public Offerings. Their systems for twenty-four hour trading will be under stress and threatened with security breaches.

Saturn in Taurus is thrifty, and this will be difficult for big banks whose income depends on credit card interest rates. Competition for new card holders who maintain a balance will be fierce. The group they will target is known as the Baby Bust. The Baby Bust will have grown into net worth mode. Sometimes called the Boomerang Gang because of their extended visits with their parents, they will have returned home to stay until their bills are paid.

Hoping for greater volume and market share, big banks will continue to merge. Boutique banks catering to the older Baby Boomers will flourish. Midsized banks will be in a squeeze. The savings rate will climb slowly even though interest rates won't move. Gambling and spectator sports will have money problems in the United States. Fewer customers will equal fewer revenues. The casinos will suggest building more establishments, and the sports teams will want more television coverage. The problem is their wealthiest audience, the Baby Boomers, would like to participate and will find interactive games more fun. Local governments will not be amused since not enough "sin" tax revenue will be collected.

The Babyboomlet will be in grammar school, and their Baby Boom parents and grandparents will not be happy with the quality of education. Expect change. Home schools, charter schools, and online schools will be popular. More innovative, small educational publishing businesses will start to appear.

Uranus in Aquarius

The year 2000 is the year for the White Elephant Skyscraper Sale. Without big nonprofit institutions to occupy their corridors, these monsters of steel will either fall of their own accord or implode.

Voluntary associations will be unable to support staff due to low membership. People will meet online and will not need a central location to publish newsletters or membership lists. National volunteer and professional organizations will cut back staff since they will only need someone to maintain their web sites.

Computer recycling bins were introduced in 1999. Word processors and game players will be able to buy any machine they want for micro amounts of cash. Excellent used machines will be in good supply, and businesses will generously donate them to anyone willing to take them. The year 2000 will be the excuse; new operating systems the reason.

The birthday of the computer is February 1, 1946, so it stands to reason that these little machines should dominate the world while Uranus is in Aquarius. Many of you will look at the cheap new portable variety that phones, faxes, pages, and accesses the Internet while you're camping on Mount McKinley.

Manufacturers of cables, satellites, miscellaneous wires, and switches will have difficulty keeping up with demand. The installers of electrical services are usually contract laborers, and they will be able to command even higher wages.

Instant education will be the keyword. A labor shortage in technology will worsen, and anyone who promises to learn will be sent to high-tech classes. Retired shop teachers will be asked to hang up their fishing poles and return to the classroom. Signing bonuses will be offered nationwide to first-year teachers.

The old ways won't work, and that will include volunteer organizations. Networking will always be important, but potential young members will not want to take the time to gossip over lunch or dinner when they could be gossiping over the Internet. Individuals will change their goals, and professional organizations will need to change theirs. Groups organized to educate, protect, or discipline their members will suddenly disappear. New groups with unusual and specialized interests will form. Fraternal organizations with charities to support will have difficulty retaining members. Charities will give up. Trade unions will be antique.

We will demand more control over our own lives. Professionals will demand a fair contract and fair payment for their skills. Labor will work when they need funds. Service workers will set their own hours. Grandma and Grandpa will have jobs. The work ethic will definitely change.

Neptune in Aquarius

Neptune made a brief visit to Aquarius from January 30 to August 24, 1998. Neptune reentered Aquarius on November 28, 1998, and will stay until 2012. Its last visit to Aquarius was in 1830, before it was discovered by astronomers with newfangled telescopes.

In Aquarius, Neptune expects us to learn new ways of doing business. Just as the steam engine changed communications in the 1830s, new operating systems with liquid crystals will dissolve further the geographic barriers between people. The old rules of marketing, demographics, and statistics won't work because it will be impossible to pigeonhole people.

The Neptune in Aquarius transit will help us solve our dependence on material possessions. Garbage will be the result. Disposal will occupy creative city planners and politicians and will be everyone's local issue.

This year, corporate boards will save a lot of money by letting their stockholders vote on the Internet. This is the paradigm shift in action. People will begin to learn that they need not feel voiceless. In the 1830s, European governments began to recognize the power of the individual. Now the individual will take that power seriously.

Worldly success will begin to take on a new definition. Dress for success will go out of fashion, and communication skills will be emphasized. The ability to use written and spoken English will become a status symbol. *Pygmalion* and *My Fair Lady* will return as a popular myth.

Expect to see all sorts of new-style universities. The current institutions will be too rigid to accept the new style of learning and, for that matter, the new type of student. Ivy League educational institutions will go the way of the gray flannel suit.

News organizations will report the theft of intellectual property. Artists will have to create new methods of cashing in on their talents.

You can expect that most creative work will continue to be computer generated until the year 2012. Film and videotape will become antique. Equities tied to the new mediums will thrive.

The lesson of this Neptune transit will be "Throwing money at a problem solves nothing. It just extends the life of the problem." The real issue will be the greed of the problem solvers.

This is the year we will see many new trends among institutions. The hospital concept will dissolve, and doctors will find their offices on wheels. There will be a population shift. Folks will find the countryside attractive and cities inconvenient. Professional folks will have to follow.

On December 31, 2000, most of us will just go to yet another outrageous party to celebrate the official beginning of the twenty-first century.

Pluto in Sagittarius
Achieving an Economic Manifest Destiny

Pluto has been in Sagittarius since November 11, 1995. The first few degrees of the transit have found the law, education, and philosophy in need of transformation. Eager for headlines, reporters will report scandals concerning endowments and charitable trusts. The ease of distributing desktop-published material will also expose scientific breakthroughs that were of no economic benefit to industry and therefore left in the archives.

The power of the U. S. dollar and the necessity of competing with the Eurodollar will determine corporate dividends. Both economies will be forced to keep interest rates low to prevent all-out economic war.

Pluto rules billions and trillions of dollars. Sagittarius rules the traveling person. Money will move abroad and into the second and third worlds. Multinational corporations will multiply and continue extending their reach. Managers of manufacturing enterprises in remote locations will become more powerful than the local governments there. The law of economics will rule. It will be in the best interest of commerce to maintain peace; allowing good employees to shoot it out will not be good business. Expect fewer revolutions and assume terrorism will wane.

The twenty-first century will start with a new kind of corporate tribalism. Independent contractors will merge their small cadres of labor in order to develop products. Creative marketing groups will sell the products, and multinational corporations will keep the supply lines flowing.

That is why the Dow Jones Industrial Average will top 12,000 by year's end. The gray flannel suit will become extinct; sweats will rule.

Winter Solstice
Capricorn Ingress—December 22, 1999, 7:45 am GMT

The Capricorn ingress predicts the year ahead for government and business. It will be a celebration for all of us—the survivors of tabloid-mania and the millennial Apocalypse prophets. Have fun. Send e-mails to your friends. Preparation begins for the millennium. This can take the form of marriage or business partners. The idea is sharing knowledge. The feminine principle will be emphasized. Aggressive behavior will go unrewarded.

Let's talk about the important stuff—money. People will be on the move, and fresh air will be the excuse. This means companies that provide trucks, such as U-haul and Ryder, will make money. Hardware stores, home centers, and furniture stores will do very well. Of course, real estate companies and builders will continue earning good money.

Folks will demand changes in local, state, and federal taxes. The elected officials of these taxing bodies will campaign on their ideas for reform. Listen carefully. The top local issue will be garbage.

Once again, foreign affairs will be an issue. The media will point out the ability of terrorists to sabotage our computer systems.

Labor will feel fairly compensated for their work. People will be healthy and won't have much interest in health care issues.

February Eclipse
February 5, 2000, 1:02 pm GMT, Washington D.C.

Yes, the market will be in for another correction. It may even dip down to the moving average. When the Dow stays even for several days in a row, it is time to buy. The New York exchanges will be the only game in the global village. The dollar will rule. Study hard and spend your cash on good companies with excellent price earnings ratios. They will exist.

There will be a crisis in government-supported institutions. Some of the directors will make headlines by pocketing unearned money. Scandals with charities, welfare, and hospitals will result in changes to the laws concerning them. There will be a media circus that people will find boring.

Stress and long hours will catch up with workers. The millennium bug will give everyone the flu. Most folks will claim the two-week variety because they need is some time off.Snowy winter weather will be another factor keeping wage earners at their home computers. While recovering, they will review their personal budgets and determine their priorities for spending in the year ahead. A lower Dow will bring out the bargain hunters, and the financial pages on the Internet will be busy.

That's it, folks. Oh I forgot, Americans will be overemployed. Finding census takers will be a monumental challenge since even retired folks will have full-time jobs. The post office might know where everyone lives, but they will be busy delivering online orders.

Aries Ingress
March 20, 2000, 7:35 am GMT, Washington D.C.

In this ingress, farmers will have problems getting into their fields. New equipment financing will be a problem in the winter, and it will be difficult to find calm weather to plant the crops. Many will throw up their hands and let the banks and insurance companies deal with it. Gardeners will be perplexed by erratic weather. Agriculture will be a major problem all year. Weather is the problem, though we will still have a surplus to export.

Commodities traders will send the prices on their contracts higher. Industries dependent on a supply of American grains will anticipate higher costs for their raw materials. The stockholders will be concerned about the earnings of these older New York Stock Exchange companies. This will keep the Dow trading in a narrow range of about 10,000–10,500 points for the next few months.

The real news will be the NASDAQ, which will enforce newly tightened rules for electronic trades and international speculation. Confidence will be the key. A lot of patents previously bought by big energy corporations will expire. Entrepreneurs will scan the government's patent site on the Internet looking for great ideas to help Americans become energy self-sufficient. Automobile companies will sell bigger and faster vehicles.

The Great Mutation
May 28, 2000, 4:04 pm GMT

Jupiter and Saturn will be conjunct at 22 degrees Taurus for the United States in the Ninth House. It has been called Techumasa's revenge, and that means that the president elected in 2000 will pass on to the next world unless modern medicine saves him, as was the case with Ronald Reagan.

Now that that's out of the way, let's get to the important stuff like what the conjunction will mean for business. Jupiter represents the sales of a corporation, and Saturn represents the finances. When they are in the same sign, they get in each other's way. This, then, will be a very difficult time to start a company or to initiate an Initial Public Offering. Wait a few months and let Jupiter move into Gemini. Investors should let their money earn interest in a savings account. The housing market in the U. S. will be at its peak. It will be a good time to sell real estate holdings, but if you are purchasing real estate be sure you absolutely love what you are buying. You will keep the property a long time, because there will be changes in tax incentives that encourage you to keep the property.

Management methods, advertising styles, communications equipment, international transport, and international law are in sharp focus. In other words, they will be a mess and changes will be absolutely necessary. This means a lot of adjustments in the labor market associated with these sections of industry. Headhunters and temporary agencies will profit. The labor shortage will continue with the focus on clerical skills of all kinds.

The Dow will slow its upward trend, and there will be fewer mergers and acquisitions to stimulate speculators. Price earnings ratios will improve, and people will begin to discuss dividends as a key factor in adding a stock or mutual fund to their portfolios.

Cancer Ingress
June 21, 2000, 1:48 am GMT

Banks and financial institutions will be in abundance. Everyone will want to borrow your money to lend to someone else. There will be a lot of money coming in from foreign sources. Commodities traders will be mystified by outrageous weather patterns. All this will add to the summer volatility of the markets.

Diplomacy will be successful, and trade negotiations with other nations will be easily accomplished. Rebuilding the systems for the electronic transfer of funds will dominate the financial news. Industries providing these machines and services will thrive.

Families will work together for the good of the family, pooling their resources and successfully negotiating to move the work place into their homes. Builders and building supply companies will continue to benefit from this trend. Office supply outfits will be busy filling orders and selling equipment. With so much work to do, families will forgo vacations.

Expect full employment. People with professional work will be well-compensated. The biggest problem will be that everyone is too busy.

Cancer Eclipse
July 1, 2000, 7:21 pm GMT

In the United States, it will be a visible eclipse. The weather will remain clear on the day of the eclipse until the next eclipse on July 31.

The markets will be at a high. For the next six months they will be extremely volatile. If you have nerves of steel, stay invested. If you are the nervous type, turn off the electricity. This eclipse will cause the markets to dip and recover for about two weeks. They do that so folks can go bargain hunting.

Mergers and acquisitions will suddenly be everywhere. U.S. corporations who buy out foreign companies will be out stalking the globe looking for any likely candidate—companies in South America the favored group. Travel, airlines, shipping, publishing, and sporting equipment will be the favored industries for acquisition. The U. S. dollar will be the currency of choice, but foreign funds will be best invested in the areas where they are earned. That means that New York Stock Exchange companies will be anxious to add to the value of their corporations with capital assets and not too anxious to please stockholders with dividends. TV and movies will lose their appeal. People will prefer the radio and will start to buy new music styles and attend concerts and festivals. Low prices for home computers will keep the office supply market busy. Indulgent grandparents will have found a new way to spoil their grandkids. Expect the software designers and Internet software dealers to benefit.

Leo Eclipse
July 31, 2000, 2:25 am GMT

Investors with a nervous stomach might as well go on a long trip where there is no electricity. The markets will take another dip. This time it will last longer because the day traders and hedge fund managers will also be on vacation. They will come back about the first week in September, giving the bargain hunters plenty of time to study and find some great deals.

Foreign markets will show some improvement, and the risk takers will move in that direction. That means they will take their New York Stock Exchange profits and put them to work overseas.

Stockholders will be pretty upset with the lack of consideration by corporate America; they were hoping to reinvest dividends. Equities related to financial services, beauty supplies, apparel, and candy will look like good values.

No matter how you look at it, the marketplace will be very exciting. Forget baseball and politics—reading the ticker tape will become the national pastime. Online trading will become a personal game with the individual in control of his or her own savings.

The Olympics will be controversial because of corporate sponsorship. Improved equipment will be featured at the games, sending many new customers to the stores. Sporting equipment manufacturers will improve their bottom lines.

The reality is that people will enjoy playing sports. Many will be getting too old for their favorite games and will search for new ones.

Libra Ingress
September 22, 2000, 5:29 pm GMT

This will be the most outrageous market since 1940 except that there won't be a world war supporting American industry in the background. Dow Industrial averages won't make sense with the overall economy very strong. The NASDAQ averages will be strong, and emerging markets will interest the investing public.

The Sun of the New York Stock Exchange is in Taurus, and it's Midheaven is at 28 degrees Taurus. Saturn will transit 28 degrees Taurus on July 27, 2000. Jupiter has crossed its Midheaven on the way to Gemini in June. Saturn will retrograde to 29 degrees Taurus on October 16, 2000, and will go direct sometime in February 2001.

The New York Stock Exchange and the Dow Jones averages love to fall in October. Even an election year will not put Humpty Dumpty back together again. The media will blame the Federal Reserve for raising interest rates. The Democrats will say the Republicans forced the Federal Reserve Bank to raise interest rates. Is this a fearless prediction, or should I bet a dollar that the Dow will reach 9,000 points again? Mercury will be stationary retrograde that day, so if you still own some stocks just put them in your safe deposit box, study hard, and research some buying opportunities. Hang on to your cash until February 2001. You'll have the opportunity to go bargain shopping then.

This will be a tight labor market with a lot of training still being given to employees. The Babyboomlet will be eligible to work, but studies will keep them out of their traditional part-time service jobs.

Folks will be extremely upset about privacy issues. Politicians won't know how to fix the problem. Incumbents will be in and out of favor. The media, looking to increase revenues, will proclaim every race a toss-up, which will be true. Most people will be too busy working, decorating their homes, or studying to pay attention. The public will be apathetic because they know the job of president is losing its appeal. There will be a change in the administration; however, it is Congress and the Federal Reserve who will really be in charge.

Traveling Americans will be concerned about the dollar and some of its problems abroad with duplicating machines. The yen and the Euro-dollar will be strong and will therefore suffer the same challenge.

Capricorn Ingress
December 21, 2000, 1:37 pm GMT

The elections will give Washington D.C. moving companies a lot of business. Women will command good salaries.

The years 2000 and 2001 will be the top of the market for land values. Farms will be the first to notice a fall in price per acre. Wise farmers will find ways to subdivide their acreage and thereby decrease any debt load acquired from thin harvests. They will be competing with land-rich insurance companies and mortgage banks for customers; therefore, clever marketing will be required.

The stock market will recover, heading toward 13,000 points on the Dow because international corporations will finally report good profits from their foreign operations. Bargain hunters will have made their selections, and some will take the plunge.

Garbage will remain a hot issue during the year ahead. Companies promising to deal with it will profit. Hot commodities will be iron, copper, wheat, and corn. Once again the Russians will have a lot of gold, oil, lumber, and labor. The prices for these commodities will remain in a narrow trading range until the Russians pay their debts, but that won't be for a long time, folks.

Capricorn Eclipse
December 25, 2000, 5:22 pm GMT

This eclipse will be visible in the United States. Everyone will be in a great mood, and the holiday will take on the flavor of an additional Thanksgiving. Americans will be very grateful for their bonuses and pretty houses. Politicians will be exhausted, and the new administration will be busy sending out party invitations.

Fortunately, this will be a cardinal eclipse and a holiday, and its effect on the markets will be fleeting. Pension and profit sharing money will sit on the sidelines until February, but the market will be set to rise after the first of January. Americans will keep the shopkeepers busy and the profits high. Computer and electronics vendors will sell their wares faster than they can manufacture them.

Grandparents will be on the road laden with educational gifts for the kids. Make your reservations early. Airplane ticket prices will be at a premium. Fuel prices will follow the weather. It promises to be a dry, chilly winter.

This will be the season for parties. Plan a few outrageous ones, but be thrifty and careful, and go for it. Happy twenty-first century!

Charting a Lunar Career Course

Alice DeVille

How do you feel about your career? Are you happy with it or ready to jump ship? Is your need for stability keeping you dry-docked? Or have you wondered what it would be like to anchor yourself in a new port?

Work climates affect beliefs about future incomes. When conditions are fairly predictable, you usually feel more secure about your career and the prospect of upward mobility. With a steady paycheck, you are more open to taking risks that help you move up the ladder of success. But when employment downturns occur, you may question whether your hard work matters and worry about what you would do if you lost your job.

We live in a constantly changing world. New times may call for reconsideration of work priorities. If so, it's time to be proactive in seeking a more satisfying career opportunity. As one door closes, are you confident that you have the talent and experience to open a new one? Maybe it's time to consider the options.

Did you know that knowledge of astrology can help you make meaningful career choices? Your interests are no accident of fate given the powerful blueprint that was mapped out at the moment of your birth. Your chart reveals the identity and initiative in your personality. Understanding your chart opens a path for maximizing your potential. If you don't already have an astrology chart, you may want to consult a professional astrologer or have Llewellyn's Computerized Astrology Services construct one for you. (See coupon in the back of this book.)

In a chart, no single planet drives your career. The relationships among planets and their places in your chart influence your overall confluence of skills, motivation, and goals. Although opinions and interpretations among astrologers vary on this topic, no one career is depicted in a birth chart. Any chart holds an array of options. This is one reason why individuals make multiple career changes in a given lifetime.

Choices are important in career charting. The way you feel about the way you make your money is also important. In fact, your response to the work you do is largely emotional, a lunar trait in astrology. Although the Moon is said to be receptive and passive, it is a major security driver in

terms of career. If you have been complaining about lack of career fulfillment, this article might change your life.

Workplace Woes

First off, before you begin analyzing your astrological makeup, you should
take a close look at your work place. The following questions may give
you clues about possible job dissatisfaction.

Are you feeling unusually restless in regard to your career and feeling
little job satisfaction ?
Has job security been a concern?
Are coworkers leaving the work environment?
If so, are the departures related to lack of interest?
How about downsizing or budget cuts?
Have you ever thought you might be next?
Do you work too many hours at the expense of your personal time?

If you are self-employed:

Has there been a change in the demand for your services or products?
Are you paid less than what you think you are worth?
Do you know of any innovative practices that might significantly
increase your business?
Are there ways you could show a better profit margin?
Do you work too many hours at the expense of your personal time?

If you answered yes to many of these questions, take the following
self-assessment test. This tool may help you make better choices for fulfilling your career goals and bringing more balance into your life.

Career Transition Assessment

How do you describe what you do for a living?
A Job
A Career
My Life Work

Have you assessed the growth possible in your job?
Yes   No

Are you satisfied with the long-range outlook?
Yes   No

Are your personal and professional goals well defined?
Yes   No

Do you like your job/career/life work?
Yes   No

Have you ever thought about leaving, not because you were
having a bad day, but because you know your talents belong in
another structure?
Yes   No

If there were no obstacles, how would you earn your income?

Is it possible to achieve this goal in your present work environment?
Yes   No

Have you ever thought you might be in the wrong position?
Yes   No

If you answered yes to the previous question, what has held you
back from leaving or transferring?

If you are self-employed, have you noticed a steady increase in
the volume of your business since you started it?
Yes   No

Have you made adjustments in your business practices?
Yes   No

Are you bored by fifty percent or more of your work?
Yes   No

Have you felt "burned out" or "stuck" in your present job?
Yes   No

Do you feel you have plateaued?
Yes   No

How is your health? Have you missed more work or reported late more often than usual in the last six to twelve months?
Yes   No

Any job-related accidents or injuries?
Yes   No

How are your relationships with others? Are you (a) clashing, or (b) relating to the people in your work and home environment?

Do you feel the work environment is (a) open and sharing, or (b) closed and detached?

Do you volunteer for more work and increased responsibilities, or do you hold back?

Have reorganizations affected your career, forcing you to take lateral or alternative assignments rather than opening doors for promotion opportunities?
Yes   No

With these changes, have you thought about your "fit" in the new work environment?
Yes   No

What have you done in the last six months that really made you feel needed, creative, and energized? (Consider your primary work place first, then other interests and hobbies.)

Study your answers. What do they tell you about the level of interest you have in your work or your business? Are you reaching your potential? If you seem to be falling short of your expectations, you should chart a course through the signs and take a look at what the Universe has to offer. Remember that your astrology chart has 360 degrees arranged in twelve departments of life. You shape your destiny by choosing the best array of attributes depicted by the planets in their signs.

If you are at the stage of life where you want to earn money for work you love, the following descriptions will help. These twelve descriptions will help you identify and guide your chosen career path, though you'll benefit most if you know your Moon, Ascendant, and Sun Sign, as well as the signs in which you may have clusters of planets. Read each appropriate personal passage and consider what they offer. In the natural Wheel of Life, Aries is the first sign. See how the descriptions fit the pattern of your chart, especially as it relates to your personal career choices.

The Moon in Your Career Destination
Desire—Aries

While the Moon is in Aries, it is tempted by an array of possibilities. This is the Moon's Desire phase. A self-directed explorer, the Aries Moon craves challenges that spark passion and offers an opportunity to shine. Enthusiasm in the workplace occurs only when it is chosen work accompanied by rapid salary increases.

Accelerated advancement is a primary goal on the Aries Moon's lunar career wish list. Aries always wants to win. Competitors may label this Moon "pushy," but it does not matter. The Aries Moon sees nothing wrong with hogging the floor and eliminating "inferior" challengers by quickly showing them the door.

No stranger to risk-taking, the Aries Moon easily brainstorms solutions to any problem. With a well-developed sense of identity, goals are ever-present for assertive Aries. He usually knows exactly what he wants, but if not Aries reverts to job-hopping. Some Aries Moons also can get fired for being too bold.

For instance, one Aries Moon told me pursued graphology as a hobby. Coworkers would frequently ask him to analyze their handwriting and he complied. One day the executive secretary of the organization brought him some anonymous samples to interpret. This Aries Moon pored over the pages of script and examined the pressure that was applied by the writer's pen. He concluded that the person was secretive, shy, impatient, strong-willed, inconsiderate, frugal, emotionally aloof, and probably needed glasses. Naturally Aries wanted to see the signature that went with the author. Can you imagine his surprise when he learned the sample he sent her belonged to the recently installed CEO of the company? And she had shared it with her new boss! From that day forward Aries was a frequent visitor to the front office. He was called on the carpet when the least thing went awry. A month later the ax fell, and from that day forward Aries Moon swore off analyzing anonymous specimens.

The career objective of Aries Moons is to find work they truly love. Females often enjoy doing what others consider a "man's work," where there is a spirit of challenge. While craving new adventures is second nature to Aries, this Moon wants successful experiences and full credit for carrying out the mission. A number of career paths fit the Aries drive for autonomy in the work environment, including: outdoor work, such as forest and range management, land surveying, logging, parks management, and fire fighting; pioneering fields, such as medicine, space travel, computer technology, entrepreneurial businesses, or anything requiring physical stamina; all positions involving the construction industry; the military, law enforcement, and sports; anything where skill with tools and instruments is required, such as the auto industry, mechanical repair work, piloting, racing, and vehicle sales; careers where precision in hand and eye coordination count.

An appropriate Aries career motto might be: I like being *first* in my field and I aim to be the *best*.

Dedication—Taurus

Inspired by the desires of the Aries Moon, the Taurus Moon takes stock of her assets, fully intending to develop the best of her Venusian traits. This is the Moon's Dedication phase, yet unlike Aries, Taurus' style is not to plunge into the job market. Taurus Moon wants to polish herself before she makes her mark in the world, and while she searches for the perfect career, she'll often sacrifice her essence in less meaningful work.

The security-conscious psyche of the Taurus Moon requires her to be methodical in depositing her paycheck in the bank. This is the first part of her plan to become educated and find the big career. If she is artistically inclined, she may earn some of that income by selling her paintings, performing weekends with a musical combo, or offering her hand-tooled jewelry at craft fairs. Unless the Taurus Moon has low self-esteem, she may enjoy working two jobs if this gives her a chance to perfect her skills. She is devoted to going a long way if it will improve her abilities.

While most Taurus Moons have a balanced approach to investing in their future, a few have been known to go overboard. A Taurus Moon client told me she once had to file for bankruptcy because she was addicted to bookstores. She felt compelled to enter every bookstore she passed in search of that vital piece of information that would change her destiny. Consequently she maxed out her credit cards on every self-help book and weekend seminar she could find, and paid out half of her monthly income servicing this debt. Eventually she surrendered her library to the repo man and was left with the contents of just one bookcase.

The lunar Dedication phase represents a planned commitment to a higher expression of self and a prosperous living standard. The Taurus Moon has a need for security that could mean infrequent job changes except in the case of a significant raise in salary. She does not like layoffs and may go into a slump if an unexpected job loss occurs. While the value barometer varies among individual Taurus Moons, the quest for competency in a satisfying career is their inner drive. The following jobs may offer the Taurus Moon some of the satisfaction they seek: occupations that call for coordination and steadiness, such as physical or massage therapists, aromatherapists, beauticians and makeup artists, acupuncturists; shop trades, such as butchering, baking, owning a shop, grocery work; any field of design, such as interior decoration, wedding or events planning, scenery and set design, graphic art, drapery, flower arranging; work relating to music, such as composing, performance, writing and teaching music; any kind of money management, such as banking, loan collection, brokerage, economics in general; land use issues, such as conservation, environmentalism, farming, geology, herbalism, and preservation.

Taurus has *high hopes* of finding her life work by listening to the *music in her soul*.

Discovery—Gemini

With the Taurus Moon's grounding in mind, Gemini Moon takes the opposite approach and searches the local job market for advancement. This is the Moon's Discovery phase, so Gemini Moon wants to show his curiosity, gathering as much information as he can. News reportage, in fact, is one of the best uses of Gemini Moon's thirst for knowledge.

On the other hand, the ever-present nervous energy of the Gemini Moon needs constant stimulation and a change of scenery. Sure, Gemini's mobility helps him gather facts about companies and broadcast his successes, but he remains dissatisfied in more conventional jobs.

This Moon sign rates high in the cooperation department and gets along well with peers. Gemini Moons can dazzle prospective bosses during interviews with a storehouse of knowledge. One Gemini Moon client of mine is called "Walking Encyclopedia" by his boss. Though he'll hang around the water cooler to exchange gossipy tidbits, Gemini still gives his full attention to the work load. He leads by walking around, waving his hands, and releasing his nervous energy. Gemini Moon gets acquainted with his crew, finds out what motivates them, and makes a mental note of this information. If he's exceptionally curious—and most of them are—he may analyze their handwriting or ask for their full names so he can check their numerology data for compatibility in the work place.

The Discovery phase of the Moon requires constant learning and many outlets for Gemini Moons to test physical and mental dexterity. They are known for their ability to do many things at once and to keep on the move. Unchallenging jobs lead them down the path of procrastination. Many Gemini aspirants opt to use their clever communicating style to voice their opinions through their writing, speaking, or debating skills.

Positions that provide a smorgasbord of tasks and educational opportunities appeal to Gemini Moons, including: careers in office work, such as typing, stenography, secretarial work, editing, management; the teaching profession and related careers, such as librarianship, information technology, language work, seminar running; fast-on-your-feet work, such as food service, courier work, service station attending, mail delivery; transportation-related careers, such as travel agenting, piloting, flight attendance, driving or trucking, railroad work, auto sales and manufacturing; work in the telephone industry, such as operator work, phone installation, customer service, and sales; such facilitor work as talk show hosting, emceeing, and serving as a panelist.

As Mercury-ruled Gemini thrives on duality, he may snag a new employment opportunity with his signature guarantee: *two for the price of one*.

Delivery—Cancer

In Cancer, the Moon moves into the more action-oriented Delivery phase. While easily retaining Gemini's discoveries, the Cancer Moon wants to blend this knowledge with intuition and divination. The mantic arts are a part of Cancer Moon's approach to the career quest, as well as a preference for people-oriented environments.

Cancer Moon has a need to nurture, so she often selects professions that cater to personal or social services. Whether she is delivering babies or an order of canapes, Cancer Moon's first priority is the needs of others. With her astute business sense, Cancer Moon is also queen of home-based businesses, but while not all Cancer Moons are domestic engineers, they do seek the emotional security of a nurturing work environment. They bond warmly with coworkers and treat employees like family. Protection is important to Cancer, so if a work mate is in jeopardy she'll put on her Joan of Arc suit and go to work.

On the other hand, this Moon wants no grumpy clients blaming her for something that is amiss. Be sure she is paid well and is supercharged by the work, or Cancer can get crabby. Her biggest downfall is she may spend too much time in the trenches, since she is committed to meeting deadlines and insuring the delivery of quality products. She'll be fine if she "takes five" and shares a few jokes with her coworkers.

Just to give one example, on a quiet Friday before a major holiday one Cancer Moon I know found herself alone in her private office. She heard voices in the adjoining main room and went out to investigate. To her shock she found two strangers looking in her boss's suit jacket, while he was on another floor attending a meeting. The strangers were carrying shopping bags stuffed with equipment they had raided from other offices. Cancer Moon approached these men, and, on sheer instinct, she jumped right out of her shell, startled the thieves, and demanded to know what they were doing. When one said they were looking for the personnel office, Cancer Moon snapped, "Well, you won't find it in my boss's jacket. Now put that wallet back." She must have sounded tough because the bandits dropped their booty and ran for the door. Cancer stalked them, yelled for help, and backed them right into the arms of building security. When she caught her breath and filled out an incident report for the building owners, she realized the high cost she might have paid by confronting these thieves.

Some career options that comfort the Cancer Moon include: home and hearth industries, such as home design and decorating, real estate, housekeeping, appliance and building maintenance, and innkeeping; imagination-oriented fields, such as writing, theater, poetry, training or curriculum designing, inventing, publicizing, comedy; personal care services, such as day or nursery school, beauty and health care, cafeteria work, teaching, psychology, therapy, charity work and fund raising; any entrepreneurial enterprises; any intuitive art, such as astrology, healing, metaphysics, tarot; green-field work, such as farming, flower cultivating, gardening, herbalism, and horticultural work.

Cancer cruises up the *Moon river* looking for career direction, has *breakfast at Tiffany's* to chew on the options, and picks the life work that illuminates her spirit while she keeps the *home fires burning*.

Demonstration—Leo

After all that nurturing, the Moon moves into the Demonstration phase with attention-seeking Leo. Always on the lookout for an appreciative audience, Leo Moon loves to demonstrate how dramatically he delivers the goods acquired in Cancer. This Moon wants the floor, or maybe a red carpet, to mark his place in the world.

In a workplace, Leo Moon needs a sunny space and a lighthearted playing field in order to do his best work. Though he cuts a dashing figure in his business suit, Leo wants to have some fun along the way. He also thrives on delegating authority and recruiting bit players to get the job done. Though his bossy roar may be intimidating, his lion heart is warm and appreciative of the end results. The limelight is his domain, but he'll insist that others get their share of the wealth too.

Hardly the coy one, Leo Moon will always step up to the plate when a decision needs to be made, and he expects such decisiveness in coworkers. If stuck in a wishy-washy workplace, Leo Moon makes all efforts to leave the job. Mediocre output is the signature of an unhappy or unmotivated Leo. Look closely—it could be that Leo has no boss to impress or no opportunity for promotion in a job. This go-getter thrives on approval and appeasing authority figures.

As an example of this need to demonstrate their usefulness, I know one Leo Moon who always needed something to chew on and so took up collecting fancy toothpicks. As irritation with his thankless job mounted, he began to paste one toothpick a day on his scheduling board. Before long the board was full. It became a talking piece for coworkers, but no one had the courage to ask the feisty Lion what it meant. It took a curious new person to uncover the board's meaning. Leo Moon was counting the days to retirement using toothpicks to mark off the days. The more irritated he became, the more time he spent on his hobby—and he only had five years to go.

If you are a discontented Leo Moon on the prowl for a new job, you may consider these career options: any motivational work, such as coaching, den mothering, guidance or camp counseling, physical education, marketing, seminar directing, teaching or tutoring; management roles, such as directing, heading or owning a business, school administration; anything to do with games or gizmos, such as carnival or circus performance or promotion, toy manufacturing or sales, game designing, sports equipment manufacture; jobs in the limelight, such as acting, art, astrology, costume or clothing design, singing, dancing, comedy, mimicry, modeling, promotion, film-making, theater; the uniformed services where ranks of general, admiral, major, or captain are possibilities, such as the military, police, shore patrol, lifeguarding, professional sports, airline, and the field of mass transportation.

The *sunshine* of Leo's life is hearing the *roar* of the crowd applauding his performance at *center stage*.

Precision/Discernment—Virgo

Impressed by the demonstrative sparkle of Leo, the Moon sets out to find Precision during this phase as Virgo is constantly fine-tuning the skills she is going to sell in the job market. Diligent Virgo Moon prepares for a career by first polishing all the trophies in her showcase, and then inspecting her current credentials. Finally, she critiques her superwoman performance and decides she really is the best candidate for the job. After all, who wouldn't want such a winning dynamo on their side?

At home in a maze of detail, Virgo seeks outlets that let her unravel and dissect the guts of the organization. Virgo Moon loves to feel needed in her workplace, and so she often chooses careers in the medical profession. Fortunately, Virgo Moon's traits—precision, diligence, speed, efficiency, coordination, and cleanliness—make her perfect in the operating room or around a hospital. She feels secure in a position that lets her carry a contact beeper and feel indispensable. Service-oriented and at home in government or corporate circles, Virgo Moon will do whatever it takes to keep the machines rolling, even if it she does it all backstage. With her workaholic and perfectionist mindset, she expects to put in extra hours.

On the downside, Virgo Moon can argue when told she is doing too much. Virgo likes the trenches and saving the day in a pinch. If you're thinking of reorganizing the work space, repositioning Virgo Moon won't be easy. She'll feel like a displaced person if you tinker with her quarters and move any one of her prized plants. Virgo Moon also likes solid financial remuneration, or she will take to sulking, perhaps even missing work.

While hypochondria is a Virgo trait, the next example exposes an unusual twist to creative avoidance. One Virgo Moon contracted "Chronic

Funeralitis" and missed a considerable amount of work over a prolonged period of time. He was a government worker disgruntled by his low pay and visibility. He "buried" the same father three times while employed by three different supervisors. When it wasn't his father's turn for interment, he routinely asked for funeral leave for paying respects to siblings, cousins, uncles, aunts, and best friends. The jig was up when two of the supervisors had coffee together and Virgo's name came up. The new supervisor expressed concern for the employee who was on leave to bury his father only weeks after burying his childhood friend. Let's hope Virgo Moon bought stock in the cemetery and negotiated a discount on the plots.

If you're a Virgo Moon hoping to find a meaningful challenge in your workplace, you may want to check out the following Sixth House careers: analytical wizardry, such as accounting, bookkeeping, budget management, program analysis, consultancy, research; communications fields, such as editing, fact finding, publishing, puzzle development, technical writing; detail work, such as cartography, calligraphy, dental work, embroidery, graphology, lens fitting and grinding, librarianship, sign painting, tentmaking, tile setting; health and medical fields, such as surgery or general practice medicine, dietitian, massage, pathology, personal training and therapy, pharmacy work, veterinarian work, animal breeding; maintenance management, such as cleaning and housekeeping, delivery services, dry cleaning, haircare, house sitting, pet walking, postal work, pest control, road paving, food service; mechanical and assembly work, handywork and repair (all products and industries), and computers.

Virgo's niche as the world server gives new meaning to the phrase, *service with a smile*.

Deliberation—Libra

Stimulated by Virgo's discriminating taste and quest for excellence, the Moon moves into its Deliberation phase. Libra Moon weighs the evidence carefully before selecting a life path. When Libra is fired up for a career change, he does his homework. Libra looks at the pros and cons of the offers on the table before he picks the one that gives him the greatest balance in his life and in his checkbook.

Since Libra Moon's cosmos is the social scene, he seeks a society where interpersonal connections complement his outgoing lunar nature. His knack for rubbing elbows with the elite makes networking an effective job hunting strategy. He is well aware of who is who and makes it a point to find out where they'll be so he can campaign hard in his people-pleaser role. With his charm and persuasiveness, Libra Moon often wins those coveted interviews.

At the same time, Libra loves the concept of downtime—not out of laziness, but because of a passion for hobbies. He puts lots of effort into these diversions, which for him are crucial investments. He doesn't have to worry if he loses his job, since any one of these pastimes might become a rewarding career.

One Libra Moon I met was known for hosting weekend cookouts. Libra Moon's guests loved his savory ribs, T-bones, and chicken breasts, especially for the quality of his barbeque sauce. Naturally everyone wanted his secret, but the Libra Moon held out knowing his creation was headed for bigger things. Each weekend, he refined the sauce by experimenting with ratios of Worcestershire, Red Hot, and chili powder until he had finally outdone himself. Afraid that he would procrastinate, his spouse ordered cases of jars and designer labels to boost his incentive to get his product to the public. Soon his work lined the shelves of specialty stores and gourmet grocers. Now he has a spot on the shopping channel hawking fifteen varieties of sauce and telling viewers he has never met a taste bud he couldn't please, and while Libra Moon stirs his secrets into the sauce, his business manager wife steers the viewers toward their checkbooks.

If you're a harmony-seeking Libra Moon, you'll like this inspiring list of job possibilities: adjudicative work, such as appraisal, conflict management, diplomacy, law, insurance, marriage counseling, mediation, negotiation, refereeing and umpiring; anything to do with athletics, such as baseball, equipment supply, golf, health club management, football, racket sports, coaching, running, sporting goods sales and service, sports medicine; human resource work, such as counseling, customer service, benefits specialist work, workplace evaluation, headhunting, hiring, recruitment, motivational psychology, test administration; meeter-and-greeter work, such as innkeeping, concierge work, hosting, meetings management, hotel management, public relations, reception, restaurant work, sales; work relating to pageantry of various sorts, such as beauty contest and parade organizing, cake decoration, catering, flower arranging, fund raising, choreography, photography or videography, political campaign management, set design, special events coordination, souvenir sales and design, wedding planning; work with partners, such as civic association work, co-owning of a company, network marketing, teamwork promotion, staff or workforce management, and variety or talk show cohosting.

Libra presents a balanced *slice of life* when he *meets the press* and offers stock options in his *recipe for success*

Depth—Scorpio

Fortified by Libra's Deliberation phase, the Moon is ready to dig in under Scorpio's influence. The Scorpio Moon's career destiny is Depth, and she is dead serious about getting to the bottom of her work pile or the source of your problems. Scorpio Moon needs opportunities to deal with issues that demand attention. This resourceful Moon will audit your books, your time reports, or your wardrobe if you choose her as your personal shopper.

Skilled in assets management and tasks that give her the access and responsibility of minding other people's money, Scorpio Moon is very scrupulous and thorough. She knows every account you ever opened and can calculate your debt ratios faster than you can get out your credit cards. If you fudge details, she will nail you every time, and in the process she may even teach you a few of her own sleight-of-hand tricks.

Scorpio Moons are lunar magicians who do their best work in quiet environments without benefit of a team. They really don't want anyone looking over their shoulder and stealing their thunder. Excessive meetings and constant interruptions drive them loony. Make no mistake in judging their controlled outer reserve, however, Scorpio Moons thrive on competition. They are known for seizing the right moment and making a play while their less-prepared rivals don't know what hit them.

A positive Scorpio Moon is full of inner power and resiliency. She seeks outlets that reward risk. At the same time, some Scorpio Moons are short on self-esteem and have to use seduction to snag jobs. They can, at times, go after potential bosses by displaying their anatomical charms. Of course, people are becoming more aware of this problem, but there are still a few individuals who remain in the dark ages.

Such was the case of a Scorpio Moon who worked at a large financial organization. Bored with her job duties, she was ready to call on the Phoenix to lift her up the career ladder. She began by arriving early at the office to search for an "early bird" with hiring authority. Eventually, she found her mark and seduced him with ease. She sat on the chair near his desk and fed his ego by commending his intense dedication to his work. Soon she became a daily fixture in his office, changing his entire morning routine. Instead of briefings and conferences with his staff, these two had coffee, Danish, bagels, and small talk from 7–9 am. Within two weeks her mark contacted her boss to request assistance on a project. Though the project fooled no one, before long the Scorpio Moon became a lunchtime fixture, too. By now she was sitting on his desk dangling her lovely, long legs. I'd tell you the rest but this is a family annual and Scorpio might silence me for spilling the beans.

While most Scorpio Moons are too insightful to resort to the above tactics, here are some ways they make a living: excavation work such as archeology, construction, deep-sea diving, dentistry, drilling, grave or tunnel digging, mining, quarry work, subway conducting; beauty and cosmetic fields, such as beautician work, cosmetic surgery, denture fitting, dermatology, ear piercing, esthetics, healing, make-up, optics; anything in the mass media, such as bumper sticker and button supply, campaign organizing, all facets of advertising, news broadcasting, TV show promotion, commercial analysis, polling, statistics; explorative and investigative work, such as deep space astrology, detective work, medicine, exploratory surgery, gossip-column writing, hypnotherapy, photojournalism, proctology, psychology, regressionism, research, spying, tax collection; the sensual art of seduction, such as exotic dancing, adult book and magazine selling and writing, lingerie design, underwear modeling, tattoo art; waste management fields, such as trash collection, junk dealing, sewer work, garbage disposal manufacturer and installation, plumbing, diaper manufacture and disposal, colonics therapy, and the tissue industry.

While the *harvest Moon* hovers over her mysterious destiny, Scorpio spices up her life work with a round of *double jeopardy*.

Dissemination—Sagittarius

Impassioned by the Depth of the Scorpio Moon, the Sagittarius Moon fills his portfolio with a sampler of exotic wares. Be advised: he is outwardbound with excitement, and there's no way to stop him. Sag Moon has already mastered his talents in numerous ports of call and relishes the thought of seeding new territory with his vagabond wisdom.

In the Moon's Dissemination phase, Sagittarius hits the trail in search of many uses for his inspirational bounty. And at the same time, Sag Moon navigates the road of opportunity to fill his ever-questing mind with unique perspectives on culture, philosophy, or human relations. His adventures are marketable commodities. And while many Sagittarius Moons are frequent travelers and often need to charge their optimistic batteries, others get their exercise by keeping the phone wires burning.

Sag Moon thrives on the gift of gab and runs record telephone marathons to keep his vocal cords in shape, especially when he is restless.

That said, you should know that the Sagittarius Moon is the epitome of commitment, but he needs a very stimulating job to keep him focused. Otherwise he drifts off course in a sea of boredom. As a project-oriented sign, Sagittarian also needs a generous expense account to cover the cost of getting the show on the road. Sagittarius will grab every hand who donates and make sure their deep pockets are emptied to fund his activities. Salesmanship savvy is Sag Moon's most infamous quality.

One Sagittarian Moon client of mine had a flair for generating enthusiasm. Accordingly, he was assigned a large project at work to loosen the organization's fuddy-duddy management style, cut through the perpetual red tape, and stimulate a creative work climate. The regular structure of the company, however, was used to doing things as they had always done according to the rule books. So Sag Moon, thinking that the organization could not alter its philosophy if the people were not comfortable with innovation, organized some celebrations revolving around the theme of change. After all, he knew something dramatic had to happen. So Sagittarius invited a world champion boxer and his wife to one of the project's kickoffs where they offered the group some tips on the importance of having a good support team in their corner. Attendees at this event were surprised to find the champ had a phenomenal memory for their names and was graciously signing autographs for everyone. The stuffy old walls of the institution crumbled under the laughter, and today none of the "old guard" is still in the organization.

If you're a Sagittarius Moon with a desire for a new career, consider one of the following options: legal defense fields such as affirmative action and consumer advocacy, civil rights activism, legislation and lobbying, political analysis, public defense, sociology, victims' rights advocacy; education fields such as behavioral modification, college or university administration, curriculum development, teaching, interactive and video-based training; spotlight-seeking fields such as activism, diplomacy, acting, lawmaking, oration, politics, preaching, pop music; travel and related industries, such as chambers of commerce, cruise ship work, languages, interpreting and translation, travel and tourism, commodities trading; fields relying on word wizardry, such as advertising, broadcasting, publishing, teaching, writing, publishing; fields relating to the spirit-world, such as spiritual ministry and guidance, philosophy, the priesthood and other religious orders, theology, shamanism, and soul traveling.

Sagittarius explores the *road less traveled* and finds a *rainbow of opportunity* on his way.

Determination—Capricorn

Since buoyant Sagittarius Moon broke new ground in the Dissemination phase, Capricorn Moon is now planning to take over the reins of her high-visibility career. In this Determination phase, Capricorn Moon conquers all obstacles and wears regal robes of authority. She sets the standard for accomplishing goals and rules the world from her Regent's throne. This Moon has built her reputation on her professionalism and organizational skills. As a model worker, Capricorn Moon amasses awards and citations during her career path.

The inspired Capricorn Moon rises to the occasion of meaningful work and sets her sights on building monuments of achievement. Her avenues may be academic, corporate, or public, and she is known for holding down two or more jobs at a time. Some Capricorns throw in full-time schooling as well and manage to get top grades.

Capricorn Moon seeks glory in her work and moves any rocks in the road on her way. Normally, she is ambitious and responsible. She salutes a pile of work and then works her way through it. There are times, however, when she's forced to take a mediocre role in her work. On these occasions, Capricorn Moon needs inspiration to get her moving. Some Capricorn Moons have intimidating mothers who fill this role. A few male Capricorn Moons have both a wife and mother who call the shots.

When Capricorn Moon's self-esteem is threadbare, she will become restless. When constraint erodes her effectiveness, she wilts on the vine. If Capricorn feels trapped, she becomes gloomy and somber. In any of these cases, Capricorn Moon will eventually pull out her strategic thinking cap and campaign for a new direction for herself.

Such was the case of a government statistician with a bossy mother and a demanding wife. He really hated his job, but a sense of responsibility had been hammered into his head. Capricorn Moon longed for a job where he could rub elbows with people, and he felt unfulfilled and listless. His rope was getting shorter, and he knew he was dying of boredom. One day, when his wife and mother were safely out spending his money, he left work early to execute a daring plan. The Moon packed his bags and joined a traveling circus, never to be heard from again!

If you're determined to make new marks in the world, Capricorn Moon, take a look at the following options: fields of management, such as accounting services, budget allocation and management, customer service, debt management, economics, human resources, international banking, strategic planning; disciplinary fields, such as academic discipline, budget trimming, drill sergeant work, the court system, law enforcement, exorcism, parenting, police work, preschool teaching; work in the public spotlight, such as acting, ballet dancing, military combat, commercial spokesmanship, government and politics, sports, television work; technical work, such as change agent work, contracts, field research, leasing or purchasing, museum curation, real estate, lab research; any role at the top of an organization, such as administration, chairmanship, executive work, government or industry leadership, ruling or owning a kingdom or barony, owning an enterprise, the papacy or presidency; various trades and professions, such as arborist work, bagpipe playing, beekeeping, bone medicine, chiropractic, dentistry, geology, gymnastics, jewelry work, lobbying, masonry, marching, music, stone-cutting, and wall building.

Capricorn stands on her *laurels* in the field of *accomplishment* and takes her bows from the top of the *pyramid*.

Distance—Aquarius

While applauding Capricorn's perseverance and determination, Aquarius concentrates on shifting the shape of his life work. He has lofty aspirations for freeing himself from the scrutiny of bosses and supervisors. That is, self-employment electrifies the Aquarian Moon's circuitry.

The Moon in Aquarius has a complex psyche driven by internal longings for change. Sometimes disruptive early life experiences affect his security needs and influence the erratic nature of his career path. As a maverick who likes to make his own hours, Aquarius Moon will break the mold of previous generations. This is one dynamo who can really get focused on his work and forget that people who care want to hear from him. His internal spirit craves independence and unique experiences. Aquarius Moon relishes being different and showing his eccentric streak.

Aquarius Moon looks to better the world for all humanity. Superman is his role model, so he is interested in liberating all the people on the planet. At the same time, Aquarius has to replenish his supply of cosmic awareness quite often. While he is a maverick who likes to make his own hours, he also leads a quite impersonal life. Some folks call him "The Clam" because Aquarius crawls into an intellectual shell to dissect emerging global conditions while ignoring the demands of his personal world.

Aquarius relates to associates who understand his need to detach until he gets the job done. His understanding falters when acquaintances need more predictable schedules than suits his pick-up-and-go style. The altruistic Moon feels at home with groups and affiliates whose missions include an all-out commitment to enterprises that promise a better way of life. This Moon is a crackerjack problem-solver. He is also an advocate for advanced technology, an agitator of public consciousness when the underprivileged are mistreated, and a willing confidante when a friend needs to let off steam—since he is perfectly capable of leaving without taking the troubled soul's baggage home with him.

You should also be aware that Aquarian Moons can be practical jokers. Their penchant for having fun often stems from dissatisfying job conditions. They have a perverse sense of humor, and anything goes when it surfaces in the work place. For instance, a sociologist with an Aquarian Moon had an interest in how people behaved in elevators when conditions were abnormal. She enlisted the aid of a colleague to record the comments of people leaving the elevators after she demonstrated some pretty weird behavior while they were passengers. For instance, when she entered the elevator she would: a) face the rear and keep riding up and down ignoring the promptings of riders; b) wait till an elevator was filled and then turn around and stare into the eyes of the closest person to see how long it would take before they looked away; c) start sobbing profusely when the elevator filled up and watch everybody move to one side to avoid her. When she wasn't riding elevators, Aquarius Moon would test the observations of her staff by wearing two different shoes, wearing her sweater inside out, or leaving a realistic-looking box of plaster of Paris chocolates on her desk to see who would take the bait. Is she still rattling around the musty old elevator cage today? Of course not—she made a quantum leap onto the Internet and is now surfing the cosmic tides.

If you'd like to try out something new in the job market, check out these eclectic options associated with the Aquarius Moon: work as an affiliate or sidekick, such as silent backing, board or city council membership, delegate work, internships, networking, philanthropy, political party work; consulting fields, such as advising, computer technology, human resources, engineering, industrial psychology, strategic and urban planning, sociology, venture capital, workforce management; cosmic crafts fields, such as astrology, space exploration and management, metaphysics, air travel, science fiction writing, visionary work; imaginative industries, such as radio, TV, music recording, hair salon work, special effects engineering, sound stage management, inventing, electrical work, auto design, computer technology, science and mathematics, software and interactive video production, website design; large-group management, such as airline staffing, beauty school operation, club or resort management, leading a cult, dance troupe management, musical groups and chorus conducting, promotion and coordination of conventions, expos, seminars, and workshops; facilitation and teaching in all forms, sports clubs, athletics, coaching, training; not-for-profit work, such as charity fundraising, defending and protecting the public domains and interest groups, politics, leading a fan club, directing development.

Aquarius takes every *step* on a rendezvous with the *brotherhood of light* for all *eternity*.

Diversity—Pisces

After the distance of the Aquarius Moon, the Pisces Moon makes a different statement. With such a variety of experiences, this goddess of versatility needs a wide berth to accommodate her creative umbrella. The Moon in Pisces is in her Diversity phase, and at work Pisces Moon sets her dreamy eyes on opportunities to impress her boss with her many talents.

On the positive side, Pisces Moon at work offers new perspectives to established routines and puts pizzazz into major initiatives. She is a shining example of a job candidate with endless useful assets. Using positive thinking, Pisces Moon leads all of us to success. She willingly lends a hand to work mates and shows compassion to staff members struggling with their work. The big downfall is that less enterprising coworkers often take advantage of her help. Pisces Moon can find herself with duties that put extra pressure on her work load. While she may not complain initially, she can acquire a martyr complex. Then you'll hear her singing the Work Place Blues. Pisces Moon must set boundaries to avoid this trap, or else seek the intervention of a supportive employer.

In a confining job, Pisces Moon checks out and at best produces marginal work. Even during a good day's work, she'll be looking over her shoulder and watching the creeping clock. This Moon deplores drudgery, yet ironically she can spend far too much time on the details of her work and so trap herself. Some Pisces Moons get so fogged out with frustration they don't see the clear signals that are telling them to leave the position. When Pisces says she is "chained to her desk" or "feels like she is in prison" in describing her work conditions, be a friend. Gently suggest that it may be time to work on her resume. Assure her there is a more appreciative outlet for her diverse talents and she deserves a new lease on life.

Pisces Moons like money but they are not particularly fond of overtime. They'll dutifully work when a crisis occurs, but they resent a steady diet of add-on hours. An industrious Pisces Moon with good work habits shared an office with a coworker who had poor workload management skills. The coworker often missed deadlines, and the boss asked the Pisces to stay late and help out. Invariably the requests for overtime occurred on Thursdays, the designated bowling night for Pisces Moon's league. One Thursday in late spring, she decided she'd have none of it this time—as her bowling team was in a tournament. Pisces Moon heard her boss telling the harried coworker in the hallway that he would get her some help, so she hid under her desk and held her breath until the coast was clear. Although she was elated that she helped her team win every game of the tournament, her real victory was in escaping deadline limbo. In case you're wondering, the Pisces Moon sent our her resume the next week.

Positions that offer travel, multidimensional tasks, and a change of scenery appeal to Pisces. Here are a few jobs that fit the bill: fields in the fashion industry, such as jewelry, shoes, or clothing design, modeling, make-up manufacturing, high-fashion photography; caregiving fields, such as urgent care, nursing, medicine, physical therapy, radiology, anesthesiology, innoculation, lab work, geriatric or pediatric medicine, prison rehabilitation, psychotherapy, Meals-On-Wheels organizing nutrition, volunteer services; "dream" works, such as acting, film production or directing, dancing or choreography, movie publicity, screenwriting, set design, stage managing, writing of the macabre, greeting card verse writing, art, interpretive dance, dream analysis, magic, regression therapy, poetry or romance writing; anything to do with the intuitive or spiritual realm, such as astrology, clairvoyance, intuitive arts, medium or mentalist work, mind reading, palm reading, psychometrics, rune or tarot reading, meditation instruction, spiritualism and psychic work; leisure time trades, such as amusement park work, bartending, boutique or souvenir shop work, celebrity tour guidance, costume industry work, music, party organizing; water industries, such as filtration systems, fish and seafood, hot tubs, oceanography, spas, scuba diving, spring water, swimming, swimsuit design, dinner and luxury cruise work, aquariums, swimming pools, surf and sand products; coastal resorts; lighthouse work.

Pisces hears *celestial music* and puts her *heart and soul* into dancing her way to the *stars*.

The Moon is Your Public

Dorothy Oja

Being your own boss can do wonders for the ego. Are you considering therefore turning your favorite interest, passion, or hobby into a business? You may be in luck. Current planetary placements in Aquarius and Sagittarius indicate that individuality is culturally and socially very trendy. Following your particular creative take, your way of doing it better or first, is supported by the cycles of the times.

Astrology shows us that each beginning has an unique energy. This initial energy is carried through the life of your venture and developed in your ongoing relationship to your venture. Therefore, one of the best ways to ensure a good beginning is to cast the chart for the start of your business. With this road map in hand, you can chart the course of your business ups and downs, struggles and opportunities for the full and complete life of your business.

This article will tell you how to use astrological knowledge to make your business better and more unique than the one down the street; in other words, how to get the edge on the competition. Creating a successful business is not easy, but this guide, and the guidance of an experienced professional astrologer, will give you a good start.

In the chart of your brand new business, you should pay particular attention to the Moon. It is the Moon that represents your public, those clients who will frequent your store or call upon the service you are marketing. The Moon also represents the general atmosphere of your establishment, as well as the people who will be drawn to your environment. This is because the Moon is the fastest moving planetary body, moving into a new sign every two-and-a-half days. Due to its speed and its intimate relationship to the Earth, it governs daily life and the continuous shifting emotional conditions we all experience. In this way, the Moon gathers, absorbs and embodies the emotional environment of any moment, which is then expressed through the moods of people in general.

Because of these factors, the Moon details the prevailing conditions necessary for you to succeed in the marketplace. Astrology is an essential tool for you in choosing the tone or the energy that best describes the

nature of your business and the mood or needs of the clients you will attract. If you work effectively with the energy pattern described by the Moon in your business chart, you will find that you will be more successful with your clients. They will come to you aware of the type of care and treatment you can offer them as indicated by the Moon. Paying attention to the Moon in your business chart allows you to be in alignment with the energy pattern that was established at the time you began your business. By expressing the full and positive potential of the Moon in your business chart, and by being aware of the negative factors and minimizing them and their effects, you will be in harmony and in a more perfect relationship to the natural order.

Remember, however, that we are discussing a general tone or energy expressed by the Moon's position; the Moon's effect is also modified by any other planets it connects with. These connections can be found in the complete chart for the start of your business. You should have your personal astrologer explain all the Moon's relationships to other planets in the chart. While you're at it, have your astrologer go over the complete chart and all its potential implications. Success awaits you—if you use the time cycles that are favorable for your business. Below, you can explore the specific ways the Moon works with your business.

Aries

If you start your business during an Aries Moon, your public will be looking for something new and progressive, stimulating and adventurous. Your business will be characterized as enterprising and cutting edge; therefore, you should strive to be clear about the identity you want to convey for your product or service. People who want fast service or delivery will also appreciate warmth and personalized attention. If you take the time to provide individualized service to your customers, people will tend to return because they will instinctively feel that they are treated as important and special. The positive, spirited attitude and atmosphere that is encouraged by the Aries Moon will also stimulate customers and make them want to return. The Moon in Aries is perfect for lighting a fire under others and getting them moving. In the end, this Moon will demand that you keep your business current and stay on top of all the latest innovations in your field. This is the Moon of the pioneer and the adventurer.

The energy in your place of business is likely to be lively and stimulating. Some possible businesses suitable for the Moon in Aries include: active or risk-taking businesses such as sports, cars, racing, adventure, travel, personal achievement programs, promotion, image consulting, or simply a new venture of any kind; businesses or products that require a fast turnover, such as a cafe or sandwich shop; businesses where you might be demonstrating an activity or craft.

> Fire is the element connected with Aries. Fiery Aries represents spirited and active participation in life and a life-enhancing expression.
>
> **Focus on:** Active customer participation in testing of your product or service. A sense of humor.
>
> **Keywords:** Courage and excitement.
>
> **Pitfall:** Arrogance or impatience could be the downfall of your business. You may often be busy but you must take time to cater to your customers.

Taurus

If you have a Taurus Moon for the opening of your business, your customers will want value for their money. At the same time, they will be willing to pay a higher price for items of high quality. Although Moon in Taurus customers are always looking for dollar value, they are willing to spend more if they think the product or service is one of exceptional or long-lasting quality. You should arrange a Taurus Moon business environment to express a feeling of substance and worth. This is not to say that you should be showy or create a false impression, but simply that your atmosphere should be comfortable and sustaining. Accouterments of the earth and natural items and plants will greatly enhance this environment. Be sure to use calm colors. Red can agitate and is not a compatible energy with ruled-by-the-bull Taurus. If you have a storefront, you should choose your location carefully. Furthermore, any major changes can affect your clientele negatively. Your customers will look for stability, steadiness, and reliability from you. They will be greatly pleased if you play soothing music as they look over your merchandise, and if you provide comfortable and accessible seating. Having sales twice a year will bring people to your store to purchase your high-quality products or service. And some of your best and most loyal customers would probably appreciate a hug.

Some possible businesses suitable for the Moon in Taurus include: art, jewelry, fabrics, gourmet foods, fine wines, collectibles, antiques, plants, or any service or business dealing with finances. Consulting in this area could involve securities analysis, financial planning, money management, tax preparation, and any advice that helps people maintain stability and security.

> Earth is the element connected with Taurus. Earthy Taurus represents endurance, dependability, and tangible quality.

> **Focus on:** Providing a comfortable atmosphere that is not rushed, with well-crafted items and thorough good service. If you are handling food, make sure it is always fresh and well-presented.

> **Keywords:** Pleasing and calm.

> **Pitfall:** High prices, poor-quality goods, unreliability, or a disheveled atmosphere.

Gemini

A Gemini Moon loves options. With a Gemini Moon for the start of your business, your customers will want variety and whimsy. Stock your store with many items, so that your customers have unlimited choices. Moon in Gemini businesses naturally have several important things going on—two or more services under the same umbrella, for instance, or diverse products. Be sure to chat with your customers and keep things friendly, as this Moon thrives on communication. People will ask you questions concerning your business and will be upset if you don't have the answers. Be prepared to answer the question your client is asking and don't veer off on a tangent. If you provide a service, clarity in communication is vital. Be sure of what you are offering and what the fees are. Do not confuse your clients with a contradictory fee structure or conflicting prices. And be sure to keep up with the latest information on your product or service. Your customers tend to have inquiring minds.

Some possible businesses suitable for the Moon in Gemini include: anything to do with writing, such as a newspaper or magazine, a card shop, a paper store, a gift shop, a service that distributes any type of information, a place for young people to gather, or services that bring people together; languages, think-tanks, the visual or graphic arts, a resume-writing service, advertising and sales, and computer-related services; any business that has to do with the element of air and wind, such as a bike shop or travel agency.

Air is the element connected with Gemini. Airy Gemini indicates attention to the visual, and people skills are of utmost importance.

Focus on: Keeping a clear and steady flow of communication with your customers. Be friendly and willing to answer questions. Find out who your customers are and learn all about them.

Keywords: Curious and friendly.

Pitfall: Gossip, nervousness, or scattered energy. Although these qualities can be part of the Gemini Moon energy, they will not endear you to your clients in the long term, as they breed mistrust and anxiety.

Cancer

Starting a business during a Moon in Cancer will give it a homey appearance to the public. This could mean a family business, for instance. You should be sure to concentrate on the basic human needs of your customers by taking care to have some comfortable chairs and drinking water on hand. If you have a store, make sure the temperature is neither too hot nor too cold. Your public will judge your business based on the overall comfort they feel at your place of business while interacting with your employees. The attitude of the people who sell your product or service will be what wins you repeat customers. Provide samples and discounts, and some candy or other munchies if you have a retail store. If you are a home-based service business, extra customer service will make your business thrive. You should think of your customers almost as extended family. It's that added touch that makes all the difference.

Some possible businesses suitable for the Moon in Cancer include: catering to the basic needs of people with food, housing, clothing, collectibles, antiques; anything that brings comfort to people, such as elder services, child care, housecleaning, psychological counseling, real estate.

> Water is the element connected with Cancer. Watery Cancer indicates sincerity, comfort, and the ability to listen to the your clients and their concerns. Being real is important to your customers!

> **Focus on:** Cultivating a natural, solicitous, sincere attitude and a comfortable environment.

> **Keywords:** Listening and nurturing.

> **Pitfall:** The tendency to give away the store, to be too saccharine and ready-to-please, and lack of firmness.

Leo

If the Moon is in Leo for the start of your business, your ego will be invested in the business. You are likely to feel proud of your business to the world and the benefits it has to offer, and you will impart a very distinctive personal flavor to your business. In fact, your personal style will be evident in every aspect of the way the business is run. At the same time, Leo is about entertainment, creativity, and warm spontaneity, so you should strive to make your establishment fun and upbeat. Leo energy is one of heart, so your customers will expect a certain generosity of spirit from your business. Celebrate your business often through promotion and fun occasions and events for people to participate in. Leo is also leadership, so you can be sure that once you are committed to your particular service, you can become a leader in your field. And if you operate your business from a heartfelt place, you will have a steady stream of loyal customers.

Some businesses suitable for the Moon in Leo include a wide range of services or products, such as: party planning, clothing and accessories, toys, costumes, and any services for children; performance and theatre arts, and entertainment of various kinds; matters-of-the-heart counseling services; and any type of life-enhancing training, such as leadership training, self-esteem building, or other such personal-support services.

> Fire is the element connected with Leo. Fiery Leo indicates warmth, loyalty, love energy, steadiness, and creative spirit.

Focus on: Drama. Make any contact with your service or business an event your customers will remember.

Keywords: Laughter and radiance.

Pitfall: A proud, stubborn, egotistical attitude will turn your customers off. Treat your customers like royalty, but make sure your attitude is one of service.

Virgo

With the Moon in Virgo for the chart of your business opening, your business will require a clean environment. You should therefore make an effort to keep things in order—in your store's appearance, stock, and records. Fortunately, this Moon has a built-in work ethic, much like an experienced craftsperson. Precision and refinement are the Virgo archetype. A continual striving for improvement and attention to detail will bring business to your door. The little details will delight your customers—the small, quality extras that will insure that your product succeeds in the marketplace. It will clearly show your customers that you have taken time and that you care about them. If you follow these impulses, people will come to know you as reliable and well-organized. Customers will also expect you to give them extra help when explaining your product or service. This is a hands-on Moon, after all; efficiency and skill, as well as on-time delivery, are other ways you can charm your potential repeat customers.

Some potential businesses suitable for the Moon in Virgo include: natural foods, vitamins, pharmacy, fitness club, cleaning, business management services, time management, editing, troubleshooting, quality-control, stress reduction, relaxation, mechanical services and supplies, computer repairs and servicing, skills assessment, training, funeral services, accounting, health-care services, and services that most people don't want to perform; also any highly skilled craft such as instrument-making and repair, woodcarving, metal, or marble craft.

Earth is the element connected with Virgo. Earthy Virgo favors real skills, practical services that perform as advertised, and tangible results.

Focus on: Offering the best service policy around. A bit of humility, too, will go a long way.

Keywords: Hard work and integrity.

Pitfall: A picky or irritable, self-righteous attitude will make people avoid your store like the plague.

Libra

When your business begins with the Moon in Libra, its chief concern will be human relations. In this case, above all you should remain fair to your customers and always give them the benefit of the doubt. What you definitely don't want are grumbling customers who will complain about you or your services to their friends and acquaintances. It makes business sense to smooth ruffled feathers as soon as possible, to refund money to dissatisfied customers, to strive to deliver the best product always. You are your own best advertising. At the same time, you might consider consulting a feng shui expert before opening your establishment to the public to create the physical arrangements necessary for maximum harmony, both visually and with respect to the energy flow. A Libra Moon indicates a perceptive group of customers who will quickly scan your environment for a pleasing atmosphere. If you have this, they will want to return. Affiliations, associations, and endorsements will favorably impress your public and encourage them to try your goods or services.

Some possible businesses and occupations suitable for the Moon in Libra are: temporary employment agencies, dating services, advertising, graphic design, fashion, accessories, jewelry, hair and beauty products, headhunting, law, art dealing, a modeling agency, photographic services, human resource management, image and color consulting, promotion of any kind, and public relations.

Air is the element connected with Libra. Airy Libra indicates the need for space and range of movement. The primary concerns of a Libra are the quality of relationships and communication with others.

Focus on: Harmonious people relationships and service with a smile. Remember, the customer is always right! You should offer a satisfaction-guaranteed policy.

Keywords: Consideration and compromise.

Pitfall: A judgmental, snobbish attitude; insincerity, indecision.

Scorpio

If the Moon is in Scorpio when you start your business, people may find you somewhat inscrutable or politically motivated. People will try to tune in to the vibes or energy they detect from your business environment. If the energy is supportive, enriching, and empowering, they will most likely return. Your potential customers will definitely be interested in knowing exactly what you stand for or where your loyalties lie. The Scorpio Moon is interested in in-depth analysis, and your customers will expect you to have some important knowledge to share with them. Persistence and focus are the foundation stones of Scorpio. You would be well served if you get to know your customers, their likes, dislikes, and personal preferences. Cater to that knowledge and you will discover repeat customers who sing your praises. Profundity works for the Moon in Scorpio, so go ahead and bowl them over with your incisive perceptions and dramatic transformations.

Some possible businesses and occupations suitable for the Moon in Scorpio include: anything to do with money or money lending, buying and selling stocks or products, stock-market analysis, occult services, lobby or campaign management, politics, fundraising, management in general, physicians, psychotherapy, counseling, recycling, detective work, transformative services, or renovation work.

> Water is the element connected with Scorpio. Watery Scorpio indicates that understanding, analyzing, and processing information are an important aspect of this sign's energy.
>
> **Focus on:** Supporting the empowerment of your customers in any way you can. If you can increase the quality of their lives, they will be repeat customers for a long time. Share the wealth in whatever form you can.
>
> **Keywords:** Strategy and empowerment.
>
> **Pitfall:** Focusing too much on money, being too intense with or pressuring your customers.

Sagittarius

If you have the Moon in Sagittarius when you begin your business, make the environment fun and fill it with novel information. Offer experiences that people can learn from and feel enriched by. You may do this by giving away free samples or providing brief demonstrations about your product or service. Choices, options, and opportunities are the staples of the Sagittarius energy. This Moon wants to feel enhanced or improved as a result of your product or service. A wide range of possibilities, elements of fun, or experiences that stretch boundaries are important to Sagittarius. Involve your customers in the understanding of what they can gain from the use of your product or service. And always maintain a positive outlook while offering plenty of encouragement.

Some possible businesses suitable for the Moon in Sagittarius include: adventure travel, legal advocacy, education services or products, teaching, philosophy and systems of thought and belief, sports and sports equipment, yoga, hiking and outdoor activity, ethnic studies, religious products and services, storytelling, balladeering, comedy and joke writing, all sorts of creative crafts, personal shopping services, and anything that give people more time and gives you an adventure.

Fire is the element associated with Sagittarius. Fiery Sagittarius indicates a brisk pace. It needs feelings of generosity and magnanimity in order to function well. It is easily engaged, enthusiastic, and spirited.

Focus on: Being positive and philosophical about the ups and downs of your business. People will want to be inspired by your infectiously engaging attitude.

Keywords: Expansion and judgment.

Pitfall: Promising more than you can reasonably deliver; exaggeration or misjudgment.

Capricorn

Starting a business during a Capricorn Moon means that there will be a no-nonsense and determined quality to the enterprise. This could be good or bad depending how you use the quality. A business-like attitude could help engender respect among your customers, and dependability, reliability, and consistency will make your customers happy. You will find that your clients often seem demanding, but that is because they want and expect the best quality and service possible. There is a methodical and sometimes slower pace to the Capricorn Moon because of its strong interest in a good foundation. If the work is worthy, the results will be strong and lasting. However, Capricorn Moon energy must work hard to fit in to a community at large if it is to remain a positive one. You can encourage this by joining local business associations, fostering relationships with other merchants, and remaining visible among the consumers. Credibility is expected, so be sure to let people know of your accomplishments, professional associations, and affiliations.

Some businesses suitable for the Moon in Capricorn include: accounting, marketing, business management, office management, quality control, consulting, time management, woodworking, ceramics, glasswork, sculpture and other highly skilled crafts, framing and housing foundation work, and working with precious gems, coins, and clocks.

> Earth is the element associated with Capricorn. Earthy Capricorn indicates a need to have tangible proof that the business is achieving its desired results. Capricorn regularly measures progress toward goals.

> **Focus on:** The consistency of your product or service and reliability.

> **Keywords:** Integrity and stability.

> **Pitfall:** Being too rigid, uncompromising, or negative. The past does not necessarily equal the future!

Aquarius

Zany is the word for a business that begins with a Moon in Aquarius. This enterprise will be seen as eccentric and fun. Your business will most likely have a unique flavor to it, or perhaps will be an old idea with a certain undefined pizzazz or twist added. You may present things in a highly individualistic and creative way, and in fact you probably will view life from an extraordinary angle. People will find your humor and honest communication refreshing; they will appreciate that you tend not to put on airs or pretend to be what you're not, and they will notice the alert and intelligent energy in all that you do. It would be to your advantage to stay on the cutting edge and on top of the latest developments in order to satisfy your customers' demand for such novelties. The Moon in Aquarius is open to the future. Your customers will relish your obvious appreciation of human idiosyncrasies and your easy ability to laugh at yourself.

Some possible businesses suitable for the Moon in Aquarius include: anything having to do with animals, humanitarian causes and human rights, advocacy work, anything dealing with the development of the individual enhancement and freedom of lifestyle, unpopular causes and social protest, technology and science, computers and video games, exotic foreign travel, a travel agency, or import-export services.

Air is the element associated with Aquarius. Airy Aquarius is alert, friendly, highly intelligent, soberingly truthful, and eccentric.

Focus on: Friendliness, acceptance, and having a humanitarian attitude laced with a good dose of humor.

Keywords: Truthfulness and individuality.

Pitfall: Being too distant, inaccessible, extreme, or blunt.

Pisces

If you have chosen to start your business during a Moon in Pisces, your public will expect you to be sensitive and kind. In fact, you may tend to draw overly sensitive people to your establishment. Make sure you can deliver some form of empathy, concern, or compassion for these folk. Focus on the traits most in keeping with a Pisces Moon: refinement, inspiration, healing, magic, artistry, and a soulful environment. Your clients will appreciate soft music or beautiful art in your place of business. If your environment provides a relaxing, poetic oasis from the rigors of the daily grind, people will surely flock to your store and want the soothing quality you provide through your product or service.

Some possible businesses suitable for the Moon in Pisces include: all things dealing with artists, such as running an artists' cooperative, art galleries, an art supply store, teaching art, and art therapy; any support services that enhance the quality of life, such as mental health services, substance abuse services, women's shelters, any services for the neglected portions of society; any service that provides behind-the-scenes support, such as theatrical staging, spiritual or mystic studies, products, and services, perfumes and aromatherapy, massage, Reiki, tarot, fine secondhand clothing, masks or theatrical props and goods, goods or services that deal with ritual, symbolism, and spiritual work; finally, any rare and hard-to-find products, specialty goods and services, such as music, poetry, dance, a video store, movies or film, fine lingerie or other fashion accessories, and rare book stores.

Water is the element associated with Pisces. Watery Pisces pays attention to the emotions and feelings, and has a mystical sensibility and artistic temperament.

Focus on: the sensitivities and subtleties of your clients, and the unspoken unity and harmony among them.

Pitfall: Being too flaky or undependable. Neglecting your business.

Keywords: Dreaminess and imagination.

There you have it. With this rough guideline in hand, you can begin to create your business plan. A complete astrological assessment by a competent astrologer is always your best option. An astrological map will guide you through the time cycles and development of yourself and your new business. Remember, however, that these are mere guidelines and cannot explain every possible business or business arrangement. You can, however, peruse the Moon positions and discern which qualities of expression form the best match between your disposition and the characteristics of your business. It will give you an excellent place to start and then you can proceed from there. Above all, enjoy what you have created.

The 27 Nakshatras, or Moon Mansions, of Vedic Astrology

Ronnie Dreyer

O ne unique feature of vedic astrology—the astrology of India—is the use of the Nakshatras in interpreting natal horoscopes and setting up electional charts. Before the twelve signs of the zodiac were introduced into Vedic culture, Indians divided the stars into the 27 Nakshatras. More commonly known as lunar—or Moon—mansions, the 27 Nakshatras are fixed star groupings which each span 13 degrees 20 minutes of the zodiac.

According to Vedic lore which appeared as early as 1,000 bce in the Atharvaveda, the last of the four books comprising the Holy Vedas, the Moon God needed one lunar month—about twenty-seven or twenty-eight days—to travel through each of his resting places, or mansions. Alternate legends depicted these asterisms as His twenty-seven amorous wives who each shared His bed during one lunar cycle. Just as the names and meanings of the zodiacal signs are derived from the shapes the stars seem to form, so were the 27 lunar mansions named and defined according to their astral patterns. Although lunar mansions have been utilized by Chinese, Egyptian, and certain French astrologers, the usage of the Nakshatras were never fully assimilated into Western astrology. Today, holidays, weddings, birthdays, and other rituals and celebrations are marked off on the Indian lunar calendar according to Nakshatra.

Because the Moon—not the Sun—is the most important luminary in vedic astrology, the Janma Nakshatra, or mansion in which your natal Moon is placed, is said to define your personality and direct your life. To calculate the Janma Nakshatra, first convert your natal Moon from its tropical to sidereal position by subtracting the ayanamsa—the difference between the equinox and 0 degrees of Aries—for your birth year. We can see how this works in the pop star Madonna's chart.

Table 1: Krishnamurti Ayanamsa

Year: Ayanamsa
1900 22°22'
1910 22°31'

1920	22°39'
1930	22°47'
1940	22°54'
1950	23°04'
1960	23°12'
1970	23°20'
1980	23°29'
1990	23°37'
2000	23°46'

According to the tropical zodiac, Madonna's Moon is positioned at 11 degrees Virgo, 32 minutes. If we subtract the ayanamsa for her birth year, 1958, of 23 degrees, 10 minutes, we arrive at a sidereal Moon position of 18 degrees Leo, 22 minutes. Purvaphalguni, then, is Madonna's Janma Nakshatra (see Table 2). According to the description of Purvaphalguni, sexuality rightly plays a large part in Madonna's motivations and success. The dispositor, or ruler of Leo—the sign in which the sidereal Moon is placed—is the Sun, and the ruler of Purvaphalguni is Venus. This means that a combination of Sun (ego and ambition) and Venus energies drive the course of her life.

As with Western astrology, each of your planets and your ascendant are defined by the Nakshatras in which they are located. By converting all your planets and your ascendant to their sidereal positions, you can see how each Nakshatra affects them. It is interesting to note too that many signs which correspond to the Nakshatras have similar definitions. Taurus and Krittika, for instance, are both gluttonous and stubborn.

> **Moon in Aswini (horse's head)**—You love luxury and the company of people. Intelligent and quick-witted, you are adept at whatever you attempt.

> **Moon in Bharani (vulva)**—You have a strong constitution and high vitality. You are even-tempered, productive, dependable, ethical, and honest.

> **Moon in Krittika (razor)**—Ambitious, proud, and intelligent, you want recognition for your achievements. You are sensual and must temper your indulgences.

> **Moon in Rohini (chariot)**—You are honest and stubborn. Attractive, charming, and artistic, you usually get what you want. Your insecurity causes you to be jealous and possessive.

Moon in Mrigsira (deer's head)—You are highly capable and a quick learner. You can also be impressionable, bashful, and much too easily influenced.

Moon in Ardra (teardrop)—Though highly intelligent, diplomatic, and free from emotional excess, you can be insensitive, temperamental, cunning, and aloof.

Moon in Punarvasu (arrows)—You are generous, loyal, sympathetic, and patient. You are often plagued by low vitality, however, and are reclusive and religious.

Moon in Pushya (cow's udder)—Like the milk-cow, you are nurturing, protective, and reliable. Generous and popular, you can be judgmental and controlling.

Moon in Aslesha (serpent)—Self-involved, secretive, and ruthless, you are also ambitious. A complex personality, you have charisma and can be argumentative.

Moon in Magha (palatial room)—Magnanimous and charming, you also love to be in charge. You can have a tendency to become too dictatorial and controlling.

Moon in Purvaphalguni (bed or sofa)—Sensual, attractive, and passionate, you love luxury. You are also fit, sexual, and can be impulsive and abrasive.

Moon in Uttraphalguni (bed)—Intelligent, compassionate, and amiable, you live a full life. You are ambitious, capable, and will live a comfortable life.

Moon in Hasta (palm of your hand)—You are clever, determined, arrogant, and unreliable. You make a good counselor but must deal with your own addictions.

Moon in Chittra (jewel)—Elegant, attractive, and artistic, you want recognition for your achievements. You are also studious, introspective, and seldom reveal your innermost thoughts and feelings.

Moon in Swati (coral)—Your cold and reserved persona conceals a sensitive and spiritual side. You are also compassionate and have a good head for business.

Moon in Vishakha (potter's wheel)—Your outgoing personality and ability to focus make you successful, but you are also jealous and difficult to work with.

Moon in Anuradha (lotus)—You are organized, a good manager, persevering, and loyal. At some point in your life, you may find it difficult to settle down.

Moon in Jyestha (talisman)—Though outwardly cheerful, you actually consider yourself a loner. At times you are moody, but your attention to ethics earns others' respect.

Moon in Moola (lion's tail)—You are intelligent, good-natured, and dynamic, but at times terribly opinionated and dissatisfied with your earnings.

Moon in Purvashadya (fan)—Resourceful and magnanimous, you are a devoted partner and friend. Sometimes you are self-centered and excessive.

Moon in Uttrashadya (elephant's tusk)—You are reliable, well-liked, and helpful, but your altruism can be overshadowed by your introspective nature.

Moon in Shravana (ear)—An excellent teacher, you are passionate about learning, traveling, and meeting new people. You are also extremely charitable.

Moon in Dhanishta (drum)—Skilled in music and the arts, you are creative and optimistic. You are also ambitious enough to rise to the top of your field.

Moon in Shatbisha (circle)—Though ethical, persistent, and good-natured, you can be impulsive, quick to judge, and argumentative.

Moon in Purvaphadrapada (cot's front legs)—You are often moody and hard on yourself and others. On the other hand, you are very passionate and eloquent.

Moon in Uttraphadrapada (cot's back legs)—You are magnanimous, persevering, and well-spoken, and you often succeed by sacrificing your own needs for other's.

Moon in Revati (fish)—Attractive and athletic, you are responsible and love doing things for others. You like comfort and will work hard for it.

The planet ruling your Janma Nakshatra determines the planetary period, or Mahadasa (Sanskrit for "great age"), in which you were born; the remaining planetary periods then follow in sequential order. Because Venus rules Madonna's Janma Nakshatra, the Mahadasa of her birth was Venus, followed by the periods of Sun, Moon, and Mars. She is presently in a North Node period.

Muhurta, or Electional Charts

The Janma Nakshatra can also be utilized for casting electional, or event, charts. Electional astrology—known in Sanskrit as "Muhurta"— is the art of choosing the most favorable date and time to initiate activities. To find the appropriate date for an event, people commonly consult the Panchang, or Indian almanac, which lists the precise time and date the Moon enters and leaves each of its mansions. The following Nakshatras are recommended for specific events.

Arrange your first date or sexual encounter for whenever the Moon transits Rohini, Mrigsira, Magha (except the first quarter), Uttraphalguni, Hasta, Svati, Anuradha, Mula (except the first three hours), Uttrashadya, Uttraphadrapada, and Revati (except the last quarter). The most favorable Nakshatras for getting married are Sravana, Rohini, Pushya, Uttraphalguni, Svati, Anuradha, Mula, Uttrashadya, Shatbishta, Uttraphadrapada, and Revati. Aswini, Mrigsira, Punarvasu, Chitra, and Dhanishta are less favorable yet still acceptable. Baby naming, an extremely important ceremony, will be auspicious if done when the Moon is in Aswini, Rohini, Mrigsira, Punarvasu, Pushya, Nagha, Uttraphalguni, Hasta, Svati, Anuradha, Uttrashadya, Shravana, Dhanishtha, Shatbishaa, Uttraphadrapada, and Revati.

Buy residential land when the Moon is in Aswini, Rohini, Mrigsira, Punarvasu, Pushya, Uttraphalguni, Hasta, Svati, Anuradha, Uttrashadya, Shravana, Dhanishtha, Shatbisha and Uttraphadrapada. Purchase land or property when the Moon is in Aswini, Rohini, Mrigsira, Punarvasu, Pushya, Uttrashadya, Shravana, Shatbisha, and Uttraphadrapada. A purchase of residential or professional space should close when the Moon passes through Rohini, Mrigsira, Uttraphalguni, Uttrashadya, and

Uttraphadrapada. Remodelling and renovating should not take place in Krittika, Pushya, Magha, Purvaphalguni, Hasta, Mula, or Revati.

Any long-term activity—starting a business, getting married, buying a house, purchasing a car—should occur when the Moon occupies Rohini, Uttraphalguni, Uttrashadya, or Uttraphadrapada. Plan your business trips when the Moon travels through Aswini, Mrigsira, Pushya, Punarvasu, Hasta, Anuradha, Mula, Shravana, Dhanishtha, or Revati. Don't begin any professional or pleasure trips when the Moon is in Bharani, Krittika, Ardra, Aslesha, Vishakha, Purvashadya, or Purvaphadrapada.

File lawsuits when the Moon is in Aswini, Rohini, Mrigsira, Pushya, Uttraphalguni, Hasta, Chitra, Anuradha, Dhanishtha, or Revati. You may consult an attorney at any time, but the actual court filing and serving of papers should only take place in these particular Nakshatras.

Do something enjoyable and creative when the Moon transits Aswini, Mrigsira, Pushya, Hasta, Chitra, Anuradha, and Revati. Such activities can include going to the movies, walking on the beach, taking a vacation, or making love. Buy a new vehicle or start planting a garden during Punarvasu, Svati, Shravana, Dhanishtha, or Shatbisha.

The most effective days for engaging in spiritual activities occur as the Moon travels through Ardra, Aslesha, Jyeshtha, and Mula. These include embarking on a spiritual journey, starting religious instruction, performing rituals and other ceremonies. Meanwhile, it is best to conduct routine or mundane activities when the Moon is in Krittika or Vishakha.

When the Moon is in Pushya, you can go wrong in any activity except marriage. By the same token, it is not recommended that you begin any new projects when the Moon is in Bharani, a Nakshatra which is considered quite unfavorable. Even though there are other factors involved in preparing an event chart, starting with the most appropriate Nakshatra will lead you in the right direction.

For more information on vedic astrology, consult these recent books on the subject:

DeFouw, Hart. *Light on Life: An Introduction to the Astrology of India.* London: Arkana Books, 1996.

Dreyer, Ronnie Gale. *Vedic Astrology: A Guide to the Fundamentals of Jyotish.* York Beach: Samuel Weiser Inc., 1997.

Harness, Dennis, Ph.D., *The Nakshatras: The Lunar Mansions of Vedic Astrology.* Twin Lakes: Lotus Press, 1999.

2000 Moon Mansion Dates

Mansion	Degrees	Ruler	Date Moon Enters Nakshatra
Aswini	0 Aries 00–13 Aries 20	South Node	Jan 14, Feb 10, Mar 8, Apr 5, May, May 29, Jun 26, Jul 23, Aug 19, Sep 16, Oct 13, Nov 9, Dec 7
Bharani	13 Aries 20–26 Aries 40	Venus	Jan 15, Feb 11, Mar 9, Apr 6, May 3, May 30, Jun 27, Jul 24, Aug 20, Sep 17, Oct 14, Nov 10, Dec 8
Krittika	26 Aries 40–10 Taurus 00	Sun	Jan 16, Feb 12, Mar 10, Apr 7, May 4, May 31, Jun 28, Jul 25, Aug 21, Sep 18, Oct 15, Nov 11, Dec 9
Rohini	10 Taurus 00–23 Taurus 20	Moon	Jan 17, Feb 13, Mar 11, Apr 7, May 5, Jun 1, Jun 29, Jul 26, Aug 22, Sep 19, Oct 16, Nov 12, Dec 10
Mrigsira	23 Taurus 20–6 Gemini 40	Mars	Jan 18, Feb 14, Mar 12, Apr 8, May 6, Jun 2, Jun 30, Jul 27, Aug 23, Sep 20, Oct 17, Nov 13, Dec 10
Ardra	6 Gemini 40–20 Gemini 00	North Node	Jan 18, Feb 15, Mar 13, Apr 9, May 7, Jun 3, Jun 30, Jul 28, Aug 24, Sep 20, Oct 18, Nov 14, Dec 11
Punarvasu	0 Gemini 00–3 Cancer 20	Jupiter	Jan 19, Feb 16, Mar 14, Apr 10, May 8, Jun 4, Jul 1, Jul 29, Aug 25 Sep 21 Oct 19, Nov 15, Dec 12
Pushya	3 Cancer 20–16 Cancer 40	Saturn	Jan 20, Feb 17, Mar 15, Apr 11, May 8, Jun 5, Jul 2, Jul 30, Aug 26, Sep 22, Oct 20, Nov 16, Dec 13
Aslesha	16 Cancer 40–0 Leo 00	Mercury	Jan 21, Feb 18, Mar 16, Apr 12, May 9, Jun 6, Jul 3, Jul 31, Aug 27, Sep 23, Oct 21, Nov 17, Dec 14
Magha	0 Leo 00–13 Leo 20	South Node	Jan 22, Feb 19, Mar 17, Apr 13, May 10, Jun 7, Jul 4, Jul 31, Aug 28, Sep 24, Oct 22, Nov 18, Dec 15
Purvaphalguni	13 Leo 20–26 Leo 40	Venus	Jan 23, Feb 19, Mar 18, Apr 14, May 11, Jun 8, Jul 5, Aug 1, Aug 29, Sep 25, Oct 22, Nov 19, Dec 16
Uttraphalguni	26 Leo 40–10 Virgo 00	Sun	Jan 24, Feb 20, Mar 19, Apr 15, May 12, Jun 9, Jul 6, Aug 2, Aug 30, Sep 26, Oct 23, Nov 20, Dec 17
Hasta	10 Virgo 00–23 Virgo 20	Moon	Jan 25, Feb 21, Mar 20, Apr 16, May 13, Jun 10, Jul 7, Aug 3, Aug 3 1, Sep 27, Oct 24, Nov 21, Dec 18
Chitra	23 Virgo 20–6 Libra 40	Mars	Jan 26, Feb 22, Mar 21, Apr 17, May 14, Jun 11, Jul 8, Aug 4, Sep 1, Sep 28, Oct 25, Nov 22, Dec 18

Mansion	Degrees	Ruler	Date Moon Enters Nakshatra
Svati	6 Libra 40–20 Libra 00	North Node	Jan 27, Feb 23, Mar 22, Apr 18, May 15, Jun 12, Jul 9, Aug 5, Sep 2, Sep 29, Oct 26, Nov 23, Dec 20
Vishakha	20 Libra 00–3 Scorpio 20	Jupiter	Jan 1, Jan 28, Feb 24, Mar 23, Apr 19, May 16, Jun 13, Jul 10, Aug 6, Sep 3, Sep 30, Oct 27, Nov 24, Dec 21
Anuradha	3 Scorpio 20–16 Scorpio 40	Saturn	Jan 2, Jan 29, Feb 26, Mar 24, Apr 20, May 18, Jun 14, Jul 11, Aug 7, Sep 4, Oct 1, Oct 28, Nov 25, Dec 22
Jyeshtha	16 Scorpio 40–0 Sagittarius 00	Mercury	Jan 3, Jan 30, Feb 27, Mar 25, Apr 21, May 19, Jun 15, Jul 12, Aug 8, Sep 5, Oct 2, Oct 29, Nov 26, Dec 23
Mula	0 Sagittarius 00–13 Sagittarius 20	South Node	Jan 4, Jan 31, Feb 28, Mar 26, Apr 22, May 20, Jun 16, Jul 13, Aug 10, Sep 6, Oct 3, Oct 31, Nov 27, Dec 24
Purvashadya	13 Sagittarius 20–26 Sagittarius 40	Venus	Jan 5, Feb 2, Feb 29, Mar 27, Apr 24, May 21, Jun 17, Jul 14, Aug 11, Sep 7, Oct 4, Nov 1, Nov 28, Dec 25
Uttrashadya	26 Sagittarius 40–10 Capricorn 00	Sun	Jan 6, Feb 3, Mar 1, Mar 28, Apr 25, May 22, Jun 18, Jul 16, Aug 12, Sep 8, Oct 5, Nov 2, Nov 29, Dec 26
Shravana	10 Capricorn 00–23 Capricorn 20	Moon	Jan 8, Feb 4, Mar 2, Mar 30, Apr 26, May 23, Jun 19, Jul 17, Aug 13, Sep 9, Oct 7, Nov 3, Nov 30, Dec 28
Dhanishtha	23 Capricorn 20–6 Aquarius 40	Mars	Jan 9, Feb 5, Mar 3, Mar 31, Apr 27, May 24, Jun 21, Jul 18, Aug 14, Sep 10, Oct 8, Nov 4, Dec 1, Dec 29
Shatbisha	6 Aquarius 40–20 Aquarius 00	North Node	Jan 10, Feb 6, Mar 4, Apr 1, Apr 28, May 25, Jun 22, Jul 19, Aug 15, Sep 11, Oct 9, Nov 5, Dec 2, Dec 30
Purvaphadrapada	20 Aquarius 00–3 Pisces 20	Jupiter	Jan 11, Feb 7, Mar 5, Apr 2, Apr 29, May 26, Jun 23, Jul 20, Aug 16, Sep 13, Oct 10, Nov 6, Dec 4, Dec 31
Uttraphadrapada	3 Pisces 20–16 Pisces 40	Saturn	Jan 12, Feb 8, Mar 6, Apr 3, Apr 30, May 27, Jun 24, Jul 21, Aug 17, Sep 14, Oct 11, Nov 7, Dec 5
Revati	16 Pisces 40–0 Aries 00	Mercury	Jan 13, Feb 9, Mar 7, Apr 4, May 1, May 28, Jun 25, Jul 22, Aug 18, Sep 15, Oct 12, Nov 8, Dec 6

Lunar Nakshatra Entry Times

January		February	
Svati	Dec. 31, 5:25:46 am	Purvashadya	Feb. 2, 2:32:00 am
Visakha	Jan. 1, 7:51:24 am	Uttrashadya	Feb. 3, 5:24:08 am
Anuradha	Jan. 2, 10:38:33 am	Shravana	Feb. 4, 7:57:17 am
Jyeshta	Jan. 3, 1:38:21 pm	Dhanishta	Feb. 5, 10:06:43 am
Mula	Jan. 4, 4:42:50 pm	Satabisha	Feb. 6, 11:49:57 am
Purvashadya	Jan. 5, 7:45:17 pm	Purvaphadrapada	Feb. 7, 1:06:08 am
Uttrashadya	Jan. 6, 10:40:01 pm	Uttraphadrapada	Feb. 8, 1:55:23 pm
Shravana	Jan. 8, 1:21:55 am	Revati	Feb. 9, 2:18:18 pm
Dhanishta	Jan. 9, 3:46:06 am	Ashwini	Feb. 10, 2:15:38 pm
Satabisha	Jan. 10, 5:47:44 am	Bharani	Feb. 11, 1:48:26 pm
Purvaphadrapada	Jan. 11, 7:22:00 am	Krittika	Feb. 12, 12:57:45 pm
Uttraphadrapada	Jan. 12, 8:24:25 am	Rohini	Feb. 13, 11:45:22 am
Revati	Jan. 13, 8:51:24 am	Mrgasira	Feb. 14, 10:13:47 am
Ashwini	Jan. 14, 8:40:46 am	Ardra	Feb. 15, 8:26:26 am
Bharani	Jan. 15, 7:52:41 am	Punarvasu	Feb. 16, 6:27:51 am
Krittika	Jan. 16, 6:29:11 am	Pushya	Feb. 17, 4:23:42 am
Rohini	Jan. 17, 4:34:50 am	Ashlesha	Feb. 18, 2:20:44 am
Mrgasira	Jan. 18, 2:16:01 am	Magha	Feb. 19, 12:26:36 am
Ardra	Jan. 18, 11:40:42 pm	Purvaphalguni	Feb. 19, 10:49:32 pm
Punarvasu	Jan. 19, 8:57:50 pm	Uttraphalguni	Feb. 20, 9:37:51 pm
Pushya	Jan. 20, 6:17:06 pm	Hasta	Feb. 21, 8:59:22 pm
Ashlesha	Jan. 21, 3:48:24 pm	Chitra	Feb. 22, 9:00:26 pm
Magha	Jan. 22, 1:41:31 pm	Svati	Feb. 23, 9:45:02 pm
Purvaphalguni	Jan. 23, 12:05:32 pm	Visakha	Feb. 24, 11:13:40 pm
Uttraphalguni	Jan. 24, 11:08:13 am	Anuradha	Feb. 26, 1:22:29 am
Hasta	Jan. 25, 10:55:12 am	Jyeshta	Feb. 27, 4:03:09 am
Chitra	Jan. 26, 11:29:13 am	Mula	Feb. 28, 7:01:44 am
Svati	Jan. 27, 12:49:27 pm	Purvashadya	Feb. 29, 10:08:59 am
Visakha	Jan. 28, 2:51:06 pm		
Anuradha	Jan. 29, 5:25:41 pm		
Jyeshta	Jan. 30, 8:22:04 pm		
Mula	Jan. 31, 11:28:07 pm		

March		April	
Uttrashadya	Mar. 1, 1:05:23 pm	Satabisha	Apr. 1, 4:00:06 am
Shravana	Mar. 2, 3:40:39 pm	Purvaphadrapada	Apr. 2, 4:56:02 am
Dhanishta	Mar. 3, 5:46:17 pm	Uttraphadrapada	Apr. 3, 5:09:35 am
Satabisha	Mar. 4, 7:17:45 pm	Revati	Apr. 4, 4:44:59 am
Purvaphadrapada	Mar. 5, 8:14:19 pm	Ashwini	Apr. 5, 3:48:56 am
Uttraphadrapada	Mar. 6, 8:38:13 pm	Bharani	Apr. 6, 2:29:46 am
Revati	Mar. 7, 8:33:39 pm	Krittika	Apr. 7, 12:55:46 am
Ashwini	Mar. 8, 8:05:40 pm	Rohini	Apr. 7, 11:14:53 pm
Bharani	Mar. 9, 7:19:36 pm	Mrgasira	Apr. 8, 9:33:57 pm
Krittika	Mar. 10, 6:19:56 pm	Ardra	Apr. 9, 7:58:16 pm
Rohini	Mar. 11, 5:10:33 pm	Punarvasu	Apr. 10, 6:31:35 pm
Mrgasira	Mar. 12, 3:54:26 pm	Pushya	Apr. 11, 5:16:07 pm
Ardra	Mar. 13, 2:33:50 pm	Ashlesha	Apr. 12, 4:13:01 pm
Punarvasu	Mar. 14, 1:10:36 pm	Magha	Apr. 13, 3:22:56 pm
Pushya	Mar. 15, 11:46:39 am	Purvaphalguni	Apr. 14, 2:46:37 pm
Ashlesha	Mar. 16, 10:24:23 am	Uttraphalguni	Apr. 15, 2:25:32 pm
Magha	Mar. 17, 9:07:09 am	Hasta	Apr. 16, 2:22:09 pm
Purvaphalguni	Mar. 18, 7:59:30 am	Chitra	Apr. 17, 2:39:54 pm
Uttraphalguni	Mar. 19, 7:07:02 am	Svati	Apr. 18, 3:22:40 pm
Hasta	Mar. 20, 6:36:08 am	Visakha	Apr. 19, 4:34:07 pm
Chitra	Mar. 21, 6:33:19 am	Anuradha	Apr. 20, 6:16:33 pm
Svati	Mar. 22, 7:04:18 am	Jyeshta	Apr. 21, 8:29:51 pm
Visakha	Mar. 23, 8:12:57 am	Mula	Apr. 22, 11:10:19 pm
Anuradha	Mar. 24, 10:00:01 am	Purvashadya	Apr. 24, 2:10:00 am
Jyeshta	Mar. 25, 12:22:09 am	Uttrashadya	Apr. 25, 5:16:53 am
Mula	Mar. 26, 3:11:12 am	Shravana	Apr. 26, 8:16:17 am
Purvashadya	Mar. 27, 6:14:44 pm	Dhanishta	Apr. 27, 10:53:07 am
Uttrashadya	Mar. 28, 9:17:30 pm	Satabisha	Apr. 28, 12:54:27 pm
Shravana	Mar. 30, 12:04:04 am	Purvaphadrapada	Apr. 29, 2:11:24 pm
Dhanishta	Mar. 31, 2:21:24 am	Uttraphadrapada	Apr. 30, 2:40:09 pm

May	
Revati	May 1, 2:21:26 pm
Ashwini	May 2, 1:20:20 pm
Bharani	May 3, 11:44:58 am
Krittika	May 4, 9:44:43 am
Rohini	May 5, 7:29:49 am
Mrgasira	May 6, 5:10:13 am
Ardra	May 7, 2:54:59 am
Punarvasu	May 8, 12:51:49 am
Pushya	May 8, 11:06:43 pm
Ashlesha	May 9, 9:43:46 pm
Magha	May 10, 8:45:15 pm
Purvaphalguni	May 11, 8:11:56 pm
Uttraphalguni	May 12, 8:03:38 pm
Hasta	May 13, 8:19:50 pm
Chitra	May 14, 9:00:06 pm
Svati	May 15, 10:04:29 pm
Visakha	May 16, 11:33:18 pm
Anuradha	May 18, 1:26:54 am
Jyeshta	May 19, 3:44:52 am
Mula	May 20, 6:25:08 am
Purvashadya	May 21, 9:23:03 am
Uttrashadya	May 22, 12:30:53 pm
Shravana	May 23, 3:37:48 pm
Dhanishta	May 24, 6:30:55 pm
Satabisha	May 25, 8:57:05 pm
Purvaphadrapada	May 26, 10:44:53 pm
Uttraphadrapada	May 27, 11:46:22 pm
Revati	May 28, 11:57:58 pm
Ashwini	May 29, 11:20:35 pm
Bharani	May 30, 9:59:15 pm
Krittika	May 31, 8:01:39 pm

June	
Rohini	Jun. 1, 5:37:12 pm
Mrgasira	Jun. 2, 2:56:30 pm
Ardra	Jun. 3, 12:10:09 pm
Punarvasu	Jun. 4, 9:28:24 am
Pushya	Jun. 5, 7:00:34 am
Ashlesha	Jun. 6, 4:54:44 am
Magha	Jun. 7, 3:17:16 am
Purvaphalguni	Jun. 8, 2:12:38 am
Uttraphalguni	Jun. 9, 1:43:14 am
Hasta	Jun. 10, 1:49:33 am
Chitra	Jun. 11, 2:30:26 am
Svati	Jun. 12, 3:43:29 am
Visakha	Jun. 13, 5:25:32 am
Anuradha	Jun. 14, 7:33:02 am
Jyeshta	Jun. 15, 10:02:17 am
Mula	Jun. 16, 12:49:16 pm
Purvashadya	Jun. 17, 3:49:21 pm
Uttrashadya	Jun. 18, 6:56:42 pm
Shravana	Jun. 19, 10:04:00 pm
Dhanishta	Jun. 21, 1:02:23 am
Satabisha	Jun. 22, 3:42:00 am
Purvaphadrapada	Jun. 23, 5:52:56 am
Uttraphadrapada	Jun. 24, 7:26:25 am
Revati	Jun. 25, 8:16:09 am
Ashwini	Jun. 26, 8:18:59 am
Bharani	Jun. 27, 7:35:34 am
Krittika	Jun. 28, 6:09:31 am
Rohini	Jun. 29, 4:07:20 am
Mrgasira	Jun. 30, 1:37:32 am
Ardra	Jun. 30, 10:49:29 pm

July	
Punarvasu	Jul. 1, 7:53:32 pm
Pushya	Jul. 2, 5:00:00 pm
Ashlesha	Jul. 3, 2:18:59 pm
Magha	Jul. 4, 11:59:51 am
Purvaphalguni	Jul. 5, 10:10:56 am
Uttraphalguni	Jul. 6, 8:58:52 am
Hasta	Jul. 7, 8:28:13 am
Chitra	Jul. 8, 8:41:02 am
Svati	Jul. 9, 9:36:34 am
Visakha	Jul. 10, 11:11:26 am
Anuradha	Jul. 11, 1:19:56 pm
Jyeshta	Jul. 12, 3:54:51 pm
Mula	Jul. 13, 6:48:17 pm
Purvashadya	Jul. 14, 9:52:24 pm
Uttrashadya	Jul. 16, 12:59:48 am
Shravana	Jul. 17, 4:03:35 am
Dhanishta	Jul. 18, 6:57:13 am
Satabisha	Jul. 19, 9:34:30 am
Purvaphadrapada	Jul. 20, 11:49:21 am
Uttraphadrapada	Jul. 21, 1:36:04 pm
Revati	Jul. 22, 2:49:33 pm
Ashwini	Jul. 23, 3:25:51 pm
Bharani	Jul. 24, 3:22:59 pm
Krittika	Jul. 25, 2:40:44 pm
Rohini	Jul. 26, 1:21:22 am
Mrgasira	Jul. 27, 11:29:15 am
Ardra	Jul. 28, 9:10:38 am
Punarvasu	Jul. 29, 6:33:15 am
Pushya	Jul. 30, 3:45:58 am
Ashlesha	Jul. 31, 12:58:29 am
Magha	Jul. 31, 10:20:27 pm

August	
Purvaphalguni	Aug. 1, 8:02:17 pm
Uttraphalguni	Aug. 2, 6:13:02 pm
Hasta	Aug. 3, 5:01:05 pm
Chitra	Aug. 4, 4:32:46 pm
Svati	Aug. 5, 4:51:38 pm
Visakha	Aug. 6, 5:57:44 pm
Anuradha	Aug. 7, 7:47:18 pm
Jyeshta	Aug. 8, 10:12:50 pm
Mula	Aug. 10, 1:04:06 am
Purvashadya	Aug. 11, 4:09:30 am
Uttrashadya	Aug. 12, 7:17:39 am
Shravana	Aug. 13, 10:18:45 am
Dhanishta	Aug. 14, 1:05:08 pm
Satabisha	Aug. 15, 3:31:28 pm
Purvaphadrapada	Aug. 16, 5:34:23 pm
Uttraphadrapada	Aug. 17, 7:11:53 pm
Revati	Aug. 18, 8:22:45 pm
Ashwini	Aug. 19, 9:06:04 pm
Bharani	Aug. 20, 9:21:18 pm
Krittika	Aug. 21, 9:07:52 pm
Rohini	Aug. 22, 8:25:59 pm
Mrgasira	Aug. 23, 7:16:38 pm
Ardra	Aug. 24, 5:42:03 pm
Punarvasu	Aug. 25, 3:45:50 pm
Pushya	Aug. 26, 1:33:09 pm
Ashlesha	Aug. 27, 11:10:42 am
Magha	Aug. 28, 8:46:28 am
Purvaphalguni	Aug. 29, 6:29:26 am
Uttraphalguni	Aug. 30, 4:12:24 am
Hasta	Aug. 31, 2:55:32 am

September		October	
Chitra	Sep. 1, 1:57:09 am	Anuradha	Oct. 1, 12:18:36 pm
Svati	Sep. 2, 1:41:19 am	Jyeshta	Oct. 2, 1:56:01 pm
Visakha	Sep. 3, 2:12:31 am	Mula	Oct. 3, 4:13:10 pm
Anuradha	Sep. 4, 3:31:17 am	Purvashadya	Oct. 4, 7:00:46 pm
Jyeshta	Sep. 5, 5:33:37 am	Uttrashadya	Oct. 5, 10:05:16 pm
Mula	Sep. 6, 8:10:52 am	Shravana	Oct. 7, 1:10:52 am
Purvashadya	Sep. 7, 11:10:37 am	Dhanishta	Oct. 8, 4:02:05 am
Uttrashadya	Sep. 8, 2:18:31 pm	Satabisha	Oct. 9, 6:26:22 am
Shravana	Sep. 9, 5:20:31 pm	Purvaphadrapada	Oct. 10, 8:15:40 am
Dhanishta	Sep. 10, 8:05:03 pm	Uttraphadrapada	Oct. 11, 9:26:45 am
Satabisha	Sep. 11, 10:24:07 pm	Revati	Oct. 12, 10:00:36 am
Purvaphadrapada	Sep. 13, 12:13:37 am	Ashwini	Oct. 13, 10:01:15 am
Uttraphadrapada	Sep. 14, 1:32:45 am	Bharani	Oct. 14, 9:34:48 am
Revati	Sep. 15, 2:23:07 am	Krittika	Oct. 15, 8:47:38 am
Ashwini	Sep. 16, 2:47:32 am	Rohini	Oct. 16, 7:46:09 am
Bharani	Sep. 17, 2:49:27 am	Mrgasira	Oct. 17, 6:35:43 am
Krittika	Sep. 18, 2:31:37 am	Ardra	Oct. 18, 5:20:24 am
Rohini	Sep. 19, 1:56:21 am	Punarvasu	Oct. 19, 4:02:58 am
Mrgasira	Sep. 20, 1:05:12 am	Pushya	Oct. 20, 2:45:05 am
Ardra	Sep. 20, 11:59:12 pm	Ashlesha	Oct. 21, 1:27:52 am
Punarvasu	Sep. 21, 10:39:24 pm	Magha	Oct. 22, 12:12:32 am
Pushya	Sep. 22, 9:07:20 pm	Purvaphalguni	Oct. 22, 11:00:56 pm
Ashlesha	Sep. 23, 7:25:39 pm	Uttraphalguni	Oct. 23, 9:56:01 pm
Magha	Sep. 24, 5:38:21 pm	Hasta	Oct. 24, 9:02:02 pm
Purvaphalguni	Sep. 25, 3:51:03 pm	Chitra	Oct. 25, 8:24:24 pm
Uttraphalguni	Sep. 26, 2:10:45 pm	Svati	Oct. 26, 8:09:21 pm
Hasta	Sep. 27, 12:45:37 pm	Visakha	Oct. 27, 8:23:09 pm
Chitra	Sep. 28, 11:44:17 am	Anuradha	Oct. 28, 9:11:06 pm
Svati	Sep. 29, 11:15:06 am	Jyeshta	Oct. 29, 10:36:23 pm
Visakha	Sep. 30, 11:25:10 am	Mula	Oct. 31, 12:38:55 am

November		December	
Purvashadya	Nov. 1, 3:13:49 am	Dhanishta	Dec. 1, 8:15:59 pm
Uttrashadya	Nov. 2, 6:12:00 am	Satabisha	Dec. 2, 11:14:24 pm
Shravana	Nov. 3, 9:20:05 am	Purvaphadrapada	Dec. 4, 1:47:44 am
Dhanishta	Nov. 4, 12:22:24 pm	Uttraphadrapada	Dec. 5, 3:44:11 am
Satabisha	Nov. 5, 3:03:30 pm	Revati	Dec. 6, 4:55:40 am
Purvaphadrapada	Nov. 6, 5:10:42 pm	Ashwini	Dec. 7, 5:18:40 am
Uttraphadrapada	Nov. 7, 6:35:54 pm	Bharani	Dec. 8, 4:54:22 am
Revati	Nov. 8, 7:16:08 pm	Krittika	Dec. 9, 3:47:16 am
Ashwini	Nov. 9, 7:12:55 pm	Rohini	Dec. 10, 2:04:44 am
Bharani	Nov. 10, 6:31:35 pm	Mrgasira	Dec. 10, 11:55:45 pm
Krittika	Nov. 11, 5:19:23 pm	Ardra	Dec. 11, 9:29:57 pm
Rohini	Nov. 12, 3:44:55 pm	Punarvasu	Dec. 12, 6:57:07 pm
Mrgasira	Nov. 13, 1:56:50 pm	Pushya	Dec. 13, 4:26:32 pm
Ardra	Nov. 14, 12:03:16 pm	Ashlesha	Dec. 14, 2:06:42 pm
Punarvasu	Nov. 15, 10:11:14 am	Magha	Dec. 15, 12:04:58 pm
Pushya	Nov. 16, 8:26:21 am	Purvaphalguni	Dec. 16, 10:27:16 am
Ashlesha	Nov. 17, 6:52:45 am	Uttraphalguni	Dec. 17, 9:17:56 am
Magha	Nov. 18, 5:33:15 am	Hasta	Dec. 18, 8:39:40 am
Purvaphalguni	Nov. 19, 4:29:38 am	Chitra	Dec. 19, 8:33:33 am
Uttraphalguni	Nov. 20, 3:43:09 am	Svati	Dec. 20, 8:59:23 am
Hasta	Nov. 21, 3:14:58 am	Visakha	Dec. 21, 9:55:55 am
Chitra	Nov. 22, 3:06:32 am	Anuradha	Dec. 22, 11:21:15 am
Svati	Nov. 23, 3:19:48 am	Jyeshta	Dec. 23, 1:13:09 pm
Visakha	Nov. 24, 3:57:11 am	Mula	Dec. 24, 3:29:02 pm
Anuradha	Nov. 25, 5:01:17 am	Purvashadya	Dec. 25, 6:05:51 pm
Jyeshta	Nov. 26, 6:34:08 am	Uttrashadya	Dec. 26, 8:59:34 pm
Mula	Nov. 27, 8:36:25 am	Shravana	Dec. 28, 12:04:45 am
Purvashadya	Nov. 28, 11:06:24 am	Dhanishta	Dec. 29, 3:14:15 am
Uttrashadya	Nov. 29, 1:59:10 pm	Satabisha	Dec. 30, 6:19:07 am
Shravana	Nov. 30, 5:06:22 pm	Purvaphadrapada	Dec. 31, 9:09:11 am

This table is computed for Eastern Standard Time and shows the exact time and date the Moon enters each Nakshatra.

FARM,
GARDEN, &
WEATHER

Farm, Garden, & Weather

How to Choose the Best Dates

Animals and Animal Breeding

Animals are easiest to handle when the Moon is in Taurus, Cancer, Libra, or Pisces. Avoid the Full Moon. Buy animals during the first quarter in all signs except Scorpio or Pisces. Castrate animals in Gemini, Cancer, Capricorn, or Aquarius. Slaughter for food in the first three days after the Full Moon in any sign except Leo.

To encourage healthy births, eggs should be set and animals mated so births occur when the Moon is increasing in Taurus, Cancer, Pisces, or Libra. Those born during a semifruitful sign (Taurus and Capricorn) will produce leaner meat. Libra yields beautiful animals for showing and racing. To determine the best date to mate animals or set eggs, subtract the number of days given for incubation or gestation from the fruitful dates given in the following tables. For example, cats and dogs are mated sixty-three days previous to the desired birth date. See tables on pages 317–318.

Garden Activities

Cultivating

Cultivate when the Moon is in a barren sign and waning, ideally the fourth quarter in Aries, Gemini, Leo, Virgo, or Aquarius. Third quarter in the sign of Sagittarius will also work.

Cutting Timber

Cut timber during the waning Moon in an air, earth, or fire sign.

Fertilizing and Composting

Fertilize when the Moon is in a fruitful sign (Cancer, Scorpio, Pisces). Organic fertilizers are best when the Moon is waning, chemical fertilizers when the Moon is waxing. Start compost when the Moon is in the fourth quarter in a water sign.

Grafting

Graft during first or second quarter Capricorn, Cancer, or Scorpio.

Harvesting and Drying Crops

Harvest root crops when the Moon is in a dry sign (Aries, Leo, Sagittarius, Gemini, or Aquarius) and waning. Harvest grain for storage just after Full Moon, avoiding water signs (Cancer, Scorpio, Pisces). Harvest in the third and fourth quarters in dry signs. Dry in the third quarter in fire signs.

Irrigation

Irrigate when the Moon is in a water sign.

Lawn Mowing

Mow in the first and second quarters to increase growth and lushness, and in the third and fourth quarters to decrease growth.

Picking Mushrooms

Gather mushrooms at the Full Moon.

Planting

For complete instructions on planting by the Moon, see Gardening by the Moon on page 295, A Guide to Planting Using Sign and Phase Rulerships on page 301, and Companion Planting on page 319.

Pruning

Prune during the third and fourth quarters in Scorpio to retard growth and to promote better fruit, and in Capricorn to promote better healing.

Spraying and Weeding

Destroy pests and weeds during the fourth quarter when the Moon is in Aries, Gemini, Leo, Virgo, Sagittarius, or Aquarius. Weed during a waning Moon in a barren sign. For the best days to kill weeds and pests, see pages 315 and 316.

Transplanting

Transplant when the Moon is increasing and preferably in Cancer, Scorpio, or Pisces.

Weather

For complete weather forecasts for your zone for this year, see page 322.

Gardening by the Moon

Today, people often reject the notion of Moon gardening. The usual non believer is not a scientist, but the city dweller who has never had any real contact with nature and no experience of natural rhythms.

Camille Flammarian, the French astronomer, testifies to Moon planting: "Cucumbers increase at Full Moon, as well as radishes, turnips, leeks, lilies, horseradish, and saffron; onions, on the contrary, are much larger and better nourished during the decline and old age of the Moon than at its increase, during its youth and fullness, which is the reason the Egyptians abstained from onions, on account of their antipathy to the Moon. Herbs gathered while the Moon increases are of great efficiency. If the vines are trimmed at night when the Moon is in the sign of the Lion, Sagittarius, the Scorpion, or the Bull, it will save them from field rats, moles, snails, flies, and other animals."

Dr. Clark Timmins is one of the few modern scientists to have conducted tests in Moon planting. Following is a summary of his experiments:

> **Beets:** When sown with the Moon in Scorpio, the germination rate was 71 percent; when sown in Sagittarius, the germination rate was 58 percent.

> **Scotch marigold:** When sown with the Moon in Cancer, the germination rate was 90 percent; when sown in Leo, the rate was 32 percent.

> **Carrots:** When sown with the Moon in Scorpio, the germination rate was 64 percent; when sown in Sagittarius, the germination rate was 47 percent.

> **Tomatoes:** When sown with the Moon in Cancer, the germination rate was 90 percent; when sown in Leo, the germination rate was 58 percent.

Two things should be emphasized. First, remember that this is only a summary of the results of the experiments; the experiments themselves were conducted in a scientific manner to eliminate any variation in soil, temperature, moisture, and so on, so that only the Moon is sign varied. Also, note that these astonishing results were obtained without regard to the phase of the Moon—the other factor we use in Moon planting, and which presumably would have increased the differential in germination rates.

Further experiments by Dr. Timmins involved transplanting Cancer and Leo-planted tomato seedlings while the Moon was increasing and in Cancer. The result was 100 percent survival. When transplanting was done with the Moon decreasing and in Sagittarius, there was 0 percent survival. The results of Dr. Timmins' tests show that the Cancer-planted tomatoes had blossoms twelve days earlier than those planted under Leo; the Cancer-planted tomatoes had an average height of twenty inches at that time compare to fifteen inches for the Leo-planted; the first ripe tomatoes were gathered from the Cancer plantings eleven days ahead of the Leo plantings, and a count of the hanging fruit and its size and weight shows an advantage to the Cancer plants over the Leo plants of 45 percent.

Dr. Timmins also observed that there have been similar tests that did not indicate results favorable to the Moon planting theory. As a scientist, he asked why one set of experiments indicated a positive verification of Moon planting, and others did not. He checked these other tests and found that the experimenters had not followed the geocentric system for determining the Moon sign positions, but the heliocentric. When the times used in these other tests were converted to the geocentric system, the dates chosen often were found to be in barren, rather than fertile, signs. Without going into a technical explanation, it is sufficient to point out that geocentric and heliocentric positions often vary by as much as four days. This is a large enough differential to place the Moon in Cancer, for example, in the heliocentric system, and at the same time in Leo by the geocentric system.

Most almanacs and calendars show the Moon's signs heliocentrically—and thus incorrectly for Moon planting—while the *Moon Sign Book* is calculated correctly for planting purposes, using the geocentric system. Some readers are also confused because the *Moon Sign Book* talks of first, second, third, and fourth quarters, while some almanacs refer to these same divisions as New Moon, first quarter, Full Moon, and last quarter. Thus, the almanacs say first quarter when the *Moon Sign Book* says second quarter. (Refer to "A Note about Almanacs," page 10.)

There is nothing complicated about using astrology in agriculture and horticulture in order to increase both pleasure and profit, but there is one very important rule that is often neglected—use common sense! Of course this is one rule that should be remembered in every activity we undertake, but in the case of gardening and farming by the Moon it is not always possible to use the best dates for planting or harvesting, and we must select the next best and just try to do the best we can.

This brings up the matter of the other factors to consider in your gardening work. The dates we give as best for a certain activity apply to the entire country (with slight time correction), but in your section of the country you may be buried under three feet of snow on a date we say is good to plant your flowers. So we have factors of weather, season, temperature and moisture variations, soil conditions, your own available time and opportunity, and so forth. Some astrologers like to think it is all a matter of science, but gardening is also an art. In art, you develop an instinctive identification with your work and influence it with your feelings and wishes.

The *Moon Sign Book* gives you the place of the Moon for every day of the year so that you can select the best times once you have become familiar with the rules and practices of lunar agriculture. We give you specific, easy-to-follow directions so that you can get right down to work.

We give you the best dates for planting, and also for various related activities, including cultivation, fertilizing, harvesting, irrigation, and getting rid of weeds and pests. But we cannot tell you exactly when it's good to plant. Many of these rules were learned by observation and experience; as the body of experience grew we could see various patterns emerging that allowed us to make judgments about new things. That's what you should do, too. After you have worked with lunar agriculture for a while and have gained a working knowledge, you will probably begin to try new things—and we hope you will share your experiments and findings with us. That's how the science grows.

Here's an example of what we mean. Years ago, Llewellyn George suggested that we try to combine our bits of knowledge about what to expect in planting under each of the Moon signs in order to gain benefit from several lunar factors in one plant. From this came our rule for developing "thoroughbred seed." To develop thoroughbred seed, save the seed for three successive years from plants grown by the correct Moon sign and phase. You can plant in the first quarter phase and in the sign of Cancer for fruitfulness; the second year, plant seeds from the first year plants in Libra for beauty; and in the third year, plant the seeds from the second year plants in Taurus to produce hardiness. In a similar manner you can combine the fruitfulness of Cancer, the good root growth of Pisces, and the sturdiness and good vine growth of Scorpio. And don't forget the characteristics of Capricorn: hardy like Taurus, but drier and perhaps more resistant to drought and disease.

Unlike common almanacs, we consider both the Moon's phase and the Moon's sign in making our calculations for the proper timing of our work. It is perhaps a little easier to understand this if we remind you that we are all living in the center of a vast electromagnetic field that is the Earth and its environment in space. Everything that occurs within this electromagnetic field has an effect on everything else within the field. The Moon and the Sun are the most important of the factors affecting the life of the Earth, and it is their relative positions to the Earth that we project for each day of the year.

Many people claim that not only do they achieve larger crops gardening by the Moon, but that their fruits and vegetables are much tastier. A number of organic gardeners have also become lunar gardeners using the natural rhythm of life forces that we experience through the relative movements of the Sun and Moon. We provide a few basic rules and then give you day-by-day guidance for your gardening work. You will be able to choose the best dates to meet your own needs and opportunities.

Planting by the Moon's Phases

During the increasing or waxing light—from New Moon to Full Moon—plant annuals that produce their yield above the ground. An annual is a plant that completes its entire life cycle within one growing season and has to be seeded each year. During the decreasing or waning light (from Full Moon to New Moon), plant biennials, perennials, and bulb and root plants. Biennials include crops that are planted one season to winter over and produce crops the next, such as winter wheat. Perennials and bulb and root plants include all plants that grow from the same root each year.

A simpler, less accurate rule is to plant crops that produce above the ground during the waxing Moon, and to plant crops that produce below the ground during the waning Moon. Thus the old adage, "Plant potatoes during the dark of the Moon." Llewellyn George's system divided the lunar month into quarters. The first two from New Moon to Full Moon are the first and second quarters, and the last two from Full Moon to New Moon the third and fourth quarters. Using these divisions, we can increase our accuracy in timing our efforts to coincide with natural forces.

First Quarter (Increasing)

Plant annuals producing their yield above the ground, which are generally of the leafy kind that produce their seed outside the fruit. Examples are asparagus, broccoli, Brussels sprouts, cabbage, cauliflower, celery, cress, endive, kohlrabi, lettuce, parsley, spinach, etc. Cucumbers are an exception, as they do best in the first quarter rather than the second, even though the seeds are inside the fruit. Also plant cereals and grains.

Second Quarter (Increasing)

Plant annuals producing their yield above the ground, which are generally of the viney kind that produce their seed inside the fruit. Examples include beans, eggplant, melons, peas, peppers, pumpkins, squash, tomatoes, etc. These are not hard and fast divisions. If you can't plant during the first quarter, plant during the second, and vice versa. There are many plants that seem to do equally well planted in either quarter, such as watermelon, hay, and cereals and grains.

Third Quarter (Decreasing)

Plant biennials, perennials, and bulb and root plants. Also plant trees, shrubs, berries, beets, carrots, onions, parsnips, peanuts, potatoes, radishes, rhubarb, rutabagas, strawberries, turnips, winter wheat, grapes, etc.

Fourth Quarter (Decreasing)

This is the best time to cultivate, turn sod, pull weeds, and destroy pests of all kinds, especially when the Moon is in the barren signs of Aries, Leo, Virgo, Gemini, Aquarius, and Sagittarius.

Planting by Moon Sign

Moon in Aries

Barren and dry, fiery and masculine. Used for destroying noxious growths, weeds, pests, and for cultivating.

Moon in Taurus

Productive and moist, earthy and feminine. Used for planting many crops, particularly potatoes and root crops, and when hardiness is important. Also used for lettuce, cabbage, and similar leafy vegetables.

Moon in Gemini

Barren and dry, airy and masculine. Used for destroying noxious growths, weeds and pests, and for cultivation.

Moon in Cancer

Very fruitful and moist, watery and feminine. This is the most productive sign, used extensively for planting and irrigation.

Moon in Leo

Barren and dry, fiery and masculine. This is the most barren sign, used only for killing weeds and for cultivation.

Moon in Virgo

Barren and moist, earthy and feminine. Good for cultivation and destroying weeds and pests.

Moon in Libra

Semifruitful and moist, airy and masculine. Used for planting many crops and producing good pulp growth and roots. A very good sign for flowers and vines. Also used for seeding hay, corn fodder, and the like.

Moon in Scorpio

Very fruitful and moist, watery and feminine. Nearly as productive as Cancer; used for the same purposes. Especially good for vine growth and sturdiness.

Moon in Sagittarius

Barren and dry, fiery and masculine. Used for planting onions, seeding hay, and for cultivation.

Moon in Capricorn

Productive and dry, earthy and feminine. Used for planting potatoes and other tubers.

Moon in Aquarius

Barren and dry, airy and masculine. Used for cultivation and destroying noxious growths, weeds, and pests.

Moon in Pisces

Very fruitful and moist, watery and feminine. Used along with Cancer and Scorpio, especially good for root growth.

A Guide to Planting

Using Phase & Sign Rulerships

Plant	Phase/Quarter	Sign
Annuals	1st or 2nd	
Apple trees	2nd or 3rd	Cancer, Pisces, Taurus, Virgo
Artichokes	1st	Cancer, Pisces
Asparagus	1st	Cancer, Scorpio, Pisces
Asters	1st or 2nd	Virgo, Libra
Barley	1st or 2nd	Cancer, Pisces, Libra, Capricorn, Virgo
Beans (bush & pole)	2nd	Cancer, Taurus, Pisces, Libra
Beans (kidney, white, & navy)	1st or 2nd	Cancer, Pisces
Beech Trees	2nd or 3rd	Virgo, Taurus
Beets	3rd	Cancer, Capricorn, Pisces, Libra
Biennials	3rd or 4th	
Broccoli	1st	Cancer, Pisces, Libra, Scorpio
Brussels Sprouts	1st	Cancer, Scorpio, Pisces, Libra
Buckwheat	1st or 2nd	Capricorn
Bulbs	3rd	Cancer, Scorpio, Pisces
Bulbs for Seed	2nd or 3rd	
Cabbage	1st	Cancer, Scorpio, Pisces, Libra, Taurus

Plant	Phase/Quarter	Sign
Cactus		Taurus, Capricorn
Canes (raspberries, black-berries, and gooseberries)	2nd	Cancer, Scorpio, Pisces
Cantaloupes	1st or 2nd	Cancer, Scorpio, Pisces, Libra, Taurus
Carrots	3rd	Taurus, Cancer, Scorpio, Pisces, Libra
Cauliflower	1st	Cancer, Scorpio, Pisces, Libra
Celeriac	3rd	Cancer, Scorpio, Pisces
Celery	1st	Cancer, Scorpio, Pisces
Cereals	1st or 2nd	Cancer, Scorpio, Pisces, Libra
Chard	1st or 2nd	Cancer, Scorpio, Pisces
Chicory	2nd or 3rd	Cancer, Scorpio, Pisces
Chrysanthemums	1st or 2nd	Virgo
Clover	1st or 2nd	Cancer, Scorpio, Pisces
Corn	1st	Cancer, Scorpio, Pisces
Corn for Fodder	1st or 2nd	Libra
Coryopsis	2nd or 3rd	Libra
Cosmos	2nd or 3rd	Libra
Cress	1st	Cancer, Scorpio, Pisces
Crocus	1st or 2nd	Virgo
Cucumbers	1st	Cancer, Scorpio, Pisces

Plant	Phase/Quarter	Sign
Daffodils	1st or 2nd	Libra, Virgo
Dahlias	1st or 2nd	Libra, Virgo
Deciduous Trees	2nd or 3rd	Cancer, Scorpio, Pisces, Virgo, Taurus
Eggplant	2nd	Cancer, Scorpio, Pisces, Libra
Endive	1st	Cancer, Scorpio, Pisces, Libra
Flowers	1st	Libra, Cancer, Pisces, Virgo, Scorpio, Taurus
Garlic	3rd	Libra, Taurus, Pisces
Gladiola	1st or 2nd	Libra, Virgo
Gourds	1st or 2nd	Cancer, Scorpio, Pisces, Libra
Grapes	2nd or 3rd	Cancer, Scorpio, Pisces, Virgo
Hay	1st or 2nd	Cancer, Scorpio, Pisces, Libra, Taurus
Herbs	1st or 2nd	Cancer, Scorpio, Pisces
Honeysuckle	1st or 2nd	Scorpio, Virgo
Hops	1st or 2nd	Scorpio, Libra
Horseradish	1st or 2nd	Cancer, Scorpio, Pisces
Houseplants	1st	Libra, Cancer, Scorpio, Pisces
Hyacinths	3rd	Cancer, Scorpio, Pisces
Irises	1st or 2nd	Cancer, Virgo
Kohlrabi	1st or 2nd	Cancer, Scorpio, Pisces, Libra

Plant	Phase/Quarter	Sign
Leeks	1st or 2nd	Cancer, Pisces
Lettuce	1st	Cancer, Scorpio, Pisces, Libra, Taurus
Lilies	1st or 2nd	Cancer, Scorpio, Pisces
Maple Trees	2nd or 3rd	Virgo, Taurus, Cancer, Pisces
Melons	2nd	Cancer, Scorpio, Pisces
Moon Vines	1st or 2nd	Virgo
Morning Glories	1st or 2nd	Cancer, Scorpio, Pisces, Virgo
Oak Trees	2nd or 3rd	Virgo, Taurus, Cancer, Pisces
Oats	1st or 2nd	Cancer, Scorpio, Pisces, Libra
Okra	1st	Cancer, Scorpio, Pisces, Libra
Onion Seeds	2nd	Scorpio, Cancer, Sagittarius
Onion Sets	3rd or 4th	Libra, Taurus, Pisces, Cancer
Pansies	1st or 2nd	Cancer, Scorpio, Pisces
Parsley	1st	Cancer, Scorpio, Pisces, Libra
Parsnips	3rd	Taurus, Capricorn, Cancer, Scorpio, Capricorn
Peach Trees	2nd or 3rd	Taurus, Libra, Virgo, Cancer
Peanuts	3rd	Cancer, Scorpio, Pisces
Pear Trees	2nd or 3rd	Taurus, Libra, Virgo, Cancer
Peas	2nd	Cancer, Scorpio, Pisces, Libra

Plant	Phase/Quarter	Sign
Peonies	1st or 2nd	Virgo
Peppers	2nd	Cancer, Pisces, Scorpio
Perennials	3rd	
Petunias	1st or 2nd	Libra, Virgo
Plum Trees	2nd or 3rd	Taurus, Virgo, Cancer, Pisces
Poppies	1st or 2nd	Virgo
Portulaca	1st or 2nd	Virgo
Potatoes	3rd	Cancer, Scorpio, Taurus, Libra, Capricorn
Privet	1st or 2nd	Taurus, Libra
Pumpkins	2nd	Cancer, Scorpio, Pisces, Libra
Quinces	1st or 2nd	Capricorn
Radishes	3rd	Cancer, Libra, Taurus, Pisces, Capricorn
Rhubarb	3rd	Cancer, Pisces
Rice	1st or 2nd	Scorpio
Roses	1st or 2nd	Cancer, Virgo
Rutabagas	3rd	Cancer, Scorpio, Pisces, Taurus
Saffron	1st or 2nd	Cancer, Scorpio, Pisces
Sage	3rd	Cancer, Scorpio, Pisces
Salsify	1st or 2nd	Cancer, Scorpio, Pisces

Plant	Phase/Quarter	Sign
Shallots	2nd	Scorpio
Spinach	1st	Cancer, Scorpio, Pisces
Squash	2nd	Cancer, Scorpio, Pisces, Libra
Strawberries	3rd	Cancer, Scorpio, Pisces
String Beans	1st or 2nd	Taurus
Sunflowers	1st or 2nd	Libra, Cancer
Sweet Peas	1st or 2nd	Cancer, Scorpio, Pisces
Tomatoes	2nd	Cancer, Scorpio, Pisces, Capricorn
Shade Trees	3rd	Taurus, Capricorn
Ornamental Trees	2nd	Libra, Taurus
Trumpet Vines	1st or 2nd	Cancer, Scorpio, Pisces
Tubers for Seed	3rd	Cancer, Scorpio, Pisces, Libra
Tulips	1st or 2nd	Libra, Virgo
Turnips	3rd	Cancer, Scorpio, Pisces, Taurus, Capricorn, Libra
Valerian	1st or 2nd	Virgo, Gemini
Watermelons	1st or 2nd	Cancer, Scorpio, Pisces, Libra
Wheat	1st or 2nd	Cancer, Scorpio, Pisces, Libra

2000 Gardening Dates

Dates	Qtr	Sign	Activity
Jan. 2, 4:32 pm- Jan. 5, 5:24 am	4th	Sagittarius	Cultivate. Destroy weeds and pests. Harvest fruits and root crops for food. Trim to retard growth.
Jan. 5, 5:24 am- Jan. 6, 1:14 pm	4th	Capricorn	Plant potatoes and tubers. Trim to retard growth.
Jan. 6, 1:14 pm- Jan. 7, 5:53 pm	1st	Capricorn	Graft or bud plants. Trim to increase growth.
Jan. 10, 4:59 am- Jan. 12, 1:48 pm	1st	Pisces	Plant grains, leafy annuals. Fertilize (chemical). Graft or bud plants. Irrigate. Trim to increase growth.
Jan. 14, 7:38 pm- Jan. 16, 10:25 pm	2nd	Taurus	Plant annuals for hardiness. Trim to increase growth.
Jan. 18, 11:01 pm- Jan. 20, 10:58 pm	2nd	Cancer	Plant grains, leafy annuals. Fertilize (chemical). Graft or bud plants. Irrigate. Trim to increase growth.
Jan. 20, 11:40 pm- Jan. 23, 12:07 am	3rd	Leo	Cultivate. Destroy weeds and pests. Harvest fruits and root crops for food. Trim to retard growth.
Jan. 23, 12:07 am- Jan. 25, 4:09 am	3rd	Virgo	Cultivate, especially medicinal plants. Destroy weeds and pests. Trim to retard growth.
Jan. 27, 12:01 pm- Jan. 28, 2:57 am	3rd	Scorpio	Plant biennials, perennials, bulbs and roots. Prune. Irrigate. Fertilize (organic).
Jan. 28, 2:57 am- Jan. 29, 11:17 pm	4th	Scorpio	Plant biennials, perennials, bulbs and roots. Prune. Irrigate. Fertilize (organic).
Jan. 29, 11:17 pm- Feb. 1, 12:10 pm	4th	Sagittarius	Cultivate. Destroy weeds and pests. Harvest fruits and root crops for food. Trim to retard growth.
Feb. 1, 12:10 pm- Feb. 4, 12:31 am	4th	Capricorn	Plant potatoes and tubers. Trim to retard growth.
Feb. 4, 12:31 am- Feb. 5, 8:03 am	4th	Aquarius	Cultivate. Destroy weeds and pests. Harvest fruits and root crops for food. Trim to retard growth.
Feb. 6, 11:02 am- Feb. 8, 7:17 pm	1st	Pisces	Plant grains, leafy annuals. Fertilize (chemical). Graft or bud plants. Irrigate. Trim to increase growth.
Feb. 11, 1:21 am- Feb. 12, 6:21 pm	1st	Taurus	Plant annuals for hardiness. Trim to increase growth.
Feb. 12, 6:21 pm- Feb. 13, 5:23 am	2nd	Taurus	Plant annuals for hardiness. Trim to increase growth.
Feb. 15, 7:45 am- Feb. 17, 9:11 pm	2nd	Cancer	Plant grains, leafy annuals. Fertilize (chemical). Graft or bud plants. Irrigate. Trim to increase growth.
Feb. 19, 11:27 am- Feb. 21, 2:21 pm	3rd	Virgo	Cultivate, especially medicinal plants. Destroy weeds and pests. Trim to retard growth.
Feb. 23, 8:58 pm- Feb. 26, 7:10 am	3rd	Scorpio	Plant biennials, perennials, bulbs and roots. Prune. Irrigate. Fertilize (organic).

2000 Gardening Dates

Dates	Qtr	Sign	Activity
Feb. 26, 7:10 am- Feb. 26, 10:53 pm	3rd	Sagittarius	Cultivate. Destroy weeds and pests. Harvest fruits and root crops for food. Trim to retard growth.
Feb. 26, 10:53 pm- Feb. 28, 7:45 pm	4th	Sagittarius	Cultivate. Destroy weeds and pests. Harvest fruits and root crops for food. Trim to retard growth.
Feb. 28, 7:45 pm- Mar. 2, 8:14 am	4th	Capricorn	Plant potatoes and tubers. Trim to retard growth.
Mar. 2, 8:14 am- Mar. 4, 6:30 pm	4th	Aquarius	Cultivate. Destroy weeds and pests. Harvest fruits and root crops for food. Trim to retard growth.
Mar. 4, 6:30 pm- Mar. 6, 12:17 am	4th	Pisces	Plant biennials, perennials, bulbs and roots. Prune. Irrigate. Fertilize (organic).
Mar. 6, 12:17 am- Mar. 7, 1:54 am	1st	Pisces	Plant grains, leafy annuals. Fertilize (chemical). Graft or bud plants. Irrigate. Trim to increase growth.
Mar. 9, 7:01 am- Mar. 11, 10:46 am	1st	Taurus	Plant annuals for hardiness. Trim to increase growth.
Mar. 13, 1:51 pm- Mar. 15, 4:43 pm	2nd	Cancer	Plant grains, leafy annuals. Fertilize (chemical). Graft or bud plants. Irrigate. Trim to increase growth.
Mar. 19, 11:44 pm- Mar. 19, 11:57 pm	3rd	Virgo	Cultivate, especially medicinal plants. Destroy weeds and pests. Trim to retard growth.
Mar. 22, 6:17 am- Mar. 24, 3:43 pm	3rd	Scorpio	Plant biennials, perennials, bulbs and roots. Prune. Irrigate. Fertilize (organic).
Mar. 24, 3:43 pm- Mar. 27, 3:51 am	3rd	Sagittarius	Cultivate. Destroy weeds and pests. Harvest fruits and root crops for food. Trim to retard growth.
Mar. 27, 3:51 am- Mar. 27, 7:21 pm	3rd	Capricorn	Plant potatoes and tubers. Trim to retard growth.
Mar. 27, 7:21 pm- Mar. 29, 4:34 pm	4th	Capricorn	Plant potatoes and tubers. Trim to retard growth.
Mar. 29, 4:34 pm- Apr. 1, 3:12 am	4th	Aquarius	Cultivate. Destroy weeds and pests. Harvest fruits and root crops for food. Trim to retard growth.
Apr. 1, 3:12 am- Apr. 3, 10:22 am	4th	Pisces	Plant biennials, perennials, bulbs and roots. Prune. Irrigate. Fertilize (organic).
Apr. 3, 10:22 am- Apr. 4, 1:12 pm	4th	Aries	Cultivate. Destroy weeds and pests. Harvest fruits and root crops for food. Trim to retard growth.
Apr. 5, 2:29 pm- Apr. 7, 4:58 pm	1st	Taurus	Plant annuals for hardiness. Trim to increase growth.
Apr. 9, 7:16 pm- Apr. 11, 8:30 am	1st	Cancer	Plant grains, leafy annuals. Fertilize (chemical). Graft or bud plants. Irrigate. Trim to increase growth.
Apr. 11, 8:30 am- Apr. 11, 10:16 am	2nd	Cancer	Plant grains, leafy annuals. Fertilize (chemical). Graft or bud plants. Irrigate. Trim to increase growth.

2000 Gardening Dates

Dates	Qtr	Sign	Activity
Apr. 16, 7:36 am- Apr. 18, 12:41 pm	2nd	Libra	Plant annuals for fragrance and beauty. Trim to in-crease growth.
Apr. 18, 2:35 pm- Apr. 20, 11:58 pm	3rd	Scorpio	Plant biennials, perennials, bulbs and roots. Prune. Irrigate. Fertilize (organic).
Apr. 20, 11:58 pm- Apr. 23, 11:47 am	3rd	Sagittarius	Cultivate. Destroy weeds and pests. Harvest fruits and root crops for food. Trim to retard growth.
Apr. 23, 11:47 am- Apr. 26, 12:42 am	3rd	Capricorn	Plant potatoes and tubers. Trim to retard growth.
Apr. 26, 12:42 am- Apr. 26, 2:30 pm	3rd	Aquarius	Cultivate. Destroy weeds and pests. Harvest fruits and root crops for food. Trim to retard growth.
Apr. 26, 2:30 pm- Apr. 28, 12:06 pm	4th	Aquarius	Cultivate. Destroy weeds and pests. Harvest fruits and root crops for food. Trim to retard growth.
Apr. 28, 12:06 pm- Apr. 30, 7:54 pm	4th	Pisces	Plant biennials, perennials, bulbs and roots. Prune. Irrigate. Fertilize (organic).
Apr. 30, 7:54 pm- May 2, 11:54 pm	4th	Aries	Cultivate. Destroy weeds and pests. Harvest fruits and root crops for food. Trim to retard growth.
May 2, 11:54 pm- May 3, 11:12 pm	4th	Taurus	Plant potatoes and tubers. Trim to retard growth.
May 3, 11:12 pm- May 5, 1:23 am	1st	Taurus	Plant annuals for hardiness. Trim to increase growth.
May 7, 2:14 am- May 9, 4:01 am	1st	Cancer	Plant grains, leafy annuals. Fertilize (chemical). Graft or bud plants. Irrigate. Trim to increase growth
May 13, 1:27 pm- May 15, 9:16 pm	2nd	Libra	Plant annuals for fragrance and beauty. Trim to in-crease growth.
May 15, 9:16 pm- May 18, 2:34 am	2nd	Scorpio	Plant grains, leafy annuals. Fertilize (chemical). Graft or bud plants. Irrigate. Trim to increase growth.
May 18, 2:34 am- May 18, 7:09 am	3rd	Scorpio	Plant biennials, perennials, bulbs and roots. Prune. Irrigate. Fertilize (organic).
May 18, 7:09 am- May 20, 7:01 pm	3rd	Sagittarius	Cultivate. Destroy weeds and pests. Harvest fruits and root crops for food. Trim to retard growth.
May 20, 7:01 pm- May 23, 8:00 am	3rd	Capricorn	Plant potatoes and tubers. Trim to retard growth.
May 23, 8:00 am- May 25, 8:07 pm	3rd	Aquarius	Cultivate. Destroy weeds and pests. Harvest fruits and root crops for food. Trim to retard growth.
May 25, 8:07 pm- May 26, 6:55 am	3rd	Pisces	Plant biennials, perennials, bulbs and roots. Prune. Irrigate. Fertilize (organic).
May 26, 6:55 am- May 28, 5:08 am	4th	Pisces	Plant biennials, perennials, bulbs and roots. Prune. Irrigate. Fertilize (organic).

2000 Gardening Dates

Dates	Qtr	Sign	Activity
May 28, 5:08 am– May 30, 10:02 am	4th	Aries	Cultivate. Destroy weeds and pests. Harvest fruits and root crops for food. Trim to retard growth.
May 30, 10:02 am– Jun. 1, 11:34 am	4th	Taurus	Plant potatoes and tubers. Trim to retard growth.
Jun. 1, 11:34 am– Jun. 2, 7:14 am	4th	Gemini	Cultivate. Destroy weeds and pests. Harvest fruits and root crops for food. Trim to retard growth.
Jun. 3, 11:30 am– Jun. 5, 11:45 am	1st	Cancer	Plant grains, leafy annuals. Fertilize (chemical). Graft or bud plants. Irrigate. Trim to increase growth.
Jun. 9, 6:58 pm– Jun. 12, 2:55 am	2nd	Libra	Plant annuals for fragrance and beauty. Trim to increase growth.
Jun. 12, 2:55 am– Jun. 14, 1:18 pm	2nd	Scorpio	Plant grains, leafy annuals. Fertilize (chemical). Graft or bud plants. Irrigate. Trim to increase growth.
Jun. 16, 5:27 pm– Jun. 17, 1:26 am	3rd	Sagittarius	Cultivate. Destroy weeds and pests. Harvest fruits and root crops for food. Trim to retard growth.
Jun. 17, 1:26 am– Jun. 19, 2:26 pm	3rd	Capricorn	Plant potatoes and tubers. Trim to retard growth.
Jun. 19, 2:26 pm– Jun. 22, 2:52 am	3rd	Aquarius	Cultivate. Destroy weeds and pests. Harvest fruits and root crops for food. Trim to retard growth.
Jun. 22, 2:52 am– Jun. 24, 12:55 pm	3rd	Pisces	Plant biennials, perennials, bulbs and roots. Prune. Irrigate. Fertilize (organic).
Jun. 24, 12:55 pm– Jun. 24, 8:00 pm	3rd	Aries	Cultivate. Destroy weeds and pests. Harvest fruits and root crops for food. Trim to retard growth.
Jun. 24, 8:00 pm– Jun. 26, 7:19 pm	4th	Aries	Cultivate. Destroy weeds and pests. Harvest fruits and root crops for food. Trim to retard growth.
Jun. 26, 7:19 pm– Jun. 28, 9:59 pm	4th	Taurus	Plant potatoes and tubers. Trim to retard growth.
Jun. 28, 9:59 pm– Jun. 30, 10:09 pm	4th	Gemini	Cultivate. Destroy weeds and pests. Harvest fruits and root crops for food. Trim to retard growth.
Jun. 30, 10:09 pm– Jul. 1, 2:20 pm	4th	Cancer	Plant biennials, perennials, bulbs and roots. Prune. Irrigate. Fertilize (organic).
Jul. 1, 2:20 pm– Jul. 2, 9:38 pm	1st	Cancer	Plant grains, leafy annuals. Fertilize (chemical). Graft or bud plants. Irrigate. Trim to increase growth.
Jul. 7, 1:47 am– Jul. 8, 7:53 am	1st	Libra	Plant annuals for fragrance and beauty. Trim to increase growth.
Jul. 8, 7:53 am– Jul. 9, 8:48 am	2nd	Libra	Plant annuals for fragrance and beauty. Trim to increase growth.
Jul. 9, 8:48 am– Jul. 11, 7:06 pm	2nd	Scorpio	Plant grains, leafy annuals. Fertilize (chemical). Graft or bud plants. Irrigate. Trim to increase growth.

2000 Gardening Dates

Dates	Qtr	Sign	Activity
Jul. 14, 7:27 am– Jul. 16, 8:55 am	2nd	Capricorn	Graft or bud plants. Trim to increase growth.
Jul. 16, 8:55 am– Jul. 16, 8:27 pm	3rd	Capricorn	Plant potatoes and tubers. Trim to retard growth.
Jul. 16, 8:27 pm– Jul. 19, 8:44 am	3rd	Aquarius	Cultivate. Destroy weeds and pests. Harvest fruits and root crops for food. Trim to retard growth.
Jul. 19, 8:44 am– Jul. 21, 7:09 pm	3rd	Pisces	Plant biennials, perennials, bulbs and roots. Prune. Irrigate. Fertilize (organic).
Jul. 21, 7:09 pm– Jul. 24, 2:44 am	3rd	Aries	Cultivate. Destroy weeds and pests. Harvest fruits and root crops for food. Trim to retard growth.
Jul. 24, 2:44 am– Jul. 24, 6:02 am	3rd	Taurus	Plant potatoes and tubers. Trim to retard growth.
Jul. 24, 6:02 am– Jul. 26, 7:01 am	4th	Taurus	Plant potatoes and tubers. Trim to retard growth.
Jul. 26, 7:01 am– Jul. 28, 8:30 am	4th	Gemini	Cultivate. Destroy weeds and pests. Harvest fruits and root crops for food. Trim to retard growth.
Jul. 28, 8:30 am– Jul. 30, 8:23 am	4th	Cancer	Plant biennials, perennials, bulbs and roots. Prune. Irrigate. Fertilize (organic).
Jul. 30, 8:23 am– Jul. 30, 9:25 pm	4th	Leo	Cultivate. Destroy weeds and pests. Harvest fruits and root crops for food. Trim to retard growth.
Aug. 3, 10:31 am– Aug. 5, 4:04 pm	1st	Libra	Plant annuals for fragrance and beauty. Trim to increase growth.
Aug. 5, 4:04 pm– Aug. 6, 8:02 pm	1st	Scorpio	Plant grains, leafy annuals. Fertilize (chemical). Graft or bud plants. Irrigate. Trim to increase growth.
Aug. 6, 8:02 pm– Aug. 8, 1:30 am	2nd	Scorpio	Plant grains, leafy annuals. Fertilize (chemical). Graft or bud plants. Irrigate. Trim to increase growth.
Aug. 10, 1:44 pm– Aug. 13, 2:43 am	2nd	Capricorn	Graft or bud plants. Trim to increase growth.
Aug. 15, 12:13 am– Aug. 15, 2:41 pm	3rd	Aquarius	Cultivate. Destroy weeds and pests. Harvest fruits and root crops for food. Trim to retard growth.
Aug. 15, 2:41 pm– Aug. 18, 12:44 am	3rd	Pisces	Plant biennials, perennials, bulbs and roots. Prune. Irrigate. Fertilize (organic).
Aug. 18, 12:44 am– Aug. 20, 8:31 am	3rd	Aries	Cultivate. Destroy weeds and pests. Harvest fruits and root crops for food. Trim to retard growth.
Aug. 20, 8:31 am– Aug. 22, 1:51 pm	3rd	Taurus	Plant potatoes and tubers. Trim to retard growth.
Aug. 22, 1:51 pm– Aug. 22, 1:55 pm	4th	Taurus	Plant potatoes and tubers. Trim to retard growth.

2000 Gardening Dates

Dates	Qtr	Sign	Activity
Aug. 22, 1:55 pm– Aug. 24, 4:59 pm	4th	Gemini	Cultivate. Destroy weeds and pests. Harvest fruits and root crops for food. Trim to retard growth.
Aug. 24, 4:59 pm– Aug. 26, 6:17 pm	4th	Cancer	Plant biennials, perennials, bulbs and roots. Prune. Irrigate. Fertilize (organic).
Aug. 26, 6:17 pm– Aug. 28, 6:55 pm	4th	Leo	Cultivate. Destroy weeds and pests. Harvest fruits and root crops for food. Trim to retard growth.
Aug. 28, 6:55 pm– Aug. 29, 5:19 am	4th	Virgo	Cultivate, especially medicinal plants. Destroy weeds and pests. Trim to retard growth.
Aug. 30, 8:33 pm– Sep. 2, 12:55 am	1st	Libra	Plant annuals for fragrance and beauty. Trim to increase growth.
Sep. 2, 12:55 am– Sep. 4, 9:08 am	1st	Scorpio	Plant grains, leafy annuals. Fertilize (chemical). Graft or bud plants. Irrigate. Trim to increase growth.
Sep. 6, 8:47 pm– Sep. 9, 9:44 am	2nd	Capricorn	Graft or bud plants. Trim to increase growth.
Sep. 11, 9:34 pm– Sep. 13, 2:37 pm	2nd	Pisces	Plant grains, leafy annuals. Fertilize (chemical). Graft or bud plants. Irrigate. Trim to increase growth.
Sep. 13, 2:37 pm– Sep. 14, 7:00 am	3rd	Pisces	Plant biennials, perennials, bulbs and roots. Prune. Irrigate. Fertilize (organic).
Sep. 14, 7:00 am– Sep. 16, 2:05 pm	3rd	Aries	Cultivate. Destroy weeds and pests. Harvest fruits and root crops for food. Trim to retard growth.
Sep. 16, 2:05 pm– Sep. 18, 7:22 pm	3rd	Taurus	Plant potatoes and tubers. Trim to retard growth.
Sep. 18, 7:22 pm– Sep. 20, 8:28 pm	3rd	Gemini	Cultivate. Destroy weeds and pests. Harvest fruits and root crops for food. Trim to retard growth.
Sep. 20, 8:28 pm– Sep. 20, 11:16 pm	4th	Gemini	Cultivate. Destroy weeds and pests. Harvest fruits and root crops for food. Trim to retard growth.
Sep. 20, 11:16 pm– Sep. 23, 2:00 am	4th	Cancer	Plant biennials, perennials, bulbs and roots. Prune. Irrigate. Fertilize (organic).
Sep. 23, 2:00 am– Sep. 25, 4:02 am	4th	Leo	Cultivate. Destroy weeds and pests. Harvest fruits and root crops for food. Trim to retard growth.
Sep. 25, 4:02 am– Sep. 27, 6:22 am	4th	Virgo	Cultivate, especially medicinal plants. Destroy weeds and pests. Trim to retard growth.
Sep. 27, 2:53 pm– Sep. 29, 10:30 am	1st	Libra	Plant annuals for fragrance and beauty. Trim to increase growth.
Sep. 29, 10:30 am– Oct. 1, 5:50 pm	1st	Scorpio	Plant grains, leafy annuals. Fertilize (chemical). Graft or bud plants. Irrigate. Trim to increase growth.
Oct. 4, 4:42 am– Oct. 5, 5:59 am	1st	Capricorn	Graft or bud plants. Trim to increase growth.

2000 Gardening Dates

Dates	Qtr	Sign	Activity
Oct. 5, 5:59 am– Oct. 6, 5:33 pm	2nd	Capricorn	Graft or bud plants. Trim to increase growth.
Oct. 9, 5:36 am– Oct. 11, 2:51 pm	2nd	Pisces	Plant grains, leafy annuals. Fertilize (chemical). Graft or bud plants. Irrigate. Trim to increase growth.
Oct. 13, 3:53 am– Oct. 13, 9:06 pm	3rd	Aries	Cultivate. Destroy weeds and pests. Harvest fruits and root crops for food. Trim to retard growth.
Oct. 13, 9:06 pm– Oct. 16, 1:19 am	3rd	Taurus	Plant potatoes and tubers. Trim to retard growth.
Oct. 16, 1:19 am– Oct. 18, 4:37 am	3rd	Gemini	Cultivate. Destroy weeds and pests. Harvest fruits and root crops for food. Trim to retard growth.
Oct. 18, 4:37 am– Oct. 20, 2:59 am	3rd	Cancer	Plant biennials, perennials, bulbs and roots. Prune. Irrigate. Fertilize (organic).
Oct. 20, 2:59 am– Oct. 20, 7:42 am	4th	Cancer	Plant biennials, perennials, bulbs and roots. Prune. Irrigate. Fertilize (organic).
Oct. 20, 7:42 am– Oct. 22, 10:52 am	4th	Leo	Cultivate. Destroy weeds and pests. Harvest fruits and root crops for food. Trim to retard growth.
Oct. 22, 10:52 am– Oct. 24, 2:30 pm	4th	Virgo	Cultivate, especially medicinal plants. Destroy weeds and pests. Trim to retard growth.
Oct. 26, 7:23 pm– Oct. 27, 2:58 am	4th	Scorpio	Plant biennials, perennials, bulbs and roots. Prune. Irrigate. Fertilize (organic).
Oct. 27, 2:58 am– Oct. 29, 2:40 am	1st	Scorpio	Plant grains, leafy annuals. Fertilize (chemical). Graft or bud plants. Irrigate. Trim to increase growth.
Oct. 31, 1:01 pm– Nov. 3, 1:41 am	1st	Capricorn	Graft or bud plants. Trim to increase growth.
Nov. 5, 2:13 pm– Nov. 8, 12:02 am	2nd	Pisces	Plant grains, leafy annuals. Fertilize (chemical). Graft or bud plants. Irrigate. Trim to increase growth.
Nov. 10, 6:12 am– Nov. 11, 4:15 pm	2nd	Taurus	Plant annuals for hardiness. Trim to increase growth.
Nov. 11, 4:15 pm– Nov. 12, 9:27 am	3rd	Taurus	Plant potatoes and tubers. Trim to retard growth.
Nov. 12, 9:27 am– Nov. 14, 11:21 am	3rd	Gemini	Cultivate. Destroy weeds and pests. Harvest fruits and root crops for food. Trim to retard growth.
Nov. 14, 11:21 am– Nov. 16, 1:19 pm	3rd	Cancer	Plant biennials, perennials, bulbs and roots. Prune. Irrigate. Fertilize (organic).
Nov. 16, 1:19 pm– Nov. 18, 10:24 am	3rd	Leo	Cultivate. Destroy weeds and pests. Harvest fruits and root crops for food. Trim to retard growth.
Nov. 18, 10:24 am– Nov. 18, 4:15 pm	4th	Leo	Cultivate. Destroy weeds and pests. Harvest fruits and root crops for food. Trim to retard growth.

2000 Gardening Dates

Dates	Qtr	Sign	Activity
Nov. 18, 4:15 pm– Nov. 20, 8:35 pm	4th	Virgo	Cultivate, especially medicinal plants. Destroy weeds and pests. Trim to retard growth.
Nov. 23, 2:33 am– Nov. 25, 10:33 am	4th	Scorpio	Plant biennials, perennials, bulbs and roots. Prune. Irrigate. Fertilize (organic).
Nov. 25, 10:33 am– Nov. 25, 6:11 pm	4th	Sagittarius	Cultivate. Destroy weeds and pests. Harvest fruits and root crops for food. Trim to retard growth.
Nov. 27, 8:57 pm– Nov. 30, 9:26 am	1st	Capricorn	Graft or bud plants. Trim to increase growth.
Dec. 2, 10:23 pm– Dec. 3, 10:55 pm	1st	Pisces	Plant grains, leafy annuals. Fertilize (chemical). Graft or bud plants. Irrigate. Trim to increase growth.
Dec. 3, 10:55 pm– Dec. 5, 9:17 am	2nd	Pisces	Plant grains, leafy annuals. Fertilize (chemical). Graft or bud plants. Irrigate. Trim to increase growth.
Dec. 7, 4:26 pm– Dec. 9, 7:50 pm	2nd	Taurus	Plant annuals for hardiness. Trim to increase growth.
Dec. 11, 4:03 am– Dec. 11, 8:48 pm	3rd	Gemini	Cultivate. Destroy weeds and pests. Harvest fruits and root crops for food. Trim to retard growth.
Dec. 11, 8:48 pm– Dec. 13, 9:09 pm	3rd	Cancer	Plant biennials, perennials, bulbs and roots. Prune. Irrigate. Fertilize (organic).
Dec. 13, 9:09 pm– Dec. 15, 10:30 pm	3rd	Leo	Cultivate. Destroy weeds and pests. Harvest fruits and root crops for food. Trim to retard growth.
Dec. 15, 10:30 pm– Dec. 17, 7:41 pm	3rd	Virgo	Cultivate, especially medicinal plants. Destroy weeds and pests. Trim to retard growth.
Dec. 17, 7:41 pm– Dec. 18, 2:01 am	4th	Virgo	Cultivate, especially medicinal plants. Destroy weeds and pests. Trim to retard growth.
Dec. 20, 8:12 am– Dec. 22, 4:57 pm	4th	Scorpio	Plant biennials, perennials, bulbs and roots. Prune. Irrigate. Fertilize (organic).
Dec. 22, 4:57 pm– Dec. 25, 3:54 am	4th	Sagittarius	Cultivate. destroy weeds and pests. Harvest fruits and root crops for food. Trim to retard growth.
Dec. 25, 12:22 pm– Dec. 27, 4:25 pm	1st	Capricorn	Graft or bud plants. Trim to increase growth.

Dates to Destroy Weeds & Pests

From		To		Sign	Quarter
Jan. 2	4:32 pm	Jan. 5	5:24 am	Sagittarius	4th
Jan. 20	11:40 pm	Jan. 23	12:07 am	Leo	3rd
Jan. 23	12:07 am	Jan. 25	4:09 am	Virgo	3rd
Jan. 29	11:17 pm	Feb. 1	12:10 pm	Sagittarius	4th
Feb. 4	12:31 am	Feb. 5	8:03 am	Aquarius	4th
Feb. 19	11:27 am	Feb. 21	2:21 pm	Virgo	3rd
Feb. 26	7:10 am	Feb. 26	10:53 pm	Sagittarius	3rd
Feb. 26	10:53 pm	Feb. 28	7:45 pm	Sagittarius	4th
Mar. 2	8:14 am	Mar. 4	6:30 pm	Aquarius	4th
Mar. 19	11:44 pm	Mar. 19	11:57 pm	Virgo	3rd
Mar. 24	3:43 pm	Mar. 27	3:51 am	Sagittarius	3rd
Mar. 29	4:34 pm	Apr. 1	3:12 am	Aquarius	4th
Apr. 3	10:22 am	Apr. 4	1:12 pm	Aries	4th
Apr. 20	11:58 pm	Apr. 23	11:47 am	Sagittarius	3rd
Apr. 26	12:42 am	Apr. 26	2:30 pm	Aquarius	3rd
Apr. 26	2:30 pm	Apr. 28	12:06 pm	Aquarius	4th
May 30	7:54 pm	May 2	11:54 pm	Aries	4th
May 18	7:09 am	May 20	7:01 pm	Sagittarius	3rd
May 23	8:00 am	May 25	8:07 pm	Aquarius	3rd
May 28	5:08 am	May 30	10:02 am	Aries	4th
Jun. 1	11:34 am	Jun. 2	7:14 am	Gemini	4th
Jun. 16	5:27 pm	Jun. 17	1:26 am	Sagittarius	3rd
Jun. 19	2:26 pm	Jun. 22	2:52 am	Aquarius	3rd
Jun. 24	12:55 pm	Jun. 24	8:00 pm	Aries	3rd
Jun. 24	8:00 pm	Jun. 26	7:19 pm	Aries	4th
Jun. 28	9:59 pm	Jun. 30	10:09 pm	Gemini	4th
Jul. 16	8:27 pm	Jul. 19	8:44 am	Aquarius	3rd

Dates to Destroy Weeds & Pests

Jul. 21	7:09 pm	Jul. 24	2:44 am	Aries	3rd
Jul. 26	7:01 am	Jul. 28	8:30 am	Gemini	4th
Jul. 30	8:23 am	Jul. 30	9:25 pm	Leo	4th
Aug. 15	12:13 am	Aug. 15	2:41 pm	Aquarius	3rd
Aug. 18	12:44 am	Aug. 20	8:31 am	Aries	3rd
Aug. 22	1:55 pm	Aug. 24	4:59 pm	Gemini	4th
Aug. 26	6:17 pm	Aug. 28	6:55 pm	Leo	4th
Aug. 28	6:55 pm	Aug. 29	5:19 am	Virgo	4th
Sep. 14	7:00 am	Sep. 16	2:05 pm	Aries	3rd
Sep. 18	7:22 pm	Sep. 20	8:28 pm	Gemini	3rd
Sep. 20	8:28 pm	Sep. 20	11:16 pm	Gemini	4th
Sep. 23	2:00 am	Sep. 25	4:02 am	Leo	4th
Sep. 25	4:02 am	Sep. 27	6:22 am	Virgo	4th
Oct. 13	3:53 am	Oct. 13	9:06 pm	Aries	3rd
Oct. 16	1:19 am	Oct. 18	4:37 am	Gemini	3rd
Oct. 20	7:42 am	Oct. 22	10:52 am	Leo	4th
Oct. 22	10:52 am	Oct. 24	2:30 pm	Virgo	4th
Nov. 12	9:27 am	Nov. 14	11:21 am	Gemini	3rd
Nov. 16	1:19 pm	Nov. 18	10:24 am	Leo	3rd
Nov. 18	10:24 am	Nov. 18	4:15 pm	Leo	4th
Nov. 18	4:15 pm	Nov. 20	8:35 pm	Virgo	4th
Nov. 25	10:33 am	Nov. 25	6:11 pm	Sagittarius	4th
Dec. 11	4:03 am	Dec. 11	8:48 pm	Gemini	3rd
Dec. 13	9:09 pm	Dec. 15	10:30 pm	Leo	3rd
Dec. 15	10:30 pm	Dec. 17	7:41 pm	Virgo	3rd
Dec. 17	7:41 pm	Dec. 18	2:01 am	Virgo	4th
Dec. 22	4:57 pm	Dec. 25	3:54 am	Sagittarius	4th

Gestation & Incubation

Animal	Young/Eggs	Gestation/Incubation
Horse	1	346 days
Cow	1	283 days
Monkey	1	164 days
Goat	1–2	151 days
Sheep	1–2	150 days
Pig	10	112 days
Chinchilla	2	110 days
Fox	5–8	63 days
Dog	6–8	63 days
Cat	4–6	63 days
Guinea Pig	2–6	62 days
Ferret	6–9	40 days
Rabbit	4–8	30 days
Rat	10	22 days
Mouse	10	22 days
Turkey	1–15	26-30 days
Guinea Hen	15–18	25-26 days
Pea Hen	10	28-30 days
Duck	9–12	25-32 days
Goose	15–18	27-33 days
Hen	12–15	19-24 days
Pigeon	2	16-20 days
Canary	3–4	13-14 days

Egg Setting Dates

Dates to be Born	Sign	Qtr.	Set Eggs
Jan. 10, 4:59 am—Jan. 12, 1:48 pm	Pisces	1st	Dec. 20—22 1999
Jan. 14, 7:38 pm—Jan. 16, 10:25 pm	Taurus	2nd	Dec. 24—26 1999
Jan. 18, 11:01 pm—Jan. 20, 10:58 pm	Cancer	2nd	Dec. 28—30 1999
Feb. 6, 11:02 am—Feb. 8, 7:17 pm	Pisces	1st	Jan. 16—18
Feb. 11, 1:21 am—Feb. 12, 6:21 pm	Taurus	1st	Jan. 21—22
Feb. 15, 7:45 am—Feb. 17, 9:11 am	Cancer	2nd	Jan. 25—27
Mar. 6, 12:17 am—Mar. 7, 1:54 am	Pisces	1st	Feb. 14—15
Mar. 9, 7:01 am—Mar. 11, 10:46 am	Taurus	1st	Feb. 17—19
Mar. 13, 1:51 pm—Mar. 15, 4:43 pm	Cancer	2nd	Feb. 21—23
Apr. 5, 2:29 pm—Apr. 7, 4:58 pm	Taurus	1st	Mar. 15—17
Apr. 9, 7:16 pm—Apr. 11, 8:30 am	Cancer	1st	Mar. 19—21
Apr. 16, 7:36 am—Apr. 18, 12:41 pm	Libra	2nd	Mar. 26—28
May 3, 11:12 pm—May 5, 1:23 pm	Taurus	1st	Apr. 12—14
May 7, 2:14 am—May 9, 4:01 am	Cancer	1st	Apr. 16—18
May 13, 1:27 pm—May 15, 9:16 pm	Libra	2nd	Apr. 22—24
Jun. 3, 11:30 am—Jun. 5, 11:45 am	Cancer	1st	May 13—15
Jun. 9, 6:58 pm—Jun. 12, 2:55 am	Libra	2nd	May 19—22
Jul. 1, 2:20 pm—Jul. 2, 9:38 pm	Cancer	1st	Jun. 10—11
Jul. 7, 1:47 am—Jul. 8, 7:53 am	Libra	1st	Jun. 16—17
Aug. 3, 10:31 am—Aug. 5, 4:04 pm	Libra	1st	Jul. 13—15
Aug. 30, 8:33 pm—Sep. 2, 12:55 am	Libra	1st	Aug. 9—12
Sep. 11, 9:34 pm—Sep. 13, 2:37 pm	Pisces	2nd	Aug. 21—23
Sep. 27, 2:53 pm—Sep. 29, 10:30 am	Libra	1st	Sep. 6—8
Oct. 9, 5:36 am—Oct. 11, 2:51 pm	Pisces	2nd	Sep. 18—20
Nov. 5, 2:13 pm—Nov. 8, 12:02 am	Pisces	2nd	Oct. 15—18
Nov. 10, 6:12 am—Nov. 11, 4:15 pm	Taurus	2nd	Oct. 20—21
Dec. 2, 10:23 pm—Dec. 3, 10:55 pm	Pisces	1st	Nov. 11—12
Dec. 7, 4:26 pm—Dec. 9, 7:50 pm	Taurus	2nd	Nov. 16—18

Companion Planting
Plant Helpers and Hinderers

Plant	Helped By	Hindered By
Asparagus	Tomatoes, Parsley, Basil	
Beans	Carrots, Cucumbers, Cabbage, Beets, Corn	Onions, Gladiola
Bush Beans	Cucumbers, Cabbage, Strawberries	Fennel, Onions
Beets	Onions, Cabbage, Lettuce	Pale Beans
Cabbage	Beets, Potatoes, Onions, Celery	Strawberries, Tomatoes
Carrots	Peas, Lettuce, Chives, Radishes, Leeks, Onions	Dill
Celery	Leeks, Bush Beans	
Chives	Beans	
Corn	Potatoes, Beans, Peas, Melons, Squash, Pumpkins, Cucumbers	
Cucumbers	Beans, Cabbage, Radishes, Sunflowers, Lettuce	Potatoes, Aromatic Herbs
Eggplant	Beans	
Lettuce	Strawberries, Carrots	
Melons	Morning Glories	
Onions, Leeks	Beets, Chamomile, Carrots, Lettuce	Peas, Beans
Garlic	Summer Savory	
Peas	Radishes, Carrots, Corn, Cucumbers, Beans, Turnips	Onions
Potatoes	Beans, Corn, Peas, Cabbage, Hemp, Cucumbers	Sunflowers

Plant	Helped By	Hindered By
Radishes	Peas, Lettuce, Nasturtium, Cucumbers	Hyssop
Spinach	Strawberries	
Squash, Pumpkins	Nasturtium, Corn	Potatoes
Tomatoes	Asparagus, Parsley, Chives, Onions, Carrots, Marigold, Nasturtium	Dill, Cabbage, Fennel
Turnips	Peas, Beans	

Plant Companions and Uses

Plant	Companions and Uses
Anise	Coriander
Basil	Tomatoes; dislikes rue; repels flies and mosquitos
Borage	Tomatoes and squash
Buttercup	Clover; hinders delphiniums, peonies, monkshood, columbines
Chamomile	Helps peppermint, wheat, onions, and cabbage; large amounts destructive
Catnip	Repels flea beetles
Chervil	Radishes
Chives	Carrots; prone to apple scab and powdery mildew
Coriander	Hinders seed formation in fennel
Cosmos	Repels corn earworms
Dill	Cabbage; hinders carrots and tomatoes
Fennel	Disliked by all garden plants
Garlic	Aids vetch and roses; hinders peas and beans
Hemp	Beneficial as a neighbor to most plants
Horseradish	Repels potato bugs

Plant	Companions and Uses
Horsetail	Makes fungicide spray
Hyssop	Attracts cabbage fly away from cabbages; harmful to radishes
Lovage	Improves hardiness and flavor of neighbor plants
Marigold	Pest repellent; use against Mexican bean beetles and nematodes
Mint	Repels ants, flea beetles and cabbage worm butterflies
Morning Glory	Corn; helps melon germination
Nasturtium	Cabbage, cucumbers; deters aphids, squash bugs, and pumpkin beetles
Nettles	Increase oil content in neighbors
Parsley	Tomatoes, asparagus
Purslane	Good ground cover
Rosemary	Repels cabbage moths, bean beetles, and carrot flies
Sage	Repels cabbage moths and carrot flies
Savory	Deters bean beetles
Sunflower	Hinders potatoes; improves soil
Tansy	Deters Japanese beetles, striped cucumber beetles, and squash bugs
Thyme	Repels cabbage worms
Yarrow	Increases essential oils of neighbors

Year 2000 Weather Predictions

Nancy Soller

The year 2000 will be an extremely cold one, as a Jupiter-Saturn conjunction will be in effect most of the year. April and May will be especially cold, since Jupiter and Saturn will not only be conjunct but parallel in declination as well. There will be widespread storms and higher than normal levels of precipitation as well due to this conjunction.

There is one blessing: the conjunction will not form until later in the year. Most of the country will have a normal winter; on the East Coast, however, Saturn, Uranus, and Venus will form a T-square on northern charts conjunct two ingress chart angles. Temperatures will be extremely chill north, and very wet and windy; the South will also be extremely wet. Alaska, however, may have above normal temperatures in winter.

A very chill, wet spring is forecast for northern portions of the East Coast. The southern Atlantic seaboard will see extremely wet weather with temperatures normal to below normal. This weather will extend inland, but areas of the Midwest just east of the Mississippi will have drier weather with normal temperatures. The Plains' temperatures in the spring will range from chill to normal, while precipitation will be slightly above normal. Areas from the Rockies to the West Coast will range from cool to above-normal in temperature, and precipitation may be a bit below normal. An extremely chill spring is forecast for most of Alaska, and there may be drought conditions in the interior of the state. Below-normal temperatures and above-normal levels of rainfall are forecast for Hawaii.

Summer will be cool and windy in northern areas of the Atlantic seaboard and in the southern seaboard. West of the Appalachians it will be wet and cool; west of the Mississippi it will be cool, dry, and windy. Some parts of the Rockies will have temperatures ranging from normal to just a little below normal with precipitation levels above normal. The West Coast will have slightly cool temperatures with precipitation normal to above normal. The Alaskan Panhandle will have drier-than-normal to normal weather, but the Alaskan interior will be chill and stormy. Hawaii will be dry, cool, and windy.

In the fall, New England will be chill, dry, and windy, but most of the rest of the country east of the Mississippi will have normal weather. Most areas west of the Mississippi will be warmer than usual and dry. The West Coast, however, will be dry and chill. Alaska and Hawaii will be cloudy and breezy, but the Alaskan Panhandle will be extremely wet.

Wind dates in the following monthly forecasts indicate that winds will range from strong to destructive.

January 2000

Zone 1: Cold, stormy, windy weather is forecast for this zone in January. The weather will be most severe north, but unusual weather is forecast for points as far south as Florida. Watch for precipitation January 1, 4, 6, 14, 19, 21, 22, 25, and 31. Watch for winds January 6, 16, 19, and 25.

Zone 2: A low barometer and abnormally damp weather is predicted both east and west. Flooding may follow in the spring. Watch for precipitation January 1, 4, 6, 14, 19, 21, 22, 25, and 31. Winds are likely January 6, 16, 19, and 25.

Zone 3: Very wet weather is forecast for areas in this zone near the Mississippi, but most of this zone will have very cold, slightly dry weather. Precipitation is likely January 2, 6, 14, 21, 24, and 26. Winds are due January 7, 16, 21, 24, and 25.

Zone 4: Seasonable temperatures and seasonable precipitation are predicted for this zone in January. Snowfall is likely January 2, 6, 12, 14, 21, 24, and 26. Winds are due January 7, 12, 16, 21, 24, and 25.

Zone 5: Seasonable temperatures and seasonable precipitation are predicted for this zone in January. Watch for precipitation January 2, 6, 12, 14, 21, 24, and 26. Winds are likely January 7, 12, 16, 21, 24, and 25.

Zone 6: A mild month is forecast for Alaska. A wet month with temperatures below normal is forecast for Hawaii. Watch for precipitation January 1, 2, 4, 6, 9, 12, 14, 17, 19, and 28. Watch for winds January 6, 9, 12, 15, 17, 19, 24, 25, 28, and 30.

February 2000

Zone 1: Very cold, windy weather is predicted for northern portions of this zone in February. There will be much precipitation both north and south. Watch for excessive precipitation February 12, 18, 27, and 28. Watch for winds February 1, 4, 6, 12, 13, 26, 27, and 28.

Zone 2: Much snow north and south is the forecast for this zone in February. Watch for excessive precipitation February 12, 16, 27, and 28. Watch for winds February 1, 4, 6, 12, 13, 26, 27, and 28.

Zone 3: Low barometric pressure and much precipitation is forecast for this zone, especially in eastern areas. Watch for much snowfall north and rain south February 5, 12, 18, and 29. Watch for winds February 6, 18, 28, and 29.

Zone 4: Seasonable temperatures and seasonable precipitation are forecast for this zone in February. Watch for precipitation February 12, 18, 20, 24, 27, and 29. Watch for winds February 4, 12, 27, 28, and 29.

Zone 5: Seasonable temperatures and seasonable precipitation are forecast for this zone in February. Watch for precipitation February 2, 12, 18, 20, 24, 27, and 29. Watch for winds February 4, 12, 27, 28, and 29.

Zone 6: Temperatures above normal are forecast for Alaska and Hawaii and both states should have less precipitation than normal. Watch for precipitation February 1, 4, 6, 14, 19, 21, 22, 25, and 31. Watch for winds February 6, 16, 19, and 25.

March 2000

Zone 1: An extremely cold month is predicted for northern portions of this zone in March. Heavy precipitation is forecast south. Watch for precipitation March 2, 6, 16, 18, 20, 26, 27, and 31. Watch for winds March 2, 5, and 31.

Zone 2: Much precipitation is forecast east. Temperatures will be normal. In the west near the Mississippi, it will be dry and warm. Watch for precipitation March 2, 4, 6, 16, 18, 20, 26, 27, and 31. Watch for winds March 2, 5, and 31.

Zone 3: A very dry, warm month is forecast east; to the west there will be more normal temperatures and precipitation. Watch for precipitation March 4, 6, 13, 18, 19, 23, 25, and 27. Watch for winds March 2, 4, 7, 9, 16, 19, 23, 30, and 31.

Zone 4: Seasonable temperatures and normal amounts of precipitation are forecast for this zone in March. Precipitation is due March 4, 6, 13, 16, 18, 19, 23, 25, 27, and 31. Winds are likely March 4, 7, 9, 16, 19, 23, 25, 30, and 31.

Zone 5: Scant amounts of precipation and seasonable temperatures are forecast for this zone in March. Watch for precipitation March 2, 3, 15, and 16. Winds are likely March 1, 3, 7, 9, 15, and 16.

Zone 6: Warm and dry is the forecast for Alaska and Hawaii this month. Watch for precipitation March 2, 3, 6, 16, 18, 20, 26, 27, and 31. Winds are likely March 1, 2, 3, 5, 7, 9, 15, and 16.

April 2000

Zone 1: A very cold month with much precipitation is forecast for northern portions of this zone. To the south it will be wet. Watch for heavy precipitation April 4, 11, 16, 18, and 23. Watch for winds April 4, 6, 9, 16, 20, 23, and 24.

Zone 2: A wet month is predicted for April with cool to mild temperatures. Watch for heavy precipitation April 4, 11, 16, 18, and 23. Watch for winds April 4, 6, 9, 16, 20, 23, and 24.

Zone 3: A cool, dry month is forecast for April. Watch for precipitation April 2, 4, 11, 15, 16, 19, and 23. Watch for winds April 2, 4, 6, 9, 15, 19, 23, and 24.

Zone 4: A cool, dry month is predicted for this zone in April. Precipitation is most likely April 4, 15, 19, and 23. Winds are likely April 2, 4, 15, 19, 23, and 24.

Zone 5: Cool and dry is the forecast for this zone in April. Watch for precipitation April 2, 11, 16, 23, 26, 28, and 29. Watch for winds April 2, 9, 16, 19, 20, 23, 24, 28, and 29.

Zone 6: Cool, rainy weather is forecast for the Alaskan Panhandle; cold, snowy weather for the rest of the state. Hawaii will be cooler and wetter than usual. Precipitation will be most likely April 2, 4, 11, 15, 19, 20, 23, 24, 26, and 28. Winds are likely April 2, 4, 6, 9, 16, 19, 20, 23, 24, 28, and 29.

May 2000

Zone 1: Very cool and wet is the forecast in the north; wet and cool is the forecast south. Watch for precipitation May 3, 6, 13, 17, and 28. Watch for winds May 3, 7, 8, 13, 17, and 24.

Zone 2: The forecast predicts for cool and wet weather east, but warmer and dry to the west. Precipitation is most likely May 3, 6, 13, 17, and 28. Watch for winds May 3, 7, 8, 13, 17, and 24.

Zone 3: Dry with seasonable temperatures east and more normal precipiation west is the forecast for this zone in May. Watch for precipitation May 6, 17, 26, 27, 28, and 30. Watch for winds May 7, 8, 9, 16, 17, 19, and 24.

Zone 4: A dry month with seasonable temperatures is forecast for this zone in May. Precipitation will be likely May 6, 17, 26, 27, 28, and 30. Winds are likely May 7, 8, 9, 16, 17, 19, and 24.

Zone 5: A dry month with temperatures from cool to normal is forecast for this zone in May. Watch for rain May 2, 3, 15, 16, 25, and 26. Winds are likely May 1, 3, 7, 9, 15, 16, and 25.

Zone 6: A cold, wet month is forecast for the Alaskan Panhandle. A cold, stormy month is forecast for the rest of the state. A cool, stormy month is forecast for Hawaii. Watch for precipitation May 2, 3, 13, 15, 17, 25, 26, and 28. Watch for winds May 1, 3, 7, 9, 13, 15, 16, 17, 24, and 25.

June 2000

Zone 1: A cool month is forecast and there will be some heavy precipitation. Look for rain June 8, 11, 12, 17, and 24. Watch for winds June 2, 9, 11, 12, and 27.

Zone 2: Wet east and dry west is the forecast for this zone. Temperatures will be cool. Look for heavy precipitation June 8, 11, 12, 17, and 24. Watch for winds June 2, 9, 11, 12, and 27.

Zone 3: Dry weather east and more normal precipitation west is the forecast for this zone this month. Temperatures will be cool. Watch for precipitation June 1, 9, 12, and 21. Watch for winds June 1, 12, and 21.

Zone 4: Dry weather is forecast for this zone this month. Temperatures will range from cool to normal. Watch for rain June 1, 9, 12, 21, and 24. Watch for winds June 1, 12, and 21.

Zone 5: Dry with cool to normal temperatures are forecast for this zone in June. Best chances for rain come June 2, 3, and 24. Watch for winds June 3, 9, 12, and 27.

Zone 6: Wet, cool weather for the Panhandle and cold, wet weather for the rest of the state is the forecast for Alaska. Below-normal temperatures and storms are forecast for Hawaii. Watch for precipitation June 2, 3, 8, 11, 12, 17, and 24. Watch for winds June 2, 3, 9, 11, 12, and 27.

July 2000

Zone 1: The July forecast for this zone is dry and cool in the north, wet with more normal temperatures in the south. Watch for precipitation July 2, 11, 15, 17, 20, 21, 24, and 28. Winds are likely July 1, 2, 20, 21, 22, and 25.

Zone 2: Expect very wet weather in the east part of this zone and normal precipitation in the west in July. Temperatures will be a bit below normal. Watch for precipitation July 2, 11, 15, 17, 20, 21, 24, and 28. Winds are due July 1, 2, 20, 21, 22, and 25.

Zone 3: Cool and dry is the forecast for this zone in July. Precipitation will be most likely July 1, 8, 11, 17, 20, 21, 24, 27, 28, and 31. Winds are likely July 6, 21, 22, 28, and 30.

Zone 4: Expect cool and dry weather in the east part of this zone with more normal weather patterns in the west. Watch for precipitation July 1, 8, 11, 17, 20, 21, 24, 27, 28, 30, and 31. Watch for winds July 6, 21, 22, 28, and 30.

Zone 5: Seasonal temperatures and seasonal precipitation are forecast for this zone in July. Watch for precipitation July 7, 8, 15, 17, 20, 24, and 30. Watch for winds July 7, 20, and 25.

Zone 6: Dry, seasonal temperatures are forecast for the Alaskan Panhandle; the interior of the state will see a very chill month with numerous storms. Hawaii will be cooler and drier than usual. Watch for precipitation July 2, 7, 8, 11, 15, 17, 20, 21, 24, 28, and 30. Watch for winds July 1, 2, 7, 20, 21, 22, and 25.

August 2000

Zone 1: Chill and dry north and wet in the south is the forecast for this month. Watch for precipitation August 6, 8, 11, 14, 16, 20, 22, 23, and 29. Watch for winds August 5, 6, 8, 10, 11, 14, 16, 20, 27, and 29.

Zone 2: The forecast for this zone calls for wet weather east and dry, cold weather in the northern Mississippi River Valley. Precipitation is likely August 6, 8, 11, 14, 16, 20, 22, 23, and 29. Winds are likely August 5, 6, 8, 10, 11, 14, 16, 20, 27, and 29.

Zone 3: Dry and cool is the forecast for most of this zone in August. Precipitation is most likely August 6, 9, 12, 16, 22, and 23. Winds are likely August 9, 11, 12, 21, and 26.

Zone 4: Eastern portions of this zone will be cool and dry. More normal weather patterns are forecast for the western portions during August. Precipitation is likely August 6, 9, 12, 16, 22, and 23. Winds are likely August 9, 11, 12, 21, and 26.

Zone 5: Less than normal seasonal precipitation and temperatures are forecast for this zone in August. Watch for precipitation August 6, 14, 20, and 31. Winds are likely August 1, 6, 9, 12, 14, 16, 20, and 27.

Zone 6: The Alaskan Panhandle will be dry and slightly cool. The interior of the state will be very cool with many storms. Hawaii will be very cool and dry. Precipitation dates include August 6, 8, 11, 14, 16, 20, 22, 23, 29, and 30. Winds are likely August 1, 5, 6, 8, 9, 10, 11, 12, 14, 16, 20, 27, 29, and 31.

September 2000

Zone 1: Cool and dry north and wet south is the forecast for this zone in September. The last week of the month will probably see more seasonable weather. Watch for rain September 3, 5, 8, 10, 14, 23, 24, 26, and 27. Watch for winds September 2, 8, 10, 14, 19, 24, and 25.

Zone 2: Wet and cool is the forecast for this zone in September. Rain is likely September 3, 5, 8, 10, 14, 23, 24, 26, and 27. Winds are due September 2, 8, 10, 14, 19, 24, and 25.

Zone 3: Dry and cool is the forecast for this zone in September. Rain will be likely on September 8, 24, 25, and 27. Winds are likely September 4, 8, 14, 24, and 25.

Zone 4: The forecast for this zone in September calls for cool and dry temperatures in the east with more precipitation to the west. Watch for precipitation September 8, 24, 25, and 27. Winds are likely September 4, 8, 14, 24, and 25.

Zone 5: A little cooler than usual with normal precipitation is the forecast for this zone in September. Watch for precipitation September 3, 8, 10, 14, 18, 20, 23, 26, and 27. Winds are likely September 8, 10, 14, and 18.

Zone 6: Normal to warm temperatures are forecast for the Alaskan Panhandle. Very cool weather with strong storms is the forecast for the interior of the state. Hawaii will be very cool and dry. Precipitation is likely September 3, 5, 8, 10, 23, 24, 26, and 27. Winds are forecast September 2, 8, 10, 14, 18, 19, 24, and 25.

October 2000

Zone 1: New England should have a overcast, wet month with temperatures closer to normal than at any time since spring. Most other areas in this zone will have a normal season. Watch for precipitation October 1, 2, 5, 11, 22, and 26. Watch for winds October 1, 2, 3, 4, 9, 12, 23, and 30.

Zone 2: Most areas in this zone will have seasonable precipitation and temperatures. Precipitation likely October 1, 2, 5, 11, 22, and 26. Winds are due October 1, 2, 3, 4, 9, 12, 23, and 30.

Zone 3: Eastern portions of this zone will have a normal season; in the west it will be dry and warmer than normal. Watch for precipitation October 8, 11, 13, 26, and 28. Winds are likely October 3, 8, 9, 13, 23, and 28.

Zone 4: Dry, warm weather is forecast for this zone in October. Precipitation would be most likely October 8, 11, 13, 26, and 28. Watch for winds October 3, 8, 9, 13, 23, and 28.

Zone 5: Dry and cool is the forecast for this zone in October. Days most likely to result in rain include October 2, 5, 12, 13, 14, 18, 20, 21, 22, and 27. Watch for winds October 2, 3, 5, 12, 13, 21, 27, and 29.

Zone 6: Extremely wet weather is forecast for the Alaskan Panhandle. Cloudy, breezy, warm weather is predicted for most of the rest of the state. Hawaii will see some cloudy, breezy, warm weather east. Precipitation will be most likely October 1, 2, 5,

11-14, 18, 20, 21, 22, 26, and 27. Watch for winds October 1, 2, 3, 4, 5, 9, 12, 13, 21, 23, 27, 29, and 30.

November 2000

Zone 1: The far north shall see overcast, breezy, wet weather; other portions of this zone will see seasonable weather. Watch for precipitation November 2, 10, 12, 15, 17, 18, and 22. Winds are likely November 2, 10, 15, 18, 22, 24, and 26.

Zone 2: Seasonal temperatures and seasonal precipitation are forecast for this zone in November. Precipitation is due November 2, 10, 12, 15, 17, and 22. Watch for winds November 2, 10, 15, 18, 22, 24, and 26.

Zone 3: Temperatures and precipitation should be normal east; in the west weather will be drier and warmer than usual. Watch for precipitation November 2, 17, 19, 22, and 26. Watch for winds November 2, 16, 22, 24, 25, and 26.

Zone 4: Warmer and drier than usual is the forecast for this zone in November. Best dates for precipitation include November 2, 17, 22, and 26. Watch for winds November 2, 16, 22, 24, 25, and 26.

Zone 5: Chill and dry is the forecast for this zone. Dates most likely to result in precipitation include November 1, 2, 4, 10, 17, 18, 25, and 26. Winds are likely November 1, 2, 8, and 18.

Zone 6: The Alaskan Panhandle will be very wet; the interior will be cloudy, windy and slightly warmer than usual. Hawaii will have some cloudy, breezy weather east. Watch for precipitation November 1, 2, 4, 10, 12, 15, 17, 18, 22, 25, and 26. Watch for winds November 1, 2, 8, 10, 15, 18, 22, 24, and 26.

December 2000

Zone 1: Cold, cloudy, breezy, and wet is the forecast north; in the south the weather will be more seasonal. Watch for precipitation December 2, 3, 4, 6, 7, 9, 11, 24, 25, and 27. Winds are likely December 2, 6, 9, 18, 22, 24, and 27.

Zone 2: Seasonal precipitation and seasonable temperatures are predicted for this zone in December. Watch for precipitation December 2, 3, 4, 6, 7, 9, 11, 24, 25, and 27. Watch for winds December 2, 6, 9, 18, 22, 24, and 27.

Zone 3: Seasonable temperatures and seasonable precipitation will be found east; western portions of this zone will be dry and a little warmer than usual. Watch for precipitation December 2, 4, 6, 9, 11, 12, 17, and 19. Watch for winds December 6, 9, 12, 15, 18, 19, and 22.

Zone 4: Dry with above-normal temperatures is the prediction for this zone in December. Best dates for precipitation include December 2, 6, 9, 11, 12, 17, and 19. Dates likely to be windy include December 6, 9, 12, 15, 18, 19, and 22.

Zone 5: Dry and very chill is the forecast for this zone in December. Dates most likely to result in precipitation include December 1, 4, 7, 9, 12, 17, 19, and 29. Dates likely to result in wind include December 1, 2, 4, 6, 9, 13, 15, and 25.

Zone 6: The Alaskan Panhandle will be extremely wet; the interior of the state will be overcast and windy. The eastern islands of Hawaii should be overcast and breezy. Dates likely to result in precipitation include 1, 2, 3, 4, 6, 7, 9, 11, 12, 17, 19, 24, 27, and 29. Wind dates include December 1, 2, 4, 6, 9, 13, 15, 18, 22, 24, 25, and 27.

2000 Earthquake Predictions

Ann E. Parker of Skokie, Illinois, has devised a way to predict the location and dates of large, destructive earthquakes. Ann notes the location of solar and lunar eclipses in the ecliptic. Then she notes that when the sign and degree of the eclipse corresponds to the sign and degree of the geodetic ascendant, geodetic midheaven, or geodetic vertex of a location on earth, a large destructive quake is likely when Mars forms a hard angle to the eclipse-point. The hard angles involved include the conjunction, square, opposition, semisquare, and sesquiquadrate. Such a quake can occur up to eighteen months before or after the eclipse. Volcanic activity is sometimes triggered instead of a quake. Any of the following events could cause an earthquake in the year 2000.

The August 8, 1998, lunar eclipse at fifteen degrees of Aquarius could result in earthquakes in California, Nevada, Montana, or Brazil. Danger dates in 2000 include January 2 and 3, February 12 and 13, March 17 and 18, April 13 and 14, June 11, 12, 16, and 17, August 24 and 25, September 16 and 17, November 4 and 5, and December 23 and 24.

The August 22, 1998, solar eclipse at twenty-eight degrees of Leo could result in a large destructive earthquake in the Marianas, New Guinea, Japan, Iran, or volcanic activity in Hawaii. Danger dates include January 18 and 19, February 29, March 1, April 10 and 11, May 1 and 2, June 29 and 30, July 8 and 9, September 14 and 15, October 16 and 17, and November 25 and 26.

The September 6, 1998, lunar eclipse at thirteen degrees of Pisces could result in large destructive quakes in Texas, Oklahoma, Mexico, the Caroline Islands, or volcanic activity in Iceland. Danger dates include January 21 and 22, February 17 and 18, March 20 and 21, May 8, 9 and 10, May 22 and 23, July 29 and 30, August 9, 10, and 11, October 7, 8 and 9, November 18, 19, and 20, and December 20 and 21.

The January 31, 1999, lunar eclipse at eleven degrees of Leo could result in a large quake in Alaska, Turkey, Iraq, Ethiopia, Bulgaria, Greece, or Poland. Danger dates include February 6, 7, and 8, March 10, 11, and 12, April 7 and 8, June 10, 11, 12, 16, and 17, August 18 and 19, September 7, 8, and 9, October 28, 29, and 30, and December 18, 19, and 20.

The February 16, 1999, solar eclipse at twenty-seven degrees of Aquarius could result in a large diastrous quake in California, Arizona, or Utah. Danger dates include January 1, 21, 22, and 23, February 28 and 29, April 8, 9, 29, and 30, July 5, 6, and 7, September 12 and 13, October 13, 14, and 15, and November 23, 24, and 25.

The July 28, 1999, lunar eclipse at five degrees of Aquarius could result in a large, disastrous quake in California, Oregon, Newfoundland, Paraguay, Pakistan, Brazil, or in volcanic activity at Mount Saint Helens. Danger dates include January 30 and 31, February 29, March 1, 30, and 31, May 21, 22, and 23, June 1, 2, and 3, August 8, 9, 10, 24, 25, and 26, October 19 and 20, and November 5, 6, and 7.

The August 11, 1999, solar eclipse at eighteen degrees of Leo could result in a large, destructive quake in Alaska, Indonesia, the Caroline Islands, Iraq, Turkey, or Poland. Danger dates include January 7, 8, 16, and 17, March 23 and 24, April 17 and 18, June 17, 18, 21, and 22, August 29 and 30, September 23 and 24, and November 9 and 10.

The January 21, 2000, lunar eclipse at zero degrees of Leo could result in a large, destructive quake in Alaska, the Philippines, China, Japan, Kenya, Italy, Hungary, and in locations around Yugoslavia. Danger dates include January 23 and 24, February 21 and 22, March 23 and 24, May 12, 13, 25, and 26, August 1, 2, 14, and 15, October 11 and 12, November 23, 24, and 25, and December 23, 24, and 25.

The February 5, 2000, solar eclipse at sixteen degrees of Aquarius could result in a large, destructive quake in California, Montana, or Brazil. Danger dates include January 3, 4, 5, 13, and 14, March 20 and 21, April 14 and 15, June 12, 13, 14, 18, and 19, August 26 and 27, September 18, 19, and 20, November 5, 6, and 7, and December 29 and 30.

The July 1, 2000, solar eclipse at ten degrees of Cancer could result in a large, destructive quake in Indonesia, China, Sumatra, Algeria, Morocco, or the Ryukyu Islands. Critical dates include January 18, 19, and 20, February 25 and 26, April 5, 6, 27, and 28, July 1, 2, and 3, September 9 and 10, October 8, 9, and 10, and November 20, 21, and 22.

The July 16, 2000, lunar eclipse at twenty-four degrees of Capricorn could result in a large, destructive quake in New England, New Brunswick, Saint Vincent Island, the Virgin Islands, Bolivia, the Dominican Republic, Puerto Rico, Brazil, or the American Northwest. Critical dates include January 15 and 16, February 10 and 11, March 15 and 16, April 30, May 1, 2, 17, and 18, July 8, 9, 31 and August 1 and 2, October 1 and 2, November 9, 10, and 11, and December 13, 14, and 15.

The July 31, 2000, solar eclipse at eight degrees of Leo could result in a large, destructive quake in Alaska, the Philippines, China, Ethiopia, Kenya, Bulgaria, Greece, Poland, Hungary, and locations in and around Yugoslavia. Danger dates include February 3 and 4, March 5 and 6, April 3 and 4, May 28 and 29, June 6 and 7, August 13, 14, 15, and 31, September 1 and 2, October 23, 24, and 25, and December 11, 12, and 13.

The December 25, 2000, solar eclipse at four degrees of Capricorn could result in a large, destructive quake on the New Madrid fault, in Nicaragua, British Honduras, El Salvador, the Fiji Islands, Iran, Turkey, Paraguay, and southeastern Europe. Critical dates include January 9 and 10, February 17 and 18, March 26, 27, and 28, April 19, 20, and 21, June 19, 20, 21, 22, and 23, September 1, 2, 27, 28, and 29, and November 10, 11, and 12.

The January 9, 2001, lunar eclipse at nineteen degrees of Cancer could result in a large, destructive quake in Indonesia, China, or the Philippines. Danger dates include January 9 and 10, February 2 and 3, March 8 and 9, April 21 and 22, May 9 and 10, July 15, 16, 20, 21, and 22, September 23 and 24, October 29, 30, and 31, and December 5 and 6.

The June 21, 2001, solar eclipse at zero degrees of Cancer could result in a large, destructive quake in Sumatra or Northern Ireland. Danger dates include January 2 and 3, February 12 and 13, March 17, 18, and 19,

April 12, 13, and 14, June 11, 12, 16, 17, and 18, August 24 and 25, September 16, 17, and 18, November 4 and 5, and December 27, 28, and 29.

The July 5, 2001, lunar eclipse at thirteen degrees Capricorn could result in a large destructive quake in Washington, D. C., Charlestown, Miami, Peru, Columbia, Iran, or Moscow. Critical dates include January 1, 2, 23, and 24, February 29, March 1, April 17, 18, and 30, May 1 and 2, July 6, 7, 8, and 9, October 15, 16, and 17, and November 25 and 26.

The December 14, 2001, solar eclipse at twenty-two degrees Sagittarius could result in a large, destructive quake in Kansas, Nebraska, western Oklahoma, Alaska, the New Hebrides, Greece, Hungary, Germany, or the Czech Republic. Danger dates include February 1 and 2, March 3, 4, and 5, April 1, 2, and 3, May 25, 26, and 27, June 4, 5, and 6, July 29 and 30, August 11, 12, and 13, October 22 and 23, and December 9, 10, and 11.

The December 30, 2001, lunar eclipse at eight degrees of Cancer could result in a large, destructive quake in Malaysia, Sumatra, or the Ryukyu Islands. Danger dates include January 15 and 16, February 24 and 25, March 31, April 1, 2, 23, 24, and 25, June 27, 28, 29, and 30, September 5, 6, and 7, October 4, 5, and 6, and November 17 and 18.

The May 26, 2002, lunar eclipse at five degrees of Sagittarius could result in a large destructive quake in California, Algeria, or Iceland. Danger dates include January 10 and 11, February 4 and 5, March 9 and 10, April 23 and 24, May 10, 11, and 12, July 16, 17, 18, 22, 23, and 24, September 24, 25, and 26, October 31, November 1 and 2, and December 6, 7, and 8.

The June 10, 2002, solar eclipse at nineteen degrees of Gemini could result in a large, destructive quake in China. Danger dates include January 28 and 29, February 28 and 29, March 28 and 29, May 20 and 21, May 31, June 1, August 7, 8, 23, 24, and 25, October 17 and 18, and December 2, 3, 4, 30, and 31.

The June 24, 2002, lunar eclipse at three degrees of Capricorn could result in a large, destructive quake in Tennessee, northwest Indiana, upper Michigan, Nicaragua, the Fiji Islands, Turkey, Rumania, or Greece. Critical dates include January 7 and 8, February 15, 16, and 17, March 22, 23, and 24, April 14 and 15, June 16, 17, 18, 20, 21, and 22, August 29 and 30, September 22, 23, 24, and 25, November 8, 9, and 10, and December 24 and 25.

Fragrant Cactus, Queen of the Night

Louise Riotte

The origins of the cactus family, according to Edward Abbey, the author of *Cactus Country*, are shrouded in mystery. Botanists theorize that the first cacti evolved from roses. They base this conclusion largely on an outstanding feature of today's cacti—their lavish, showy flowers, which closely resemble roses in shape and structure

To those who see them for the first time, cactus blossoms are incredibly beautiful. It is unbelievable that such a delicate, ethereal flower should spring from such a thorny plant. Many of us who live far from the desert can enjoy the beauty of cacti that adapt readily to pot culture. A perfect example is the tropical Queen of the Night (*Hylocereus undatus*); this native plant of Brazil is a fast-growing cactus that will develop, in time, to a large plant even in a pot. The plant is easy to grow if kept in partial shade and moderately moist. The delicate white flowers are nocturnal, and are the largest and the most beautiful flowers of all the cacti. Their blooms begin to open at sunset and, emitting a delightful vanilla aroma, reach full development during the night, withering at dawn.

The lofty organ-pipe cactus, native to Arizona's Sonora Desert, is a spectacularly fragrant night bloomer whose flowers are scattered high up on the "organ pipes," or stems, of the cactus. Meanwhile, the slender, gray-whiskered senita cactus is another night blooming species, whose lovely pale lavender flowers with white stamens are smaller but more plentiful than the organ-pipe.

Edible Cacti

But the beauty of the flowers is not their only treasure. Many cacti have provided sustenance for the people who live in the arid climes where cacti grow. The tough, homely, opuntia prickly pear, for instance, grows in sun-scorched flats and in chilly mile-high uplands, and can be found in all of the contiguous United States except Maine, New Hampshire, and Vermont. This cactus provides food for both man and beast; cattle have been known to eat it in times of drought. The flat cactus pads, called *nopales*, are delicious and succulent with a flavor somewhat similar to green beans, and they can be used in egg dishes, vegetable casseroles, or salads. I often find nopales offered in the vegetable section of the supermarket. Anne Lindsay Greer, in her book *Cuisine of the Southwest* tells how to prepare them: "All 'eyes' must be removed with a knife or potato peeler before preparation…. Wash, cut in squares, and cook in salted water. Choose unbruised cactus; smaller leaves are generally more tender. Refrigerate."

"It is no simple task," says Sigmund A. Lavine, author of *Wonders of the Cactus World*, "to describe completely the fruits of the cacti. They vary tremendously in the color, shape, and taste among the

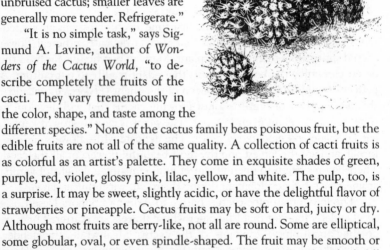

different species." None of the cactus family bears poisonous fruit, but the edible fruits are not all of the same quality. A collection of cacti fruits is as colorful as an artist's palette. They come in exquisite shades of green, purple, red, violet, glossy pink, lilac, yellow, and white. The pulp, too, is a surprise. It may be sweet, slightly acidic, or have the delightful flavor of strawberries or pineapple. Cactus fruits may be soft or hard, juicy or dry. Although most fruits are berry-like, not all are round. Some are elliptical, some globular, oval, or even spindle-shaped. The fruit may be smooth or covered with scales, woolly or have silk-like hairs, even bristles which fall off as the fruit matures.

The various opuntia cacti are the source of the most delicious fruits borne by the Cactaceae. These fruits are pear-shaped and are known popularly as "prickly pears." Care must be taken to avoid the barbed bristles when peeling the fruit, but both Native American and Mexican women do this with great ease. However, if the Apache legend is to be believed, all cacti with edible fruit were once densely covered with needle-like spines, and as a result it was very difficult to harvest cactus fruit. But Killer of Enemies, a legendary demigod who watched over the Apaches, took pity on his people and used his magic to make most of the spines vanish.

Killer of Enemies must have been very powerful for how else can one explain the lack of spines on *Opuntia ficus-indica*, also known as "Indian

Fig?" This nourishing and refreshing fruit is easy to pick and is a staple food for Indians in both North and South America. The Indian Fig is so highly regarded that it has been introduced in many subtropical countries where it is not native. Its fruit feeds livestock and pigs; moreover, the stems supply green fodder during subtropical summers when grass frequently dries up because of the excessive heat and lack of rain.

The fruit vendors of the southern states and Mexico offer other varieties of cactus fruit to their customers. The round red berries of *Opuntia streptacantha*, popularly known as "tuna cardona," are also popular. The sweet pulp of this fruit is used to make preserves and an alcoholic drink. Garambullos, the small, blue, olive-shaped berries of *Myrtillocactus teometrizans*, are also a great favorite in the Mexican markets. Generally speaking, cactus fruit has a pleasant taste, though some species have a slightly musky flavor. The cream of the crop, however, is *Echinocereus stramineus*. Its crisp, juicy, brownish-red fruit tastes like strawberries. Native American tribes in the Southwest, Central and South America prized cactus fruit, but the Pimas and Papagos of Arizona absolutely relied on cactus berries, grinding the stems of the prickly pear to extract sugar.

Other parts of cacti are also edible. Delicate cactus flowers are tasty treats when dipped in batter and fried in hot sunflower oil. Honeybees gathering pollen rummage through the golden anthers of prickly pear blossoms during the day. But when night comes moths, who take their bearings by the Moon, take over the task. And, surprisingly, certain species of cactus are pollinated by bats.

Few cactus plants have achieved more publicity than barrel cactus. Its name is appropriate; not only are these desert drinking fountains shaped like barrels but they also contain large amounts of water. Their pulp is full of moisture months after a rain. They are not, however, as desert folklore would have you believe, filled to the brim with crystal-clear water, available to save the lives of dehydrated travelers. When the top of a barrel cactus is lopped off, the plant is filled with a soft mass of storage cells that hold moisture like a sponge. The pulp yields a sticky liquid when scooped out and squeezed that is slow to evaporate and so bitter that only a desperate traveler would drink it. The barrel cactus, perhaps, has saved the lives of people only when no other moisture was available.

The pulp is cut into cubes and boiled in sugar syrup, then it is pressed through strainers and hardened. This makes a sweet delicacy called "cactus candy." Today, cactus candy is sold in many markets in the Southwest and Mexico, packaged in boxes with desert scenes.

There is still another cactus plant highly regarded by the Native Americans of the Southwest. To them, the reddish-purple, cucumber-shaped fruit of the saguaro, or *Carnegie gigantea*, is far more delectable than cactus candy. When the berries of the saguaro mature in late June, the Pima and Papago set up harvest camps and gather tons of the fruit. They eat the fruit fresh and also make preserves or jam from it. Juice extracted from the fruit is boiled down to a molasses-like syrup and stored in pottery jars. The tiny black seeds of saguaro fruit can be strained from the fruit's pulp and ground into meal that makes a kind of cake.

A saguaro will bear fruit annually for as long as three years without receiving a drop of water. Saguaro fruit is so important to the Papago people that its harvesting is celebrated each year with a festival which once marked the start of the Papago New Year. The name saguaro itself derives from the Pima word for "friend."

Maguey is the name given to several kinds of agave plants that grow in Mexico and the Southwest. The name is often used for the pulque agave, which yields a juice used to make tequila. The agave, often called the "century plant," lives fifteen to twenty years. This is quite a long time, says Edward Abbey, "if you are sitting around waiting for it to bloom." The agave flowers bloom only once during its life; it dies after flowering.

The agave's green leaves are edged with red spines that grow to nine feet long and a foot wide. In the center of the plant, the leaves fold over the agave's core and conceal the heart like an artichoke's leaves. When the moment arrives for the plant to bloom, the inner leaves part and within a few days a great tumescent stalk rises to a height of up to fifteen feet. Its blooms, loosely clustered pyramids of large, showy golden flowers at the tips of the stalks, are an impressive sight in the desert moonlight.

Once the center leaves of the agave part to bloom, you can dig up the agave, cut out its heart, roast the heart on slow coals, and enjoy a wonderful meal. It is especially delicious seasoned with the wild garlic that grows in the same area. If left undisturbed, the agave blooms for only a very short time. In a few weeks the flowers fade, the stalk dries up, and the entire plant dies. But what a glorious end. It has accomplished the purpose for which the sacrifice was made; the flowers are succeeded by pods containing the seeds for the next generation of agave.

Cactus as Lumber

Whether they are smooth or prickly, all cacti are classified as Cactaceae by taxonomists. They comprise a family of perennial herbs and shrubs furnished with aerials. But can you imagine sleeping in a "cactus herb hotel"? Few foreigners spend the night at the hotel in Cachi, a small town in northern Argentina, but visitors are amazed to discover that the hotel's beams, floors, and doorsills are cacti. And this is not a unique situation in the semiarid regions of Central and South America where lumber is scarce. Carpenters in these areas make window frames from cactus stems. Many Mexicans live in cactus-framed houses plastered with adobe. Native craftsmen fashion furniture from the woody skeletons of the saguaro.

The long, tapering, woody ribs which form tubes inside saguaro tissue and support the plants in high winds are excellent substitutes for lumber. Ranchers use these ribs to build shelters, while novelty manufacturers make planters and other small items from saguaro "wood" for the tourist trade. The Papago use the saguaro ribs for varied purposes, including roofing their houses and making cradles. Man is not the only architect to use the giant saguaros for home building. Desert woodpeckers peck holes in them for nests. Indeed, many species of birds use different types of cacti for nest building, relying on cactus thorns to help protect their young.

Cactus in Religion

Oasis dwellers in the Syrian Desert surround their property with high mud fences topped with prickly pears to keep out stray animals. Farmers in Central America enclose a fence of cacti around their fields to protect them. Natives of the island of Curacao, off the coast of Venezuela, hang their washing on cacti, using the spines for clothespins. Moreover, to the Native tribes of the Southwest United States and Mexico, cactus plants are considered sacred. The Zuni classify the cactus as one of the plants that give themselves to the people. They in fact regard certain cacti that bear edible fruit so highly that a member of the Zuni Cactus society, charged with the control of game and the curing of wounds, carries a specially beaded prayer feather when he approaches the cactus to be used in ritual whipping of a chief.

Cactus societies also flourish in other Southwestern tribes that conduct ceremonial whippings with cacti. These lashings are said to make the individual receiving them strong and brave. It is a custom also among the Hano to beat the ground with cactus stems, believing that doing so would cause the ground to freeze so their warriors would leave no tracks. Cacti, as might be expected, are ruled by Mars.

Certain kachinas, masked representatives of the ancient gods, carry cactus stems during their seasonal visits to the pueblo of Acoma. Men who rub against these stems believe they will have more vigor. The Hopi people, meanwhile, believe placing bits of cactus in the corners of their houses will cause them "to grow roots."

Cacti are also used in the sacred rites held in kivas, the enclosed chambers where pueblo-dwelling Native people conduct religious ceremonies. Many Native Americans believe that cacti make powerful "lucky pieces," often tucking a bit of cactus in their belts to insure a successful hunt. The Native people of Mexico also associate cacti with magic because of a very old legend about the first Aztec community. According to the legend, long ago the Aztec people migrated south from a cold land. Their chief deity, Mexitli, instructed their leaders to keep traveling until they came upon a great eagle perched on a cactus devouring a snake. The Aztecs spent years wandering, but one day the headmen saw the omen that Mexitli had foretold. The Aztecs built their homes at the spot where the cactus grew. Today, the community they founded is called Mexico City, after Mexitli, and the omen that led to the city's establishment appears both on the flag and the seal of the Republic of Mexico.

Peyote, Blessing or Addiction?

Long ago, many Native American tribes used *Lophocora williamsii*, a spineless, gray-green "button" cactus covered with white hairs, in religious ceremonies. Commonly called *peyote*, this native plant of northern Mexico and Texas contains mescaline, a mild, nonaddictive hallucinogen. Although the use of peyote is forbidden by law in several states, there is no federal mandate against it.

It is not known when the Indians began using peyote as a hallucinogen. But it has been established that early residents of the Southwest and Central America not only used the cactus as a medicine but also ritually ate peyote buttons to foretell the future and combat witchcraft. Peyote is not easy to find. It grows only in a few areas, and according to legends it has the ability to "sink into the ground" if those who search for it have not purified themselves.

No Native Americans consider peyote more sacred that the Huichol of Mexico. They prize peyote as a medicine, as well as for fortune-telling, amulets, and religious rites. For the Huichol, peyote was once a deer. In their peyote ceremonies, the sacred music is accompanied by deer-hoof rattles and notched deer bones that are rubbed together. The area inhabited by the Huichol does not grow the peyote cactus. Therefore, the Huichol tribe make a pilgrimage northward to the state of San Luis

Potosi to gather the cactus—a journey of forty-three days. Elaborate purifying ceremonies must be performed at the start of and during the trip.

Increasing numbers of Indians throughout the Americas have become convinced that God created peyote so that man could communicate directly with Him. As a result, the Native American Church of the United States, which combines ancient peyote ceremonies with modern religious thought, has become popular.

The Hedgehog Cactus—A Prickly Beauty

To bring a bit of the desert's beauty into their homes, the hedgehog cactus can be easily grown; its only requirement is full sunshine. Its unusual name, meanwhile, comes from the resemblance of its long spines to a porcupine's quills. At the same time, the hedgehog cactus produces some of the most beautiful flowers in the cactus family. Its blossoms are brilliant and spectacularly large in relation to the size of the plant. Few hedgehogs grow more than ten inches tall, yet their flowers may be more than three inches in diameter fully opened. One species, the comb hedgehog cactus, has blossoms five or six inches across on a plant no more than nine inches tall. The colors of its blossoms vary from white, yellow, and pale green, to a mixture of red and blue, and every shade of purple. The satiny sheen of the petals heightens the blaze of color. There is a further touch of elegance in the subtle shading from dark at the base to lighter color at the tip of the petals. One species, the golden hedgehog, is admired for its pale lavender flowers with reproductive organs of contrasting color; dark green female stigmas at the center of the flower and male stamens of a delicate whitish yellow.

Potting Mix for Cacti

Cacti do not need rich soil, but their growing medium must be porous and permeable. You can buy a mix or make your own:

2   parts sharp sand
2   parts loam
1   part old brick pounded into bits
½   part peat moss.

Also note, more cacti are killed by kindness, or overwatering, than any other factor. And care must be taken if chlorine is employed to purify the water supply in your area.

The Waves in Your Garden

Harry MacCormack

Having written previously about connections between cosmic bodies and growth and well-being here on Earth, in this article I will take you deeper into the possible nature of these connections. As gardeners, we create and extend relationships between various plants and ourselves or other humans. In this discussion we will touch on what we do when we garden: manage the very basis of the life process in continual creation. The motions of earth waters make a suitable model for understanding the life-giving energy waves in your garden.

Thomas Moore once wrote: "Relationships have a way of rubbing our noses in the slime of life—an experience we would rather forego, but one that offers an important exposure to our own depth." Gardens, as reminders of the mythic garden, are relationships enacted. Our gardens reflect our management of life forces—our health, our fertility, or lack of both. Gardens are extensions of ourselves as lovers, antagonists, and manipulators over a host of plants, microbes, minerals, and insects.

Indeed, as Jeffry Wolf Green wrote in *Pluto Volume II*, the garden of Eden story has so forcefully imprinted the three religions of Western culture that our understanding of fertility and reproduction, maleness and femaleness flounder under a heavy sense of jealousy and betrayal. When we are in relationship with plants in gardens or in the wild we carry at least some consciousness of our expulsion from the Garden.

As two-legged creatures, humans are flawed. Or as the Hopi Elders put it, we are two-hearted. We are constantly attempting to negotiate a truce between the need to socialize and the need to find integration with plants and animals outside our species. The question remains whether it is possible for humans to balance our lives and become one-hearted.

In our gardens, we see the seasonal cycles of birth and death and are made aware that the seeds for rebirthing come in the shape of death each year. Perhaps it is our fear of being part of this life process, and the fact that we can own none of it, that makes us at once curious about gardening and hesitant to risk it.

In the end, with millions of suns surrounding us, and even more millions of planets, we must recognize we are here to learn the patterns of a highly structured universe. And some of our most important lessons can come in gardening experiences.

One day about a year ago, a friend of mine who is a forest ecologist introduced me to Dr. O. Ed Wagner, a physicist. Dr. Wagner was getting some attention from forestry people because of his research regarding spacing relationships among trees and other plants. His book, *Waves in Dark Matter*, showed the existence of what he termed W-waves, so-called because they are found in trees and "wood."

W-waves are very slow moving waves that move through plants and other life-forms. In his research, Dr. Wagner reveals that W-waves are not electromagnetic, but rather directional and perhaps linked with gravitational forces. Apparently, plants and other beings act as oscillators for W-waves to move through. There is some evidence to show that plants communicate with each other by means of these waves. When under attack by insect or saw, plants go into collective defense, producing abnormal chemical reactions and displaying peak measurements in wave activity. Furthermore, Dr. Wagner has discovered that the spacing between structures on plants—that is, between nodes and branches—are quantized and repeat from plant to plant within a species. This kind of ordering occurs on a larger scale with planetary and Moon orbits and ring systems, which, according to Wagner's theories, occurs on the basis of wave energy. In fact, Dr. Wagner's measurement of W-waves show energies coursing up and down tree trunks and stems. It is likely that W-wave forces facilitate sap flow and help create the gravitropism which produces a plant's organization. W-waves apparently move through all matter and are universal.

Like gravity, W-waves penetrate matter and space and are dependent on bodies for their resonance. That is, W-waves are usually not noticeable unless moving through matter. Certain specialized shapes produce the most efficient resonances and frequency multipliers.

The presence of an apparent W-wave background field throughout the universe makes it possible for all life to be in communication. One of the functions of living creatures is to take in this wave energy from the vacuum and reenergize the waves for further transmission. Each body will do this differently and create what we know as life diversity. The manner of a body, how it communicates, is stored in its genetics and chemistry.

Seeds, eggs, stones, and so on hold the potential to communicate interactively on a cosmic level.

Gardening is relationship management. If certain plant shapes resonate differently than other plant shapes, then a host of choices are available to us as we lay out a garden. First, we need to determine, by measuring plant wave energy, how close a cabbage should be, for instance, to a tomato plant in order to encourage optimal growth. That is, we need to know how much one plant's resonant field can be effectively reduced by proximity to another resonant field—even if all other gardening techniques are normal—and we need to determine if certain plants interact more easily with other plants' resonant fields.

Some plants have vibration rates that are antagonistic or overpowering to other plants. With study we may be able to discover that some mixtures of plant vibrations prove beneficial in unexpected ways. For instance, the right mixture of plant wave energies might create patterns that would confuse harmful insects. I have observed century-old orchards where multiple species cropping works in staving off insects without the use of insecticides. Intercropping of "companion" plants in gardens may be successful at least in part because of wave interactions.

Dr. Wagner's research has involved the possible role of plant and animal shapes in resonance. Meanwhile, other researchers have focused on how various substances affect the attraction and transmission of waves. Wilhelm Reich, Carey Rheems, Dr. T. G. Hieronymus in this country, and Dr. Plocher in Germany, have all been able to capture wave energies and utilize them to bring harmony and change to various life forms. They have discovered how atomic structures are birthed, held, transported through waves and how these waves have been affected by the clay content in various soils. Their research has focused on discovering how certain sites of positive/negative action-exchange in wave activity in the soils have trasmuted to biological action, and they have studied the ways to enhance the activity in these sites. Dr. Plocher, for instance, has made certain powders that can help to heal polluted water and a host of other environmental problems. Victor Roehich has created an Intrasound powder and gel (just released in the U. S. by Tri Star) that can almost immediately balance diseased or sickened biosystems.

The energy of material matter in a crystallized form must be able to move through a body freely. Blockages can occur when matter is overpowered with energized clays or other substances. A healer's hand can bring relief to a body that is out of balance or toxic by refocusing waves. The resonate focus of energy waves through an atuned body to a body that has been corrupted is an amazing process. People with the proverbial "green thumb" also tap into wave energy. Physical change can occur through the wave communication in meditation and prayer. Agricultural-based cultures have always maintained fertility by reliance on various religious practices. Even in the face of modern, reductionist science-based agriculture, agricultural communities are usually very prayerful and aware of the physical connectiveness between all life forms.

Subtle wave energies are the basis of continual harmonious creation. As humans we are healthiest and reach our greatest potential if we align ourselves with wave interactions. Without wave sensitivity, humans can become blocked and mentally desensitized from how our physical world operates. Such disharmonies result in a pursuit of power and money rather than fertility as our human purpose.

Dr. Wagner and a host of other researchers have noticed and docu-mented the effects of disrupting wave energies on W-waves. Energy disrupting waves are usually generated by human machinery. Wagner found that plants close to a power station or transmission lines showed significantly altered W-wave patterns. Electrical lines can emit disorienting wave patterns which overpower normal subtle wave patterning. Such disorientation at cellular levels may be a primary factor in disease, cancer, AIDS, degeneration, and sudden unexplained deaths. In your garden you can see this disorientation in stunted plants (when fertility and all other factors are measured as normal), and in plants who have been attacked by insects to the point of death.

Furthermore, Dr. Wagner hypothesizes that W-waves are primary agents in Sun cycles. Our Sun is currently undergoing an energy transfor-mation; it began in 1996 and will peak in 2012 and is the result of a very long cycle of cosmic wave balancing. This prolonged event was predicted by such cultures as the ancient Tibetan, Mayan, and Hopi peoples. In fact, the Sun's current wave pattern change is so complete that many ancient calendars chose to end when the cycle completes itself in 2012.

Such very large wave occurrences transform our planet, and life forms will seek a new balance as a result. According to the wise ones of the an-cient cultures, the current proliferation of very intense storms, droughts, earthquakes are all signs that the rebalancing is already underway. These changes are a cosmic cleansing or healing, and I have seen signs of change in my garden these past two seasons as many plants have shown strange variations in growth habits and ripening.

If W-waves hold the organizing forces necessary for continuous universal creation, it may be possible to focus them or to enhance them through our human awareness and transmission capabilities. Many healers already do this, and many cultural ceremonies may assist fertility by celebrating subtle wave changes through the seasons.

Currently, there are many devices which apparently attract these wave energies so that our gardens and farms become ever more fertile and productive. I suspect that the W-waves so carefully investigated by Dr. Wagner are the same waves that have occupied the attention of some of our leading and most controversial scientists for over fifty years. Wilhelm Reich, for instance, created devices to capture and transmit what he called Orgone energy. Unfortunately, Reich's books and devices were banned in the U. S., and he was imprisoned in the late 1950s because of his work. Furthermore, Dr. T. G. Hieronymus perfected instruments to collect and measure wave energies, calling his system Cosmiculture, or the New Agriculture. Dr. Carey Rheems studied the influences of electromagnetic energy on soil and nutrition, and he developed an agricultural methodology based on his findings. His ideas, which have become known as radionics, mirror much of the research of Dr. Wagner.

For those who love the idea of being able to work with cosmic wave energies, the Hieronymus "Eloptic Energy Analyzer" allows one to study wave energies coursing through biological life forms. It operates on the principle that biological resonance can be monitored by comparing scales of healthy and unhealthy energy environments. His years of testing has allowed him to discover the perfect energy balance for soils, organic gardening systems, insect management, and water pollutant control.

Another of Hieronymus's tool, the "Cosmic Pipe," helps attract energies to a garden. The history of this type of tool may go back to the time of Stonehenge, when there was a greater reliance on cosmic energies for food production and health. The cosmic pipe device has a head that is fixed at least six feet above the soil in a garden. The head helps attract focused wave energies, and a conductor sends these energies into a resonance chamber where vials are placed. The vials are filled with whatever the gardener wishes to modulate—seaweed extract for instance, or fish fertilizer. This modulated energy then goes to the amplifier-radiator in the pipe, buried at least two feet in the ground. There the modulated energies are broadcast horizontally in a broad band out through the soil to the roots of plants, to the ionic structures of soil, or into the molecular arrangements of minerals and microorganisms. Concentrating cosmic energies in this fashion has proven to show tremendous changes in gardens and farms. Soil nutrients are balanced over a season rather than being released in peaks and valleys. Harvests have been doubled, the time of production shortened by two weeks to a month, and the sugar content of fruits has been markedly increased. Hay crops have been tripled. Chemical residues have also been removed.

Both Hieronymus and Reich found that by using a pipe device to concentrate cosmic energies, they were able to affect weather. Balance was achieved in these experiments. Reich, by aiming his pipe, was able to create storms, a practice which attracted the attention of the military and eventually led to his imprisonment and death.

Although wave technologies have become popular with a number of prominent organic, biodynamic, and sustainable growers, these same growers are also very careful to utilize the best composts, minerals, and green manuring techniques for balancing soil health. As with your own body, achieving cosmic balance requires careful diet and studied awareness. Using devices to manipulate wave energies adds a dimension to life which is not easily available in our otherwise market-driven realities. My burgeoning knowledge of W-waves and subtle energies has changed the way I move in the world.

To learn more, I suggest you read the texts I have listed in this article. And if you want to pursue your own investigations, you can purchase tools to monitor and manipulate cosmic energies at the following locations:

Hieronymus, P.O. Box 3326, Eatonton, GA 31024

Acres, USA at P.O. Box 8800, Metairie, LA 70011

Transforming Your Garden into an Old-Fashioned Retreat

Carly Wall

Do you feel your flower beds have lost the sparkle and charm you recall from your grandparents' delightful garden? From my youth I ran barefoot down a downy path under the light of the Moon. Lightning bugs swirled around me, and I inhaled the intoxicating scent of old roses, lavender, and jasmine. Grown-ups sat on the porch, talking while their glasses tinkled from the ice.

Why don't we do this anymore? Possibly because the plants have changed. We've created hybrids and plants that can't reseed themselves or live without sprays, fungicides, and sprinklers. In fact, today's plants need so much care, that gardening is rather frustrating. You can spend hundreds of dollars trying to recreate grandma's effect, but to no avail. Still, there is hope, because we know one of her little tricks: she used heirloom seeds.

Heirloom—The Only Way to Go

Because of a frightening loss of seed and plant diversity and habitat today, we stand to lose precious plant life. With the current "homogenized" growing practices, heirloom plants become ever more important.

Heirloom plants are traditional varieties of green life that have been passed down from generation to generation. As plants particularly suited to grow in certain areas, you can save yourself some trouble if you use them. In your area, there are likely one or two old gardeners who have their own special brand of flower seed, as well as the historical story behind that plant. Heirloom seeds are open-pollinated or self-pollinated, which means that pollination occurs without intervention by man. While hybrids produce consistent varieties, heirlooms are prized for their scent, flavor, and diversity. Also, heirloom plants are less apt to be wiped out by disease because of diversity, and unlike most hybrids, seeds saved from heirlooms can be planted year after year.

There are literally thousands of varieties of heirloom plants that fall into one of three categories: biennial (growing the first year, producing seeds the second); perennial (living and bearing seed year after year); or annual (flowering and maturing seeds the first year). Annuals are easiest to grow; they have a range of colors and scents, and are the easiest to collect seed from. Heirloom flowers are hardy and attract bees, butterflies, and hummingbirds to your garden. Your garden will sow itself even if you don't save the seed and replant, thus creating better gardens each year.

Different Heirloom Annual Flowers

There are so many flower varieties to choose from you may get dizzy. For cut flowers, there's the Gloriosa Daisy in yellow or mahogany, the Cleome in pink and violet, the Cosmos in sunny yellow, the Ageratum, the Woodland Tobacco, the Four-o-Clocks, the Tassel flower, and the pink Lavatera. If you pine for flowering vines, choose from the blue Morning Glory, the Hyacinth Bean, Cup and Saucer vine, Spanish Flag, the Cardinal Climber, or Love in a Puff. For a night garden, here are some fragrant flowers to choose from; Jasmine Tobbaco, Moonflower, Evening Scented Stock, Night Phlox, and Sweet Alyssum. Some of my personal favorites include amaranth, asters, bachelor buttons, borage, broom corn, calendula, cleome, cosmos, hollyhock, lamb's ears, marigolds, nasturtiums, poppies, sunflowers, and zinnias. Here are some descriptions of heirloom flower varieties from the a popular catalog:

Amaranth, Elephant Head (*Amaranthus gangeticus*)—passed down through immigrants from Germany. The reason it's called Elephant Head is that the deep-red bloom grows quite large (thirty-six to forty inches) with long protuberances that look like an elephant's trunk.

Bachelor Buttons (*Centaurea cyanus*)—An old favorite garden flower with deep marine-blue flowers growing one to two feet.

Calendula (*Calendula officinalis*)—Bright yellow-orange double flowers bloom profusely from spring until frost and have been used in medicinal salves since at least the early 1600s.

Calliopsis, Elegant (*Coreopsis tinctoria*)—An excellent cut flower, this hardy two-foot garden daisy blooms from late July until a hard frost. The three-inch diameter flowers are brilliant yellow or velveteen burgundy.

Cosmos, Mixed Seashell (*Cosmos bipinnatus* var. *cosmicos*)—From the Greek word meaning "harmony," these easy-to-grow, frilly leafed plants can be used as a companion plant with corn to deter ear worms or in the carrot patch for a touch of beauty.

Marigold, Pinwheel (*Tagetes patula*)—A two to three foot tall marigold with scallop-edged petals, radially striped maroon and yellow with golden centers. Most petals are evenly striped, some are yellow with red markings, and some are all yellow or all red.

Poppy, Cornfield (*Papaver rhoeas*)—Beautiful flowers, single and double, in red, pink, and a range of pastels. This poppy grew wild in European meadows for centuries and has been eliminated from many cornfields by herbicides.

Poppy, Mixed Carnation (*Papaver somniferum*)—A beautiful, Royal Horticultural Society variety that provides an early summer spectacle with prolific pink, red, and salmon double blooms born on strong, upright one to two foot stems.

Sunflower, Russian Mammoth (*Helianthus annuus*)—A favorite that is easy-to-grow with ten to fourteen inch, large-headed flowers and nine to twelve foot stalks. First offered by seed companies in 1880s. Edible seeds are medium-large, good for eating or for the birds.

Tobacco, Aztec (Jasmine-scented) (*Nicotiana alata*)—A profusion of fragrant, trumpet-shaped, one to two inch pale pink to white blossoms on a stem arising from a rosette of thick velvety leaves. It is especially fragrant in the evening and makes a good trap crop for potato beetles. Grows to five foot tall.

Zinnia, Mixed Beautiful (*Zinnia elegans*)—This selection is diverse in color and petal type. Grows to four feet tall and blooms through the entire summer. An easy to grow garden favorite.

Saving Seed

Saving seed used to be something that was done naturally. We know why that habit fell away, but now since we are bringing back the natural way of planting and growing, we can again regain this pleasant occupation. If you do want to save the seed, you will need to take only a few measures.

Let the plant go to seed till the seeds dry completely on the plant. On a dry day when the dew has been dried by the Sun, collect the seeds. To be sure they are moisture-free and to avoid mold, lay the seeds out on a few newspapers in a darkened room for three to five days. Pick through and collect the seeds. You'll want to store each kind of seed separately; for a pretty effect, I like to store my seeds in jelly jars with ring lids. Be sure to label and date your jars. If seeds are stored dry and well-sealed, they should keep for several years. Some seeds do require a cold dormancy period—so you should stick them in the freezer for several weeks before planting. Be sure to check the seed requirements in a seed catalog or gardening directory. Also, because storage time reduces germination, you may want to plant your self-saved seeds more thickly than normal to get a good crop.

After you've experimented with the annual heirloom flowers, you may be tempted to go on and try heirloom perennials, and then on to heirloom fruits and vegetables. You may even become a true heirloom fanatic. If so, here are a few organizations or associations, exchanges, and catalogs you may want to check out.

Organizations

Heritage Rose Foundation, 1512 Gorman Street, Raleigh, NC 27606

Historic Iris Preservation Society, 12219 Zilles Road, Blackstone, VA 23824

Thomas Jefferson Center for Historic Plants at Monticello, P.O. Box 316, Charlottesville, VA 22902

Flower and Herb Exchange, 3076 North Winn Road, Decorah, IA 52101

Heritage Seed Program, RR3 Uxbridge, Ontario L9P 1R3 Canada

Seed Exchangers, P.O. Box 10, Burnips, MI 49314

Seed Savers Exchange, 3076 North Winn Road, Decorah, IA 52101

Catalogs

Abundant Life Seeds and Books, P.O. Box 772, Port Townsend, WA 98368, (206) 385-5660

Down on the Farm, P.O. Box 184 Hiram, OH 44234

The Fragrant Path, P.O. Box 328 Fort Calhoun, NE 68023

Seeds of Change Organic Seeds, 1364 Rufina Circle #5, Santa Fe, NM 87501, (505) 438-7052

Jack-O'-Lanterns on a Moonlit Night

Carly Wall

We all know jack-o'-lanterns are a tradition on All Hallow's eve. Halloween simply wouldn't be the same without pumpkins glowing on porch rails as children look for tricks or treats. As a child, I was afraid of jack-o'-lanterns even though they grew as ordinary pumpkins in the garden all summer long. I began to suspect that the Moon had something to do with it, for the Moon is especially bright around Halloween. We set out into the dark night with only the hideous, candle-illuminated faces and the light of the Moon to lead us back home.

As I grew older, I realized it wasn't the fault of the pumpkins. We were after all, the carvers of those faces. But I did wonder just how this custom had come about. It seems an Irish story claims there was an old man named Stingy Jack too mean for heaven or hell. So upon death he was cursed to walk the earth with only the Moon to light his way. If he could find them, however, Stingy Jack carved out turnips, potatoes, or rutabagas and placed a burning coal inside. Since his name was Jack and he carried a lantern, his name became "Jack of the Lantern" or "Jack-o'-lantern." It became a tradition on All Hallow's eve, the night when spirits were loosed to walk the earth, to carve vegetables, place a light inside, and put them in windows or doors. This was meant to scare off Stingy Jack or any other spirits that might be lingering around. Eventually, pumpkins became the favored carving item in America. Pumpkins are abundant at that time of year, and they make quite an impressive lantern. That's why Washington Irving put the pumpkin in "The Legend of Sleepy Hollow." Pumpkins are the scariest vegetables around.

Growing Your Own Jack

Native Americans favored pumpkins for food, and so did the colonists. It's a holiday tradition to have pumpkin pie at the Thanksgiving table, just like the early settlers did. All pumpkin varieties have hard shells, but certain varieties are said to be best for eating or for decoration.

The eating varieties are the buff-colored sugar-pie or Dickinson, the Golden Delicious, and the Small Sugar. There's also a squash, the Lumina that resemble the pumpkin except that it is white-skinned—it makes for a ghostly jack-o'-lantern. For carving, there's the Big Max, Mammoth gold, and the huge Big Mac which grows to twenty pounds or more. Still not enough? Well, you could grow a pumpkin the size of your truck! This variety, called Dill's Atlantic Giant, was developed by Howard Dill of Ontario. Or you can grow miniature pumpkins; there's Jack Be Little or Munchkin, each growing one pound or less, or Spookie and Small Sugar

at two to eight pounds. There's also a middle ground—Jackpot, Autumn Gold, and Ghost rider—each weighing in from eight to twelve pounds.

Even with all the varieties, pumpkins are easy to grow. To do so, you must plant the seed when the ground is fairly warm and the danger of frost has passed. Place three or four seeds per hill and then thin to the best two vines. To grow larger pumpkins, select a few of the developing fruits to keep on each vine and pinch off the rest. Insects you have to watch for are the seed corn maggot, the spotted and striped cucumber beetle, the squash vine borer, the pickleworm, and squash bugs. Since bees pollinate pumpkins, you should welcome them freely to your garden. Without them, you wouldn't get much of a harvest and what little that came would be poorly shaped. To harvest, merely wait until the shell has hardened and the color of the pumpkin is good. You'll also want to harvest before the first heavy frost. Cut your pumpkin off the vine with a sharp knife, leaving a good stem on your pumpkin. That's all there is to it.

The Secret to Growing Huge Pumpkins

The secret to growing huge pumpkins is work. To start, you should prepare your seedbed by mixing fertilizer into the soil. Plant three to five seeds per mound and thin to a single plant. Three weeks after seeding apply one cup of nitrogen fertilizer around the vine. Repeat every two or three weeks. Keep your plant well-watered all summer and allow only one fruit to develop on each plant. You may want to keep straw under the fruit to keep it from staying wet and rotting against the ground.

Eating Lots of Pumpkin

After all your hard work, not only can you decoratively display part of your harvest, you can eat it too—in all kinds of delicious ways. Pumpkin is actually good for you—one cup of pumpkin contains over half of the daily recommended allowance for vitamin A. Pumpkins are also high in vitamin C, niacin, pantothenic acid, and potassium. Other minerals include phosphorus, silicon, iron, magnesium, and calcium. Pumpkinseeds are high in zinc and other minerals. Here is a collection of recipes. Pumpkins have a variety of uses, and perhaps you can serve a few of these dishes on the night you light your jack-o'-lantern!

Dried Pumpkin Flour

This flour can be used to replace a quarter of the wheat flour in recipes to add moisture, nutrition, and color. Dry the inner fruit of the pumpkin in thin slices. When dried, place in a 175°F oven for 15 minutes to pasteurize, then pulverize to a powder in a blender. You can store the flour in plastic bags or jars with tight-fitting lids.

Pumpkin Muffins

- 1¾ cups flour
- ¾ cup of sugar
- 3 teaspoons baking powder
- 1 teaspoon pumpkin pie spice
- ¼ teaspoon salt
- ¾ cup canned pumpkin
- ½ cup skim milk
- ¼ cup oil
- 2 eggs
- 3 tablespoons brown sugar

Heat the oven to 400°F. Line muffin cups with baking cups. Mix dry ingredients. In a small bowl, mix pumpkin, milk, oil, and eggs. Blend in dry ingredient mix. Spoon batter into cups and sprinkle a little brown sugar on top of each muffin. Bake for 15–18 minutes.

Pumpkin Bread—In Jars

⅔ cup shortening
4 eggs
⅔ cup of water
½ teaspoon baking powder
1 teaspoon ground cloves
1 teaspoon cinnamon
1 cup chopped walnuts
2⅔ cup sugar
2 cups canned pumpkin
3⅓ cups flour
1½ teaspoons salt
½ teaspoons allspice
2 teaspoons baking soda
8 wax paper circles, lid-sized
8 wide mouth pint jars

Pour batter into greased canning jars until half full. Place jars on a cookie sheet. Bake at 325°F for 45 minutes. Remove from oven, cover with waxed circle and seal with lids. Turn upside down to cool; can be kept a long time. To serve, shake from jar, slice and serve with whipped cream.

Traditional Pumpkin Pie

2 eggs
½ cup sugar
½ cup corn syrup
1½ teaspoons ground cinnamon
1½ teaspoons ground ginger
¼ teaspoon ground cloves
1 16-ounce can pumpkin
1½ cups evaporated milk
1 unbaked pie shell

Beat eggs and stir in rest of the ingredients. Place pie plate with unbaked crust on a cookie sheet. Pour in the pumpkin mixture. Carefully place in oven. Bake 425°F for 15 minutes, then reduce heat to 350°F and bake 45 to 55 minutes or until a knife inserted into the center comes out clean.

Pancakes Pumpkin Style

1   cup flour
2   tablespoons brown sugar
2   teaspoons baking powder
½   teaspoon pumpkin pie spice
1   cup milk
½   cup canned pumpkin
1   tablespoon vegetable oil
2   eggs

Combine dry ingredients. Beat eggs with milk, pumpkin, oil, and fold in dry ingredients. Spoon batter onto hot griddle. Flip when bubbles form. Warm maple syrup, beat in cream cheese, and pour over pancakes.

Pumpkin Cake

4   tablespoons shortening
½   cup of canned pumpkin
1¼   cups sugar
3   eggs
1   cup milk
2½   cups flour
4   teaspoons baking powder
½   teaspoon salt
1   teaspoon cinnamon
¼   teaspoon nutmeg
½   cup raisins
½   cup chopped pecans

Mix together dry ingredients. In a bowl, beat eggs, add pumpkin and shortening. Blend with dry ingredients, raisins, and nuts. Pour batter into two round, greased, and floured cake pans. Bake 375°F 30 minutes until a toothpick comes out clean. Frost with cream cheese frosting and pecans.

Pumpkin Pudding

1½   cups canned pumpkin
2   tablespoons flour
1   cup brown sugar
2   cups milk

1 cup evaporated milk
1 teaspoon cinnamon
½ teaspoon ginger
½ teaspoon nutmeg
2 eggs, beaten

Butter a cast-iron pan. Combine ingredients in pan; bake 350°F 35 minutes.

Pumpkin Cookies

½ cup shortening
1¼ cups brown sugar
1 cup pumpkin puree
2 eggs
2 cups all-purpose flour
2 teaspoons baking powder
½ teaspoon salt
½ teaspoon ginger
1 teaspoon cinnamon
1 teaspoon nutmeg
¾ cup chopped walnuts

Combine dry ingredients. In a second bowl, cream shortening and work in sugar and eggs. Stir in the pumpkin and nuts. Gradually stir in dry ingredients. Drop spoonfuls of batter on a cookie sheet. Bake at 375°F 10–15 min.

Pumpkin Candy

1 cup solid-pack pumpkin
1 cup sugar
1½ cups flaked coconut, lightly packed
½ teaspoon cinnamon
¼ teaspoon nutmeg
1 cup finely chopped pecans
1 cup crushed cereal

Mix the first five ingredients in a large heavy saucepan. Cook over medium high heat, stirring constantly for 15–20 minutes. After it thickens and forms a ball in the center as you stir, add the crushed cereal, mix well, then turn onto a buttered baking sheet. Cover loosely with foil or plastic wrap. Let cool completely. Lightly butter hands. Shape candy into balls. Roll in nuts. Cover, and store in refrigerator. Yields 30 pieces.

Old-Fashioned Roses

Touched by Love and the Lunar Rays

Carly Wall

There is an old love charm involving both a beloved flower and a certain phase of the Moon. It seems that long ago, a wandering lover could be reclaimed by gathering three roses on the night of Midsummer Eve. One of the roses had to be buried under the Moon's rays in the wee hours of that night under a yew tree, another in a freshly dug grave, and the last under the girl's pillow. A woman had to leave the rose for three nights under her pillow and on the fourth night, if she burned the flower, she would haunt her lover's dreams. The dreams would bother the lover, and he would find no peace until he came to her.

The rose has been an enduring symbol of love for centuries and in many different cultures. In particular, roses emerged in two areas—the Mediterranean and China—and from these areas, it developed and spread. Historically, the Romans became lovers of the rose, and its petals were used profusely during their festivals and banquets to scent the air. In Arabian cultures, the scent of roses was so beloved that a legend claims that perfumery owes its origin to roses. According to the legend, an Arabian princess ordered hundreds of baskets of rose petals to be thrown into her moat as a tribute to her love on her wedding day. The next day, she discovered an oily substance floating on the water's surface, and she scooped it up. She discovered that the oily substance had the heavenly scent of the rose, and the art of perfumery was born.

People have worked to create roses with distinctive scents. Damask roses were cultivated in Europe during the Middle Ages, and the *Rosa Gallica officinalis* was grown in France. Most of the old varieties have delicious and intoxicating scents. On the other hand, recent hybrids have been bred for color and show, holding little if any aroma. The breeding of roses took on a feverish hold in Europe during the end of the eighteenth century. Though people loved the Damask and Gallica varieties, cultivators developed new and popular hybrids. The Moss rose, for instance, evolved from the Damask rose at the Leyden Botanic Garden in 1720. In Paris in the mid-1800s, a chance hybrid of the European and China rose created wonderful new varieties; the Bourbon and Noisette.

While pleasure has long been the main attraction to roses, it was also discovered that there were many useful qualities to the plant. Roses have been cultivated in medicinal herb gardens as medicines. For instance, rose water was often used to keep headaches at bay. Old-time usage also called for rose vinegars to help cool fevers and inflammations, and rose decoctions helped pains in joints, eyes, ears, throat, or gums.

Cultivating Your Garden

Old species of rose require very little in the way of care. Primarily you need a spot that gets full Sun. These roses require at least six hours of Sun in order to stay healthy and bloom correctly. You should also prepare your soil and allow for adequate spacing and water. Roses need good drainage as they do not grow well where the soil stays wet.

For soil considerations, remember that rich, loamy soil can be had if you break up the ground and add good compost. Work the compost into the soil completely. Shrub roses may need more space, but the older roses are narrow and look better when clumped together (18 inches apart). Dig the holes at least 24 inches deep and place the root-ball in, making sure to cover with soil by sifting it on top. Take care to avoid any air-pockets, then soak the soil completely and tamp the ground gently.

The roses will appreciate long, slow watering on a regular basis at least for the first year. A summer drought can kill your newly planted rose. Meanwhile, a good mulching in the fall will ensure that the winter weather won't be a problem.

Getting to Know Some of the Old Roses

Old European Roses are hardy, once-blooming shrubs which bloom prolifically at the start of summer. The following is a list of some of the more common species.

> **Gallica:** These roses have pinkish, single flowers. The fragrance and form vary. Some are very heavily scented, while others are only slightly scented. Some varieties include: Belle de Crecy, Charles de Mills, Duc de Guiche, James Mason, *Rosa gallica officinalis*, *Rosa gallica versicolor*, and Tuscany Superb.
>
> **Damask:** An old group of hybrids, they are said to be a cross between the *Rosa gallica* and *Rosa phoenecia*. They are taller than Gallicas, with fountain-shaped shrubs. The flowers range in color from white to deep pink in clusters. They have a

beautiful rose scent. Some varieties are Celsiana, Ispahan, Madame Hardy, and Marie Louise.

Alba: The tallest of the old European rose groups, they have dense, dark-green foilage and the blossoms range from white to blush pink. The Albas are especially resistant to disease and hardy. Some varieties are Alba 'Maxima', Armide, Celeste, Konigin von Danemark, Maiden's Blush, Madame Legras de St. Germain, and Madame Plantier.

Centifolias: These are the Cabbage Roses hybridized by Dutch breeders in the sixteenth and seventeenth centuries. They have thorny bushes and their blooms are double and very fragrant. Some varieties of these are Fantin-Latour, Rose de Meaux, *Rosa x centifolia varegata*, and Petite de Hollande.

Moss: These roses grew from a mix between the Centifolia and Damask. They have a "fuzzy" growth that resembles moss that grows on the buds and sepals. This mossy growth has a balsam scent. They are somewhat hardy but prone to powdery mildew in warm, moist climates. The varieties include the Capitaine John Ingram, Mousseline/Alfred de Dalmas, *Rosa x centifolia*, Muscosa Alba, and Salet.

Other old varieties are related to the European rose groups but have some unique breeding characteristics.

Noisettes: These roses are early repeat bloomers that climb. They came from a cross between the *Rosa moschata* and Parson's Pink China during the first decade of the nineteenth century when they were heavily interbred with Tea Roses. They are distinctive and very old-fashioned with soft colors. They also have lovely scents. Some varieties are Blush Noisette, Bougainville, Elie Beauvilain, Gloire de Dijon, Madame Alfred Carriere, Narrow Water, and Madame Berard.

Damask Perpetual Rose: This group resembles the old European garden roses but are more bristly, and unlike the old Europeans, they can rebloom. They also have compact growth. David Austin has worked with breeding the Damask Perpetual pollen with his English roses. Some of these varieties include Blanche-Vibert, Marie de St. Jean, Miranda, Rose de Rescht, and Rose du Roi.

Bourbon: The Bourbon roses developed from a crossing between the Autumn Damask and the Old Blush China in the early part of 1819. They carry an intense fragrance somewhat like the Damask. Most grow into large shrubs, but are more susceptible to powdery mildew and blackspot than the other roses. Some varieties of these include Comtesse de Rocquigny, Louise Odier, Maggie, Madame Ernst Calvat, Madame Isaac Pereire, Mrs. Bosanquet, and Souvenir de la Malmaison.

The Crafty Rose

If you grow these fragrant roses, then you will soon be blessed with an abundance of blooms. What to do? Never let the blossoms fade away. You can use them to create all kinds of interesting things; dry the petals and add them to potpourris, create rose waters (good for aging skin as an astringent facial splash or to add to cake or cookie batter instead of vanilla to impart a delicate rose flavor), perfumes, or rose beads. Many herbal books have how-tos on all these crafts, but I'll share here a special recipe for something called a Beau Catcher—great to make on a moonlit night when you are pining for a lost love (and a lot easier than the previously mentioned love spell).

A Beau Catcher is simply a fancy name for a "rose jar." This is a moist potpourri which smells heavenly and lasts for many years. It's great to have on hand whenever you want to scent a room. Victorian ladies believed it turned gentlemen into tigers because of the scent—so it makes a great wedding shower present, or a favored gift for any occasion.

Here's how you can create your own:

Rose Jar

12	fragrant rose blooms, petals separated and placed in a quart Mason jar.
	A stone crock or large-mouthed jar with lid
	A saucer to fit inside the crock and a good sized rock
¾	cup sea salt or coarse salt
1	ounce ground cinnamon
¾	ounce ground cloves
¾	ounce ground allspice
¼	ounce ground nutmeg

$\frac{1}{2}$ ounce orris root

2   ounces lavender buds

25   drops rose essential oil

1   ounce rosewater

Allow one month for the potpourri to ripen. Make a layer of petals in the bottom of the crock. Sprinkle salt on top. Continue making layers of the petals and salt until you run out of petals. Place the saucer on top and weigh it down with the rock. Cover and set aside for 10 days. Remove the salted petals and set aside.

In a bowl combine half the cinnamon, and the rest of the spices, mixing well. Now sprinkle the spices on the bottom of the crock, layer it with petals and continue to layer. Set this aside for two weeks. After two weeks, combine the remaining cinnamon, lavender buds, orris root and mix in the spicy rose petal mixture. The petals will be limp and moist. Now add the rosewater and rose essential oil. Toss well with a wooden spoon.

Pack this mixture into a decorative glass jar with a tight fitting lid. Add ribbons or decorations if you wish. Whenever you want to scent a room, remove the cover for a half hour. As long as you keep the lid on and only open for a half hour at a time, your Beau Catcher will last for years and years. A neat way to preserve the bounty from your garden!

Old Rose and Rose Associations

The Heritage Rose Foundation
1512 Gorman Street
Raleigh, NC 27606-2919
U. S. A.
(919) 834-2591

The American Rose Society
P.O. Box 30,000
Shreveport, LA 71130-0030
U. S. A.
(318) 938-5402

The Canadian Rose Society
10 Fairfax Crescent
Scarborough, Ontario
Canada M1L 1Z8
(416) 757-8809

Theme Gardens for Kids

Deborah A. Duchon

Josey, age seven, loved spending time in the garden. Using her child-sized tools, she dug and planted and weeded alongside Mom. She tried to help her Dad spread mulch and rake leaves using a full-sized rake.

When Josey asked for a garden space of her own, her parents were thrilled. They wanted to help Josey make it a special place. Sure, they were familiar with the usual "first gardens for kids" kinds of plants. In fact, they had always encouraged Josey to have her own plants as part of the family garden. She had already raised radishes, carrots, and sunflowers from seeds and had helped bury bulbs in the fall. No, Josey's parents wanted to help their daughter plant a garden that was really special.

Luckily for Josey, her family lived in Michigan, so they could visit the famous Children's Garden at Central Michigan State University and get some ideas for their own yard. The Children's Garden has been called "The Most Magical Half Acre in America" for good reason. The gardeners there have a flair for whimsy combined with exceptional skill and an uncompromising commitment to make their half acre the best of the best. As a result of their leadership in the field of children's gardens, similar gardens are sprouting up in many other botanical gardens across the nation.

One of the most popular displays at the Children's Garden is the Pizza Garden. Within this small circle grow the plants that are needed to make a pizza: wheat to make the crust, tomato plants for the sauce, clumps of basil, oregano, thyme, parsley, garlic, onion, and fennel to spice the sauce. Some pizza toppings may be included in your garden as well: green peppers, for instance, and eggplant. The possibilities are as endless as pizza itself. And since no pizza is complete without cheese, the smart folks at Michigan State planted a circle of marigolds around the perimeter of the pizza garden to represent cheese. The children love it.

Near the entrance to the Children's Garden lies another flower bed that is easily reproduced in the home garden. These are the plants with animal names—a wonderful theme garden to spur a child's imagination. A few plants with animal names are:

- Lamb's ears, *Stachys lanata*
- Tiger lily, *Lilium lancifolium*
- Snapdragon, *Antirrhinum* spp.
- Zebra grass, *Miscanthus sinensis 'Zebrinus'*
- Turtlehead, *Chelone obliqua*
- Goatsbeard, *Aruncus dioicus*
- Elephant ears, *Colocasia esculenta*
- Butterfly bush, *Buddleia davidii*
- Foxglove, *Digitalis* spp.
- Butterfly weed, *Asclepias tuberosa*
- Pussy-willow, *Salix caprea*
- Dogwood, *Cornus* spp.

Josey's visit to the Children's Garden enouraged her to think up her own theme. All the way home, she talked about various themes. One was to plant a garden with symbolic colors. In school, she had studied flags and had become interested in the symbolism of colors. She thought it might be fun to grow flowers that had meaningful colors, such as red, white, and blue. She looked through gardening books and seed catalogs to find flowers with red, white, or blue flowers, but lost interest after a while.

Then she thought it might be fun to grow a garden with her favorite color, red. She had read through the gardening books and seed catalogs and had amassed an impressive list of red flowers. Her parents suggested that she stick with annuals so that she would get faster results:

- Salvia, *Salvia splendens* (Josey especially liked the name "Red Hot Sally")
- Zinnia, *Zinnia elegans*
- Tulip, *Tulipa* spp.
- Geranium, *Geranium* spp.

Josey also wanted to add a few plants with girls' names, because she, after all, was a girl. She went back to the gardening books and seed catalogs and was pleasantly surprised with the numbers of flowers she had to choose from. We don't know what she ended up deciding on, but here are a few that she considered:

- Shasta daisy, *Chrysanthemum x superbum*
- Black-eyed Susan, *Rudbeckia* spp.
- Lily, *Lilium* spp.
- Iris, *Iris* spp.
- Rose, *Rosa* spp. (Her parents strongly discouraged this particular choice. "Maybe when you're a little older," they advised, and suggested rose moss, *Portulaca grandiflora,* as an alternative.)
- Petunia, *Petunia x hybrida*
- Veronica, *Veronica* spp.
- Sweet William, *Dianthus barbatus* (for her baby brother)

Theme gardens have been in vogue lately They show sophistication beyond simple flower or vegetable gardens. Theme gardens, such as "herb gardens," "Bible gardens," and "Moon gardens," represent the purposeful growing of plants that have meaning beyond their use or appearance. They bring plants together that have something meaningful in common. Unfortunately, they so often have adult themes that children get left out.

Josey's experience shows that theme gardens can be just as meaningful for children as for adults, although the content of the themes generally perform at a different level. Not many articles on theme gardens would include a pizza garden or a plants-with-animal-names gardens.

On the other hand, children's theme gardens, as the products of a child's imagination, are not necessarily unsophisticated. It was interesting that Josey came up with the idea of all red flowers as an expression of her favorite color. When the all-white garden at Sissinghurst Castle in England was first introduced, it seemed the height of sophistication in a country where a profusion of floral colors was the norm in every garden ranging from the most humble cottage to the grandest castle. The fact is that children see life from a fresh standpoint, and by allowing them to indulge their imaginations they can come up with garden themes that most adults never could.

If there is a special child in your life, your may want to devote a small space in your yard for a theme garden that will excite that child's sense of wonder, just as it did for Josey.

Black Silk Soil for Your Garden

Penny Kelly

Plants need healthy soil in order to produce healthy fruit—it is almost impossible for a plant to be healthy when planted in thin, depleted soil. It takes Mother Nature thousands of years to create one inch of topsoil, so if you want good, organic food now you will have to help out a bit.

The first thing to understand is that plants typically have roots that dig their way down into the earth three to six feet or more. When they have enough healthy soil in which to do this, they will find most of the minerals and nutrients they need to produce healthy fruit that can repel pests, mildew and various kinds of rot, and support human life.

Dead soil, however, has no micro-organisms living in it and is electrically static. Plants will not be healthy in such soil. To remedy this, your best option is compost. Compost is brown gold. If possible, you should put it on your garden in the fall. To obtain it, look in the yellow pages for wholesale nursery and garden suppliers, or drive out to the country and ask around for someone who might have compost you could buy or carry off free for the hauling.

After compost, you need a good soil inoculant. An inoculant contains strains of living organisms that characterize healthy soil—bacteria, beneficial fungi, enzymes, algae, and natural yeasts. Micro-organisms greatly improve the ability of plants to get the nutrients they need from the soil. Such inoculants are available at sources such as Peaceful Valley Farm Supply—to receive a catalogue call (530) 272-4769. Follow the sdirections that come with the inoculant and spray it onto your garden and into your compost pile, preferably on an overcast day in the spring when the weather is not too cold (at least 50°F).

Another element that helps build a great soil is paramagnetic rock powder. We have used this powder on the lettuces and greens in our winter solar garden, and the plants seemed to spring up and grow several inches overnight. Rock powders help to remineralize soil and return it to a state of electrical conductivity that encourages necessary chemical reactions around plant roots. An electrically static soil discourages the free movement of nutrients through soil and into the root hairs of a plant. The result is weak, limp, unhealthy plants that produce vegetables and fruits that are soft or unsweet. Paramagnetic rock is not expensive to buy, but it is expensive to haul because it is very heavy. To find a source for rock in your area, check the Acres, U. S. A. newspaper; located at 2617-A Edenborn Avenue, in Metairie, Louisiana, 70002, (504) 889-2100.

Applying compost, spraying an inoculant, and spreading rock powder will get your garden off to a good start for the year, but soil building is much like making soup. You can put all the stuff in the pot and give it a stir, but you need to allow time for the flavors to blend. In the same way, you need to allow time for your soil amendments to react in ways that change the color, texture, smell, and composition of your soil. In the end, you want something I call "black silk"—a rich, dark soil that is loved by plants. It may take a few years to achieve this, but is well worth the effort.

In fact, what you do in between growing seasons can have a great effect on your soil. While most people, including myself, are tired of the garden by the time August rolls around, you can make a world of difference by planting a good cover crop once the garden is done. The roots of cover crops do wonderful things for soil. Clovers, alfalfa, and vetch break up clays and hardpan while fixing nitrogen. Buckwheat and oat smother weeds and grasses and add phosphorus, and rye makes sweeter vegetables. All of these cover crops add lots of green, carbon-based matter to the soil for the micro-organisms in your inoculant to eat and thrive on.

You can be picky about cover crops by cleaning out your rows as each is harvested, turning over the soil, and sprinkling a few cover crop seeds where the plants were recently growing. Or you can wait until the garden is completely done for the year, till it under and spread your cover crop over the entire thing. In the spring, you simply till the cover crop under, wait a couple of weeks for your micro-organisms to digest it, and you are ready to plant. In my experience, no other soil-building process is quite as powerful in "blending" the flavors of your soup as a cover crop.

There are dozens of ways to build excellent soil. These few practices—compost, soil inoculants, rock powders, and cover crops—have proven to be the most effective in the least amount of time. Since what you're after is top quality nutrition for yourself, you really can't afford not to build a good soil. Unhealthy plants will be subject to insects and diseases. Truly healthy fruit of any kind, from apples to zucchini, will remain fresh long after it is picked and provide healthy sustinence.

Garden Rescue Remedies

Penny Kelly

When it comes to garden rescue remedies, you shouldn't think of poisonous sprays and chemicals. Instead, to outwit persistent pests, plan your defense in advance. Once you discover garden invaders and the damage they cause, it is already too late. You can begin rescuing your garden as far in advance as when you are deciding what to plant.

When you plant crops such as peas, lettuce, spinach, cabbage, kohlrabi, and broccoli in early spring, insects may be scarce. However, there are plenty of hungry rabbits, deer, woodchucks, and other hungry creatures who would love to break their winter fast on a few tender greens. My grandmother, a serious gardener all her life, put a fence around her entire garden to keep out kids, dogs, cows, rabbits, chickens, deer, and woodchucks. For grandma, the purpose of the fence was to communicate her intent to keep animals out, and as such it did a very good job.

Physical barriers of all sorts will usually do an excellent job of keeping your young seedlings intact until they are ready for harvest. Among the items that do an excellent job of protecting your plants are row covers, tomato cages, chicken-wire tents, unused flower pots, cardboard boxes with a large stone on top, old bushel baskets, Styrofoam cups turned upside down, and a handy thing called a "rudy wrap." Row covers can be laid atop new seedlings and kept in place indefinitely. They may have to be put back in place after a strong wind or when a deer gets his hoof caught in it, but they are wonderfully discouraging not only to rabbits and deer but later to cabbage moths and other flying insects. Up to five feet in width—though you may cut them smaller—they will last for quite a few years if you carefully rinse, dry, fold, and store them at the end of the season.

To keep deer away from my tomato plants at a time when one bite will finish off a young plant, I turn my cone-shaped tomato cages upside down, putting the wide end down around the young plant the same day it goes in the ground. This seems to discourage deer—perhaps because they can't angle their heads to nip off that tender tomato. Sometimes this method discourages rabbits and cutworms, but it is not foolproof. A large Styrofoam cup with the bottom cut off and placed around the base of the

plant gives added protection. You can
turn the tomato cages right side up as the
plants get bigger.

I usually put all my cole crops—cabbages, broccoli, kohlrabi, Brussels sprouts,
cauliflower, and kale—in one long row so
I can place a chicken-wire tent over the
top of them. A length of twenty-four-
inch chicken wire folded in half and set
over the entire row does a wonderful job
of deterring almost everything that would
gladly feast on your early vegetables.
Deer, raccoons, rabbits, woodchucks, and
birds will not be able to get at the shoots.

I have used old flower pots, cardboard boxes, and bushel baskets turned
upside down and held in place with a
heavy stone. There are drawbacks, since
these must be removed each morning to
give the plants light, and put back each
evening for protection. Furthermore,

cardboard boxes can collapse under the weight of the rock, especially after
an evening storm, which can crush the plant it was supposed to protect.
These objects can also cut your plants cut off from the water they need.

Meanwhile, if you need long-term protection from deer, woodchucks,
raccoons, and rabbits after your plants are fairly large, you may choose to
invest in something called rudy wrap. Used by professionals and those
who manage orchards, vineyards, and tree farms, rudy wrap is available
from orchard or nursery supply companies. The wrap comes in several
forms—a continuous roll about thirty inches wide that you can cut to any
length, and a large, white, flexible plastic rectangle about three by two
feet with tabs on one end and slots near the other into which the tabs fit.
If your plant is small, the wrap will not stand up by itself and you will have
to use a stake to provide something for the wrap to lean on. Rudy wrap
can be pulled aside for cultivating or feeding plants. When you need serious protection for larger plants such as grapevines, fruit trees, blueberries,
raspberries, or other precious plants, you simply encircle the plant with
the rudy wrap with an opening at the top to let in rain and sunlight.

Let's say you are not bothered by animals as much as by insects, mildews, and fungus diseases. If you succumb to poisonous chemical sprays, you will have only made your problems worse and guaranteed that they continue. Why? A plant cannot be healthy in poor soil. When a plant is not healthy it will be lacking in minerals and its electromagnetic field will be both out of tune and off-color. Insects—all of whom sport antennae for the express purpose of tuning in to unhealthy EM fields—will search out and destroy these plants. Although I know of two exceptions to this rule—potato beetles will destroy a healthy potato plant in a matter of days, and horn worms will strip healthy tomato and parsley plants in no time at all—as a rule of thumb you can tell what shape your plants are in by how much damage insects do to them. The healthier your soil is, the less pests will bother your plants. The best defense against insects, molds, and mildews is to plan to create healthy, living soil.

Of course, if it is already August and you're fighting squash bugs and cucumber beetles by the thousands, and your tomatoes and zucchini are threatening to collapse from mildew and fungus, you need rescue medicine. Several things might help get you by. For the insect problem, a light mist of soap, water, garlic, and mineral oil works well. Crush three or four cloves of garlic in a quart of water and let sit overnight. In the morning strain out the garlic and add a tablespoon of dish soap or Fels Naphtha hand soap. Let the soap dissolve, then add one or two tablespoons of mineral oil and shake well. Insects don't like garlic or soap, and the oil helps the soap and garlic stick to the plant.

In a pinch, you can also use organic pesticides such as pyrethrin, rotenone, or *Bacillus thuringiensis* (Bt). The pyrethrin and the Bt work quickly and degrade quickly, especially in the sunlight. It's best to apply after 6 pm so they have a chance to work overnight. Pyrethrin is effective against a wide variety of moths, mites, beetles, and worms, while the Bt is effective mainly with moths and worms. The rotenone lasts a bit longer, is not affected by the Sun, and also works against a wide variety of garden pests. A little of these chemicals goes a long way, so use sparingly.

If mildew and fungus threaten your harvest, the first thing to do is ask yourself if you've been overwatering. Many people water because their plants don't look well, but plants most often look sickly because of a lack of minerals or nutrients. Overwatering only weakens the structure of the plant. Also, you should be aware that a light dew settles on the plants shortly after dusk, and at night the deep roots of plants bring water up from the earth. This "root tide" is more pronounced during a Full Moon, so you should cut back on water during the Full Moon. To contain some of this moisture, you can put a deep, airy layer of straw around your plants. When evening dew falls from the plants, it will be kept from the soil. This will not only insulate the plant from the nightly moisture, but it will also limit the growth of mildew fungi that live naturally in the soil. You should also remove diseased leaves and stalks from plants, and consider thinning out plants a bit so more air can move among them.

If you need a miracle to save a sick garden, there are several more drastic measures you can take. These potent healing methods can be used singly or together to rescue flowers, fruits, vegetables, or grains. The first is to apply a kind of "compost tea," made by putting a shovel-full of good living compost (filled with micro-organisms) into a five-gallon bucket of water. Let this brew steep for up to a week, then strain through cloth and spray the liquid directly on the plant using a three-gallon sprayer. If you don't have a sprayer, simply apply the tea-water generously around the base of the plant. If you don't have any compost, you can substitute with one of several herb teas. Make a decoction of either chamomile flowers, the leaves and stems of nettles, oak bark, or sage leaves. When cool, strain and spray, or use to water your plants directly.

Another option is to apply a light dusting of either bentonite clay or montmorillonite clay in the evening to the leaves and around the base of the plant. Clay is a natural healer and, for reasons that are not well-understood, it energizes plants tremendously and helps them balance themselves and eliminate diseases of all kinds. Montmorillonite is a greenish clay that can usually be purchased for around $15 for a twenty-five-pound bag. You can purchase it from pottery and ceramics suppliers, or from some nursery and garden centers, as well as from excavating contractors.

Finally, to assure that your garden will continue to grow beautifully, apply a generous side-dressing of compost after the plants have finished flowering. At this point in their lives they are getting ready to put every ounce of energy they have into producing a fruit or vegetable and you can help make sure they don't exhaust themselves and get sick from a lack of nutrients. While this can be a very labor-intensive rescue operation, the reward for doing so is more than worth the effort. When the cucumbers you picked in late September are still good two months later and you are eating garden tomatoes in your Thanksgiving salad, you will already be planning your garden rescue remedies for next year.

Special Crops from the Garden of Eden

Penny Kelly

At the start of each spring, an annual affliction occurs that every serious gardener has suffered from. It starts with a vision of the coming year's garden—a vision of Eden. The gardener dreams of the ways this year's garden will be laid out. It will will have no weeds, no mildew, no insect damage; every plant will be healthy, and will produce miraculous quantities of succulent fruits and vegetables. The gardener will float from plant to plant, every tool and necessary soil amendment magically at hand, plucking the beautiful bounty, attentive to each plant.

Each year the vision mutates a bit and expands. You begin to envision strange fruits, exotic vegetables, mesmerizing herbs. Like an addict, you know you have to grow these new essentials. You're certain they will be worth the effort, and you're convinced you can figure out what each specialty needs. Trance-like, you order the seeds and buy the plants regardless of the cost in order to keep the vision of Eden alive.

This year it was Mexican epazote, lemon grass, chocolate mint, white sweet potatoes, a new heirloom tomato, a new kind of cucumber, Thai peppers, and a couple of sweet bay plants. At the end of the season the results tallied as follows.

The epazote seeds had been in the ground exactly one day when a horrific spring storm dumped a half-foot of rain on the newly tilled, south-sloping garden. The epazote, which is supposedly very easy to grow, may have sprouted and be growing easily somewhere down by the pond which is where the seeds, along with two inches of topsoil, probably ended up. If it is down there, I haven't found it because I still have no idea what epazote looks like.

Afraid to plant the lemon grass in the garden because of its reputation for invasiveness, I therefore left it in the greenhouse. Planted in a pot on one of the raised parsley beds, the roots grew out the bottom of the pot and down among the parsley. Eventually, my husband came along, saying "Who left this here?", and picked up the pot to move it, jerking the roots out without realizing what he'd done. I found the lemon grass a few days later, in shock and half dead. I'm still hoping it will revive.

The chocolate mint also did not get out of the greenhouse, as I knew mint tends to be invasive. I gave instructions to my daughter, who was my garden manager this year, to keep it watered, and the last time I saw it, it looked pretty good. Someone must have moved it though, because by the end of the season I couldn't find it anywhere.

Meanwhile, the white sweet potatoes not only made it out of the greenhouse, they even went into the ground and grew beautifully. But since my daughter/garden manager sprained her leg in September and ended up in a cast for two months, she was unable to dig anything, and the sweet potatoes are still in the ground. I was too busy, which was why I hired her in the first place. Maybe they'll come up next year like the red potatoes, which never got out of the ground last year because there's always something that doesn't get harvested.

The heirloom tomatoes I seeded myself along with a dozen other varieties of tomato. I carefully marked them with the name and date, then urged my daughter to get them in the ground when they were ready for transplanting. She did with help from my granddaughter, who neatly removed all the identification markers as the plants went in the ground, which neatly eliminating any chance of saving seeds from them or knowing exactly which varieties were which. When the tomatoes had grown and ripened it was easy to tell which were cherry tomatoes and which were Roma, but the rest were anyone's guess.

The unique cucumber plant that was supposed to produce red cucumbers went into the ground right next to the Thai peppers so I could keep my eye on both of them at the same time. The cucumber didn't make it but the pepper plant prodigiously produced tiny, brilliantly colored purple peppers about the size of my little finger. I picked a strawberry basket full of them and took them inside to wash for dinner, thrilled that one of my new specialty crops had produced something. Making up a salad from a variety of fresh veggies, I put in about a dozen of the pretty purple peppers, tossed the whole works, and sat down with my husband to eat. We chatted in a relaxed manner as we ate until he suddenly exploded out of his chair, breathing heavily, gasping in fact, and shouting frantically as he ran toward the sink, "There's something in this salad!... What did you put in this salad? Where's the water? I need water!" Half an hour later, smoke still belching from his mouth, eyes still running like waterfalls, face bright red, breath coming in starts, he was almost able to talk. I fished one of the peppers out of the salad, cut it in half, and ran it over my tongue. It was like a slice of flame!

"I think it was the peppers," I said matter-of-factly. "Maybe I should take the rest of them out of the salad."

"How many did you put in?" he said, gasping.

"About a dozen," I replied.

I found out only later that one Thai pepper, topped, sliced, and with seeds removed, was enough to make a sixteen-quart pot of soup taste red hot. The peppers are even hotter when the seeds are left in. And speaking of seeds, I did save some from the Thai pepper, but I'm just not sure we can eat them if I grow them.

As the snow settles over the sweet bay plants—which should have been left in the greenhouse because they are tender perennials—the first seed catalogs of the new year have arrived and the vision is renewing itself. Eden is developing, though there's still a lot to learn along the way.

PERSONAL LUNAR FORECASTS

Understanding Your Personal Moon Sign

Gloria Star

To imagine the sky without the ever-changing reflections of the Moon would be unnatural. The Moon is Earth's constant companion; she has a power all her own. Her cycles control the ebb and flow of Earth's tides, yet there is something more which has led to our fascination with her. Musings about the Moon and her influences fill the art and literature of humans. Since the time of the Sumerians, written records have linked Moon cycles with changes in nature and in behavior. Do you know that you have your very own personal Moon? It is one of the features of your astrological chart and a significant indicator of how you express yourself.

Your horoscope is based on precision calculations. Astrologers can chart the positions of the Moon, Sun, and planets based on the exact date, time, and place of your birth. This detailed picture—your astrological chart—symbolizes the complexity of your whole being. You probably know about your Sun sign, which describes the ways you express your ego and your drive to be recognized—something easy to see and even easier to show to others. Your Moon tells a more intimate story, since it describes your subconscious, internalized nature. You feel your Moon.

To find the exact degree and placement of your Moon sign, you need a copy of your astrological chart based on your birth data. Since the Moon changes signs every couple of days, and planets move by degrees each minute, an accurate calculation is necessary to figure your exact set of energies. To obtain your chart, you can visit an astrologer or order your chart calculations from directly Llewellyn Chart Services by using the "Astrological Services Order Form" in the back of this book. Meanwhile, you may use the handy tables and simple calculation method described on page 62 of this book to provide a close approximation of your Moon's sign.

While you may be unfamiliar with the astrological concepts associated with the Moon, you are more acquainted with her energy than you realize. Whenever you tune into your feelings about anything, you're connecting through the energy of your Moon. Your Moon is a finely attuned internal recording device that collects, stores and assimilates everything you experience and feel. Your habits and attitudes reflect the qualities of your Moon. These messages replay themselves as your habits and attitudes, and some operate automatically. You can add more information and make alterations at any time, but since you hold these impressions at a very deep level it's not always easy to change or erase an old internal message.

Your Moon sign provides insights into your inner strengths and shows where you feel most susceptible. This energy shapes your thought patterns and stimulates your emotional nature. The attributes of your Moon sign represent the ways you absorb life experiences. As you learn more about yourself by experiencing life, you may find that by concentrating on the nature of your needs as illustrated by your Moon sign you can help provide true comfort, security, and safety in your environment. Once you're in the flow of the energy of your Moon, you carry your sense of home into every life situation.

You share your lunar energy whenever you express nurturance and support. Your Moon sign shows how you express these sentiments toward others and yourself. Whether you're male or female, your Moon indicates the way you nurture others. Your Moon portrays your archetypal feminine quality and represents your relationship with your mother, with women, and with the feminine part of your psyche. Your Sun, on the other hand, represents the archetypal male quality and illustrates your connection to men, your father, and the masculine elements of your inner self.

When you probe into the mystery of your Moon, you uncover the repository of your soul. Your Moon contains all that you have been and therefore influences all that you can become. Shining forth from within your eyes, the light of your Moon reflects the inner, soulful aspects of yourself. Your Moon represents your emotional tendencies and needs; it is the part of you which has flown to the pinnacle of ecstasy and which also remembers the true emptiness of despair. Your capacity for contentment increases when you strive to fulfill the needs defined by your Moon's sign.

This part of the *Moon Sign Book* is designed to help you understand the basic planetary cycles throughout the year 2000 that will influence you at an emotional level. Transits to your Moon stimulate change, and you may discover that some of the cycles help you reshape your life while others stimulate a desire to delve into the mystery of yourself. Astrology can show you the cycles, but you are the one who determines the outcome through your responses. By becoming open to your own needs, and by responding to the planetary energies in a way which allows you to fulfill these needs, you can experience a renewed sense of self-confirmation and a deepened feeling of personal security.

Aries Moon

Aries Moon is a pioneering soul, driven to meet life's challenges with courage and commitment. You often embrace leadership roles—forging ahead and opening formerly closed doors. Aries Moon can be a powerful force, igniting your ever-ready warrior spirit and individuality.

For you, life must be filled with excitement and adventure. You become impatient in boring situations. As a result, your relationships can suffer unless you're with others who value autonomy. If your needs are not met in an intimate relationship or if growth has reached a standstill, you'll deal with the problems directly or move on. Similarly, in your career you'll perform at your best when your passion is engaged and when you feel free to exercise your talents. You are a mover and shaker.

In a crisis, you always rise to the occasion. You often work in crisis-oriented situations such as medicine, counseling, military service, or politics. Your awareness of the importance of taking the initiative is likely to prompt you to take action when others are paralyzed by their fear or lack of conviction. At some level, however, you may never grow up, preferring instead to maintain your daring (and occasionally impertinent) drive to have what you want when you want it. While your tendency to flirt with disaster may keep you feeling alive, you can alienate others if you carry things too far, leaving the impression that you're only out to have a good time. The trick of maintaining your passion for life and love without selling out or compromising your fiery drives requires true finesse, and may only arise after time has helped you embrace a more emotionally mature outlook. Even finding the right place to make your home can take some time, but you'll feel settled when you're in a place that gives you freedom to express yourself in your own special way.

Famous Individuals with Aries Moon

Antonio Banderas, Bill Gates, Jacqueline Kennedy Onassis

The Year at a Glance for Aries Moon

While last year was filled with a series of opportunities to expand your awareness and security needs, your drives this year will require you to strike a sensitive balance between freedom and responsibility. Your special abilities and talents are gaining greater acceptance and leading to increased recognition, and, as a result, new pathways may be opening which were previously closed or even nonexistent in your life. The key concept for this year is very clearly *change*, with many changes centering around your personal desire to improve your life and lot. If you follow the impulse to break away from an existing situation, determining exactly how to execute that change without disrupting your emotional stability can be rather challenging. Some changes may require that you assume an unusual level of risk, although measuring that risk against your capabilities will help you take actions that assure growth instead of compromising your deepest needs.

The opportunities presented through the cycle of Jupiter are a mixture of the need to expand horizons which opened last year without stretching your resources too far. During the first half of the year, it's time to strengthen your security base, and then, from July through December, you'll feel ready to reach into new territory. Travel or education can become important resources in themselves, and the friends you're making along the way can have a powerful impact on your choices.

Saturn's cycle can bring some frustration and restraint if your Moon is between 10–20 degrees of Aries, although the way you handle your obligations will determine how intensively you feel those restraints. Taking on responsibilities you can readily handle will work to your benefit, but if you get in over your head, you may feel it is taking a while to get back on track. The very slow-moving transits of the outer planets—Uranus, Neptune and Pluto—have an especially significant impact on you. If your Aries Moon is from 0–1 degrees, you feel more stability while Saturn transits in sextile to your Moon during the summer and early fall of this year. If your Aries Moon is from 2–7 degrees, you experience the influence of Neptune drawing you to pursue your imaginative creativity and to develop your spirituality, and your psychic impressions may be right on the mark this year. If your Moon is from 7–14 degrees Aries, you are under the influence of Pluto transiting in trine to your Moon, marking a year of changes in the way you fulfill your needs and desires; changes in home environment may be on your agenda. You'll experience a restless excitement if your Moon is from 15–21 degrees Aries, with Uranus transiting in sextile aspect to your Moon, stimulating your sense of adventure and desire to break free of unnecessary restraints. If your Moon is 22–29 degrees Aries, you feel the need to take careful steps to bring stability to your life since Saturn is traveling in semisextile to your Moon.

Regardless of the exact degree of your Aries Moon, you'll benefit from this time of greater awareness about your deepest drives and needs. Seeing yourself and your needs more clearly allows you to make choices which ring true and make sense, stimulating your courage to make the changes which literally breathe new vitality into your emotional life.

Affirmation for the Year

My intuitive voice speaks the truth of my Higher Self. In
all things I seek guidance from the Source.

January

Jupiter's transit in Aries continues to provide an emotional lift, although it may be tempting to overobligate your time or resources or to give in to unnecessary indulgences. You're attracting support, however, and may feel that others are more open to your needs and concerns, since the beneficial cycles of both Jupiter and Venus are positively influencing your Moon. During the Lunar Eclipse on the 20th you may feel inspired to reach toward the fulfillment of a profound hope or dream, though you may bid goodbye to a situation that has been familiar and comfortable for a long time as you open to new realizations. This is an excellent month to make improvements at home and the perfect time to strengthen an emotional commitment.

February

Your exuberance continues, and although Jupiter finally leaves its transit in Aries, Mars moves into Aries this month, adding a vigorous drive to achieve satisfaction. To make the most of this cycle, open yourself to the visionary images and ideas that can shape your future during the Solar Eclipse on the 5th. While you may feel that answering the needs of the moment is sufficient, your deeper longings may call for more. At the same time, your awareness of the precious quality of remaining focused in the *now* can give you a powerful advantage, personally, and professionally. Avoiding purely selfish motives helps you maintain a clear conscience, and acting as a champion for an important cause can be exceptionally rewarding for you.

March

You may feel driven to break through the barriers that have inhibited the realization of your dreams and desires, and by joining forces with others who understand your motives you can be quite effective initiating change. Containing your impatience can be difficult from the 1st to 22nd. Your passion may be impossible to ignore, yet the choice of how you direct it makes a huge difference in whether or not you are well received. Brash arrogance can result in a negative response, while a considerate attitude coupled with courageous actions may garner the reaction you hope to receive. The challenge centers on being honest with yourself about your motivations and projecting that honesty toward others.

April

You may feel your heart brimming with love, and your ability to embrace a true sense of your personal worth helps you exhibit your most attractive qualities. Your sense that you have a new lease on life can prompt you to initiate actions during the Aries New Moon on the 4th, and you may feel especially clear about what you want. Carving out the best path may be easier after the 6th, when others may be more sympathetic to your concerns or needs. Making alterations to your personal environment fares quite nicely from the 6th to the 30th, and if you want to move you'll have the best success from the 13th to the 30th. Relationship matters reach a peak during the Full Moon on the 18th, and you may also better understand the needs and concerns of your partner.

May

Slowing down to the measured pace of others who mistrust innovation or change works to your benefit. Your ability to nudge progress works best after the 14th, when your ideas are better received and your ability to illustrate your intentions is strongest. Pushing too intensely can be disruptive and may actually interfere with the results you desire. Stating your opinions can be helpful, but stepping away from battles that are not yours to fight preserves your integrity during the Full Moon on the 18th. Objectivity improves after the 24th, when you may be able to help salvage a situation by bringing others to an understanding of fresh possibilities.

June

Set plans in motion from the New Moon on the 2nd through the Full Moon on the 16th, since this time period also coincides with cycles that enhance confidence in your abilities. Friction with others can arise after the 15th, particularly if your attitudes are interpreted as self-serving. Your drive to fulfill your desires intensifies after that time, and you may be upset if you have to wait. Taming your passions can be helpful, although you may not like postponing anything important. Progress can occur, but you may be walking on eggshells to keep certain situations under control, including your own temper. Find ways to channel these feelings; repression is not the answer. Staying physically active helps you let off steam.

July

A brewing crisis may finally reach its peak during the Solar Eclipse on the 1st. If you feel under attack, the way you respond makes all the difference in the long term; however, as a means of gauging your actions, be aware of how others react to you. The Lunar Eclipse on the 16th can bring family matters into the spotlight, and others may seem to air their grievances too frequently. Fortunately, your feelings of love and acceptance help you tolerate demanding attitudes from others, and after the 23rd you may find that they are looking to you for leadership or insights. The second Solar Eclipse on the 30th marks a time when your creative ideas are right on the mark. Trust your intuition, and follow your heart!

August

It's time to allow yourself to play, and whether you're vacationing or funneling more energy into a project near and dear to your heart, your passions will guide your actions. Allowing your romantic desires to flow freely can stimulate powerful changes in a close relationship, or if you're seeking a new love your attraction to someone can be almost overpowering. Sending the right signals without feeling too vulnerable from the 1st to the 10th can become a fascinating game, adding to the intrigue of discovering how someone feels about you. Since you do like to know where you stand, your direct nature takes over during the Full Moon on the 15th, and real progress sets the stage for love to blossom.

September

Your preference usually is to take the lead, but this month you'll make great strides by following signals from others. The joy of forging a strong bond with your partner can help you realize the value of honoring your mutual autonomy—it's almost as though you gain more freedom by supporting your partner's own independence. You may have misgivings about an agreement or an offer, but looking carefully into any obligations will help you make the best decision. If you're wary of someone else's motivations, research is much more satisfying than trying to guess what's going on. To see what's beneath the surface, turn over a few rocks after the 18th.

October

Dealing with demands from everyone else can get on your nerves, particularly if you're not getting what you want out of a situation. Your sensitivity reaches a peak during the Aries Full Moon on the 13th, and you may grow exhausted by the demands of others. However, this will also be the best time to explore the true nature of your job, relationship, or living situation to uncover the things that are really bothering you. Renovations or improvements at home may be on your agenda after the 19th, and some of these improvements may involve getting rid of a few things you no longer need or want. If you really want satisfaction, think about changing habits or attitudes so you can deal with the outside world more effectively.

November

Although you may not be seeking a battle with others, a few altercations can arise this month. You may unwittingly arouse anger in your acquaintances, especially if they have issues with you or feel threatened by you in some way. Before you determine that your only option is to wage war, seek counsel from the 1st to the 8th from a trusted advisor. During this period, your ideas may result in options that can lead to a unforeseen solution. However, stubborn attitudes from others can still be difficult to swallow during the Full Moon on the 11th, even if you are trying to be nice. Better circumstances arise with the New Moon on the 25th, when you can also be successful repairing a misunderstanding or taking actions to heal an old wound.

December

The complex nature of relationships (especially family ties) gives way to higher principles, and your desire to bring others closer together can initiate a better understanding among those who share the circle of your life. You're feeling generous, and although you may be tempted to go overboard be sure that others understands your feelings. You may accomplish most by giving more time and energy to a relationship. The stuff can be almost inconsequential, and from the 9th through the Full Moon on the 11th you'll relish the benefits of sharing an intangible experience with someone you love. While you may be in a festive mood from the 9th to the 31st, your actions or attitudes send mixed signals to someone after the 23rd. Keep your intentions clear!

Taurus Moon

Through the essence of your stable and enduring Taurus Moon, you feel profoundly connected to Mother Earth. You crave a life of serene security and make choices which feature reliability and practicality. Once the fruits of life ripen, you know how to savor the sweetness of your successes. Even if some parts of your personality are not conservative, when it comes to the big choices you're most comfortable with options which promote growth and sustain stability.

Rushing into commitment is not your style. You like to have time to make a connection, and once you've made a promise you're in for the duration. As a result, adapting to change can be difficult, especially if it seems beyond your control. Fear can overtake your confidence if you sense that your security is being compromised, and it is during such times that deciding between holding your ground through life's storms or stubbornly resisting necessary change can make all the difference.

The force which sustains you is love, and by opening your heart to the power of love you feel animated and alive. While loving others is part of this experience, simply allowing love itself to grow through your creative acts adds to your sense of wholeness. Avoiding the temptation to become possessive is your primary test, since letting go of someone or something can seem like you've lost a part of yourself. People intimate with you may understand how deeply you care, although they will also appreciate if you can allow them occasionally to make their own mistakes or follow a path you might not have chosen for them. Fortunately, your patience reminds you that personal development and evolutionary change require time, trust, and most of all—love.

Famous Individuals with Taurus Moon

Lee Iacocca, Demi Moore, Pope John Paul II

The Year at a Glance for Taurus Moon

With Saturn's influence in Taurus bringing its stabilizing force to your Moon, you may feel you're on especially solid ground this year. Further good news comes from Jupiter's transit in Taurus for the first half of 2000;

it will add opportunity, abundance, and a feeling of optimism and hope. There are some looming challenges, and you may feel that some of the things you've counted on to remain constant are changing, but adopting a responsible attitude will go a long way toward helping your cope with these circumstances.

Saturn's cycle over your Moon is indicative of a period of endings, although you may feel many of the things reaching completion seem like a natural outgrowth of your maturation. From February through July—while Jupiter travels in your Moon Sign—you will be ready to take on the world, although determining your limits can be rather difficult. If you overextend, you may later feel rather intensely burdened and resentful. Part of the problem can stem from giving in to excessive materialism or greed, but if you are attentive to your motivations you can avoid the traps associated with choices driven by such feelings. Your appetites are likely to increase, but that does not mean that they all must be satisfied at once!

The slower moving cycles have a yearlong impact. If your Moon is between 0–1 degrees Taurus, you may feel that everything is starting to move along now that Saturn is moving away from the slowdown conjunction to your Moon. If your Moon is between 2–7 degrees Taurus, you may be caught in the tendency to see things only as you want to see them, since Neptune is transiting in square aspect to your Moon. Remember to keep your emotional boundaries intact! If your Moon is between 8–14 degrees Taurus, you're making adjustments and trimming away what you no longer need while Pluto transits in quincunx aspect to your Moon this year. You may be feeling a ridiculous mixture of rebellion frustrated by the weight of your current obligations if your Moon is between 15–21 degrees Taurus, since Uranus is square your Moon while Saturn conjuncts your Moon. While you may be tempted to burn your bridges, you may also realize that you're still standing in the middle and must extinguish that torch! For those with the Moon between 22–30 degrees Taurus, Saturn is transiting in conjunction to your Moon from July to December and marks a time when it is absolutely necessary to embrace the fulfillment of your true needs in a responsible manner.

Since the rarity of Saturn transiting in your Moon sign (every twenty-eight to twenty-nine years) is the primary feature of this first year of the new millennium, you may be taking your life and your needs more seriously. All individuals with Taurus Moon are experiencing this influence, and you may find that many of the most significant decisions and actions fall upon your capable and steadfast shoulders. The manner in which you handle these circumstances sets the stage for the next several years, and it is your opportunity to find the most viable alternatives for long range growth.

Affirmation for the Year

I honor my responsibilities with a loving heart.

January

Driven by a desire to establish a firm foundation, you may feel that you're ready to buckle down and do the work required to assure that you'll have what you want when you want it. Launching a serious effort during the New Moon on the 6th helps assure a safe and steadfast beginning, although you may have some anxiety about whether or not you can sustain your concentration if you're doing things you'd prefer not to do! Any emotional conflicts in your work or your relationship can create confusion around the time of the Lunar Eclipse on the 20th, when your anxieties can undermine your sense of stability. A loving connection helps you get back on track after the 24th, and you may find it easier to set priorities.

February

Outside disruptions may force you to alter your plans, and you may find yourself in the midst of a crisis during the Solar Eclipse on the 5th as situations you cannot control will destabilize your sense of order. Since others are likely to look to you for comfort you may find yourself wondering what happened to your own support system, although your ability to handle things can keep you going when others are in need. Your rewards arise during the Full Moon on the 19th, when you have a chance to indulge your own needs for a change. Romance can blossom, although you may also be drawn into an unrealistic attraction from the 20th to 24th. These are the things that inspire poetry and music and set your muse dancing—enjoyable experiences, but not permanent!

March

If you're feeling perplexed about the mixed messages others seem to be sending, step back for a moment to review your own actions and attitudes. Those situations where you've been reluctant to commit may have come back to roost in some way. Fortunately, your care and tenderness can turn the tide when it counts, and during the New Moon on the 6th you may have a chance to make amends or begin anew. Expressing your true feelings makes a difference, and extending yourself during the Full Moon on the 19th can bring purely magical results in connecting with someone you love. You're ready to embrace, laugh, and love with gusto!

April

Evidence of your dedication and hard work comes forward and adds to your sense to accomplishment and feeling of security. Although you're making exceptional progress on the things that are near and dear to your heart, you may be procrastinating about other obligations. However, if you're dealing effectively with your priorities, you may finally be in a position to solicit the kind of assistance which allows you to move at a more satisfying rate on a project. In close relationships, it's time to demonstrate your feelings and put actions behind your pretty words. Tensions build near the Full Moon on the 18th, when a power struggle can lead to a separation. At home, moving or renovations can add the kind of elements to your personal environment that are truly comfortable. After all, one of your top priorities is a comfy nest!

May

Concentrating your energy on your top priorities may leave some things (or people) waiting for your attention, but this is a time when a large project or significant event requires most of your time and attention. Unanticipated changes can interrupt your flow after the Taurus New Moon on the 3rd, although allowing safeguards in your schedule can help diminish the effects of disruption. One thing is certain: you're definitely sure about what you want and need. During the Full Moon on the 18th you're challenged to strike a balance between your needs and the demands of others, but by adopting a loving attitude you'll go much further than you would have by focusing on your own selfish desires. A generous spirit draws greater abundance, while stubborn self-importance can rob you of the things you truly need.

June

With Jupiter and Saturn continuing their influence on your Moon you may feel unwilling to alter your course, though it's a good idea to review your obligations. If others seem less reliable, you may be the one left to solve problems. Still, it's not necessary to go down with sinking ships! Fortunately, fresh possibilities emerge, and even if you're dealing with some unfinished business after the 17th, your ability to reach a resolution helps get everything on track. By taking advantage of this time to accomplish a sense of closure, you'll put yourself in a better position to move into even more satisfying circumstances in your work and your relationships.

July

Although you may not feel that you want to set the world on fire, it is time to expand your horizons. You're likely to opt for quietly incorporating the elements which help assure long-range security. During the Solar Eclipse on the 1st you'll benefit from participating in a project where others are combining their talents toward accomplishing a mutual goal. The Lunar Eclipse on the 16th brings your attention to spiritual concerns, and a close relationship can evolve into a more transcendent experience and inspire your creativity. The final eclipse this month (a solar eclipse on the 30th) may be accompanied by a challenge to drop the things which are complete, releasing unnecessary emotional attachments.

August

This month is filled with a few contradictions. While you may feel very certain about your choices and feelings, situations in the outside world can take you away from your personal priorities. Dividing your time between personal and professional concerns, or your preferences and those of others, can be frustrating if you're not gaining any real satisfaction. Looking for the reasons you're involved with your current situation can be helpful, and, if you're unhappy, may lead you to seek something more gratifying. In a close relationship, talking about your concerns during the Full Moon can open Pandora's box. If possible, wait to introduce ideas which could alter the course of your relationship during the New Moon on the 29th.

September

If you're feeling disgusted by politics or inconsistencies of a leader, friend, or advisor, you may want to examine your own ideals! Often, external conflicts mirror inner turmoil. Other times, they provide an opportunity to see how much you've grown away from destructive behaviors or attitudes. Shared ideals can bring you closer in a relationship, and any romantic inclinations during the Full Moon on the 13th will raise your connection to a more profound level. In many ways, your sensuality opens the doorways to true alchemy—but only if you have a willing partner!

October

Contrasts between your values and those of others can challenge you to adopt a more open-minded attitude in order to grow. The differences can be tantalizing if other factors help insure the integrity of an intimate relationship. At work, maintaining your objectivity helps, and if your position or ethics differ from others you may need some finesse to determine the best ways to assert yourself. Choose your battles, since conflicts arising from the 1st through the Full Moon on the 13th may not actually involve you. Perhaps you simply need to wait on the sidelines with towels and water. After the 23rd you may have reasons to step into the action, especially if a "territorial" dispute arises.

November

Everything may seem to be building to a peak centered around the Taurus Full Moon on the 11th, and you may feel a definite change of heart. If you're in a situation which is working to your benefit and where your needs are being met, this is the time to vigorously renew your commitment. However, if your circumstances seem to be leading nowhere or a relationship is not satisfying your needs, then you are reaching a turning point. Your partner's needs may also be changing, and a healthy relationship may undergo positive evolutionary shifts. Artistic ventures, investments, and career endeavors can become more satisfying, and after the 13th you're in the perfect position to talk over future plans, although changes are most promising next month.

December

Your patience pays off from the 1st to the 9th, when your dedication and commitment can prove to be the perfect ingredients for reaching a resolution in a personal crisis or finding common ground in a relationship. Business matters improve early in the month but can run into snags from the Full Moon on the 11th through the 21st due to unclear or misleading information. Your usually realistic attitudes can give way to fanciful hopes, although you're back on track on the 23rd. For this reason it's a good idea to avoid signing long-term contracts or getting involved in situations which can affect your life for years to come. The Solar Eclipse on the 25th marks an exceptional period of new direction and personal fulfillment—all you have to do is figure out what you want!

Gemini Moon

One thing is certain: your Gemini Moon thrives on change. Always drawn to innovative ideas, the challenge to fill your mind and soul with the kaleidoscopic possibilities presented by life can keep you forever young. Trying new things helps you maintain an open mind, and sharing your ideas with others helps broaden your horizons, particularly if their background or lifestyle contrasts with your own. People or situations you consider boring garner very little of your attention. All too often, distractions can be a real problem for you, placing you in the uncomfortable position of juggling too many things at once.

Your secret desire may stem from a need to be revered for your intelligence and wisdom, but it can take some time before you feel fully appreciated. You do enjoy close relationships and may feel especially drawn to others who share your curiosities and love of freedom. If you're placed in a situation which inhibits your independence, you may shut down or withdraw your affections. Essentially lighthearted, you love to laugh and your humor frequently brings smiles to the faces of others. Heavily charged emotional situations are not comfortable for you, although your objectivity is usually welcome during trying moments, when you can keep your wits while others are foundering.

Your sense of home arises when you feel connected to people or places through shared interests and views. To feel settled and secure, you must first embrace the essence of personal freedom which is expressing your mind as it links harmoniously to higher principles and true wisdom.

Famous Individuals with Gemini Moon
Goldie Hawn, Tina Turner, Queen Victoria

The Year at a Glance for Gemini Moon

You're filled with anticipation as you enter the new millennium, and during this year you will have a chance to open new horizons. While you may feel that things are getting off to a slow start, a gradual momentum builds as your confidence, hope, and optimism grow. Your success will depend upon your ability to maintain your focus on your priorities while you explore fresh options, but if anyone can deal with this challenge, it's you.

To your benefit, Jupiter will transit in Gemini, your Moon's Sign, during the second half of the year 2000. Your preparation for this period may require that you complete some obligations before you can fully take advantage of the freedom promised by Jupiter's transit over your Moon, and from January through the end of June you may have to juggle a series of responsibilities before you're free to explore fresh horizons.

The most intriguing challenge you're facing is a long-standing one: Pluto has been transiting in the sign opposite your Moon for the last few years, and is continuing its cycle. Until the year 2009 you may feel a series of transformational changes occurring in your life, and you're learning to trust the profound insights of your intuitive voice. The two-year period when Pluto exactly opposes your Moon will be the most intensive.

If your Moon is between 0–1 degrees Gemini, Saturn will move into a stabilizing conjunction during the summer and early fall, marking a time of new responsibilities and a slower pace. If your Moon is between 2–7 degrees Gemini, you feel a sense of expanded consciousness and a more profound connection to the spiritual plane while Neptune transits in trine aspect to your Moon. You're in the intense cycle of Pluto opposing your Moon if your Moon is between 8–14 degrees Gemini, and you may decide to make a major move, drastically change old habits, or release outworn attitudes this year. The most significant period you're facing this year occurs from August through October, when balancing your changes requires honesty about your true needs.

If your Moon is between 15–21 degrees Gemini, you feel the stimulus of Uranus in trine aspect to your Moon, prompting you to take risks and allow your most profound needs to emerge as your top priority. If your Moon is between 22–29 degrees Gemini, the first two months of this year will bring fast-paced growth. After that time, you may be working to solidify your foundations so that you can take advantage of a period of strong opportunity emerging early next year.

All individuals with Gemini Moon will find this to be a year of profound personal discovery, with the planetary cycles offering a chance to understand your inner dynamics more clearly. As a result, you can fine-tune your goals with your ever-changing needs, and you may, in the process, discover the key to unmasking a higher level of awareness and insightful understanding.

Affirmation for the Year
I am eager to release old burdens so that my spirit can
fly freely!

January
While you may be eager to take advantage of the benefits of a partnership,
your independent nature is likely to prompt you to define your obligations. Agreements made now may be called into question during the late
summer, especially if you take on commitments which are difficult to
handle. For that reason, determining your limitations now can save you
headaches later! You may also decide that it's time to break free of a relationship which is draining your energy and resources. Prompted by the
Lunar Eclipse on the 20th to delve into the recesses of your spirituality for
guidance, your creativity emerges in full force and can shine through as
ingenious ideas or artistic expression. After this time, most of the obstacles inhibiting your sense of fulfillment will seem easier to handle.

February
It's time to devote attention to your inner self. Reflecting on your needs
during the Solar Eclipse on the 5th can lead you to an understanding of
your true feelings, and in many respects this can be a period of great illumination. If you're willing to make the changes necessary to revolutionize
your life, you may experience some feelings of newfound freedom and
rebirth. In order to take full advantage of these cycles, you may have to let
go of a familiar situation or move away from a relationship which is no
longer nourishing your personal growth. Simply escaping is not the
answer. You're ready for solutions, inspiration, and refreshed vitality. A
love relationship can take an unusual turn near the time of the Full Moon
on the 19th, though initiating a new attraction may not go as planned.

March
Your eagerness to keep things moving forward can stimulate positive
changes. Breakthroughs from the 1st to 12th can be uplifting in a close
relationship, and you may be willing to take a few more emotional risks in
order to find out how you truly feel about someone or something. Communication problems can run rampant, since Mercury's retrograde seems
to unleash loaded issues. Watch the signals you're sending during and
after the Full Moon on the 19th, when a flirtation can get you into

trouble if you fail to follow through on promises—spoken or implied. You may also misread others and have trouble clarifying your intentions.

April

Although it may not be true, you feel that progress is coming to a halt at present. While your desires are not slowing down, the complexity of some situations may require that you take more time to develop them and become fully aware of the impact of your thoughts, actions, and words. Initiating communication during the New Moon on the 4th helps stir a stalled situation, and developments beyond your control can force you to take a look at your priorities after the 14th. Relationships with your children can bring great satisfaction from the 6th to the 30th. You may feel more willingness to make a specific commitment, since your ability to see a situation more realistically helps you determine healthy choices.

May

Mars moves into Gemini, activating your desire to get what you want when you want it. Knowing what you want can be the problem—since you're likely to have choices that make you wish your clone was operating more efficiently! Control issues may emerge midmonth, and near the time of the Full Moon on the 18th you will feel that you have reached the end of your patience with those who seem intent on directing your life. If you've been laying the foundations, it will be easier to step onto a different path in pursuit of new horizons after the 21st. If you've been waiting for your ship to come in, it's likely to be at the docks; you need to have the right ticket—and that's what all the preparation, waiting, and testing have been about for the last few months. Fortune smiles after the 24th.

June

The Gemini New Moon on the 2nd stimulates a series of opportunities to fulfill your hopes and desires. Creative ventures are rewarding, although taking financial or emotional risks is not a good idea from the 1st to the 5th. If you're interested in pursuing a new romance, test the waters before you jump into them. Relationships, both professional and personal, are a key feature of the month, building to a climax during the Full Moon on the 16th. Although the general climate around you may be highly conservative for most of the month, the restraints are most apparent after the 22nd as Jupiter completes its cycle in Taurus. Promise to complete your obligations, since next month you'll to move into new territory.

July

The expansive energy of Jupiter moves into Gemini, initiating a period of optimism. Since this cycle will last nearly a year, you have time to take advantage of the drive to fulfill your needs. This month you're uncovering some of the elements that are in the way of experiencing joy. The Solar Eclipse on the 1st and Lunar Eclipse on the 16th are accompanied by highly charged emotional changes, triggering anxieties about your security. When you have an objective frame of mind, you realize that you've been drawn into another's issues or fears. At this point, confidence helps you forge a different path. Forward motion gains momentum after the 14th, and by the Solar Eclipse on the 30th you'll be standing on new ground.

August

Riding the wave of new inspiration, you may feel that you can meet any challenge. By embracing your spirituality, your strength and hope are restored, and you may feel that you're ready to accept responsibilities which once seemed impossible. Saturn sneaks into Gemini for a couple of months, urging you to evaluate your needs in a realistic manner. Think of this as a preview period—a time to familiarize yourself with the tests you'll face during the next two years. Listening to your intuitive voice while watching the signals from your relationships, work and outer life experiences, you may feel that you're armed with ample input to help you make the best choices. Expressing your passion for a person or an ideal during the Full Moon on the 15th, you may feel that you have a new lease on life. Enjoy it!

September

Your objectivity is at a premium, and others may rely on your judgment of a situation or may need your help to resolve highly divisive situations. Fortunately, your ability to remain impartial is boosted by several planetary cycles, although you may feel that you have your own dilemmas to resolve near the Full Moon on the 13th. Your intuition may be your most reliable resource. Loving relationships blossom, and expressing your feelings flows much more easily. However, philosophical differences can lead you to question your involvement with certain individuals, and may be the reason to bid good-bye to a situation which is counterproductive to your growth. By the new Moon on the 27th you're ready to make your feelings and intentions known, setting out on a more fulfilling path.

October

Although you may be itching to jump into a conflict-ridden situation, you may end up in more trouble in the end! You may also be feeling more emotionally vulnerable—with old issues rising to the surface in surprising ways. As a result, you may feel that you're ready to address some unfinished business of your own, even though it seems painful on some levels. If you're extracting yourself from an unhealthy situation, you may still find yourself grieving the loss, especially if you're breaking an old habit; the psyche works that way—even the things you're ready to release can prompt you to feel sad. Work toward resolutions near the Full Moon on the 13th. From that time forward, you'll feel more emotionally centered.

November

Once again, your soul seems renewed. While some complexities remain, your desire to move forward and ability to assert yourself is growing stronger. Your vision of what you want and need becomes exceptionally clear from the 4th to 17th, and you may even feel that you're experiencing a kind of precognitive awareness of what is yet to be. Measuring the risks, you may feel ready to step onto a more daring path, and if your attitudes reflect love and concern, you'll draw strong support. However, if you present an arrogant or defiant attitude, you may create unnecessary hostility. After a few adjustments, your hope returns in full with the New Moon on the 25th. Make time to enjoy your close relationships.

December

Your feelings about your partnership may be changing, so it's a good time to address your concerns. During the Gemini Full Moon on the 11th you may feel especially vulnerable to criticism, but you're also ready to share what's in your heart with someone you love. Love has a magical way of transforming your life this month, and opening your heart to the true experience of giving and receiving love can inspire the courage and conviction to follow your dreams. Your desire to create an ideal life may have seemed unrealistic, but the more you're willing to actualize it, the more amazing each day becomes. Spend time during the Solar Eclipse on the 25th determining what you need to do to bring your hopes and wishes into reality. You can let go of the past and move forward.

Cancer Moon

Because your Cancer Moon brings you into harmony with the natural rhythms of life, you are the true "moon child." Your intuitive sensibilities are powerful and allow you to understand the changing nature of life. Capable of fostering comfort and tenderness, you are the first person others seek for support. While your focus often is family and home, you sense a connection in every aspect of your life. Whether you're raising children, encouraging students, or directing a company, your style is involvement, and you prefer to draw anyone you love into your inner circle.

To thrive, you need a secure nest, filled with the people and things you love; this sustains you when the world seems menacing or cold. In matters of the heart, you need a partner who will share your passion to establish a secure home and whose ideas about family are similar to your own. Family and tradition may hold a special place in your heart, and if you're responsible for the care of others, you're likely to take it rather seriously. Participating in nourishing growth feeds your soul, but letting go when it's time for your fledglings to fly can be discomforting. Instead of feeling abandoned, celebrating another's growth can fill you and allow you to release your fears. Letting go of the past can also be painful, although once you've found a way to incorporate the wellspring of support into necessary evolutionary changes, you can weave a beautiful tapestry of life which reflects reverence for the past and hope for the future.

Welcoming the continuing cycles of change adds a vibrant confidence to your soul and can bring marvelous alterations over the course of your lifetime. Once you step into this flow, a quality of contentment shines in your eyes as you drink in the feeling of peace arising from your awareness of the timeless nature of the essence of life.

Famous Individuals with Cancer Moon

Eva Gabor, Aaron Spelling, Prince William

The Year at a Glance for Cancer Moon

You're on fairly stable ground as you enter the new millennium, although there are plenty of changes happening around you! While you may feel that some things are changing without your permission or beyond your control, other modifications may arise because you can see the value of releasing old or outworn situations. In many ways you may feel that it's time to open your life to new possibilities. Before you can fully welcome the new, however, you must make room for it.

The clarifying cycle of Saturn is transiting in a solid and supportive aspect to your Moon this year, marking this as a time to take a serious look at your needs, priorities, and responsibilities. You may finally be ready to alter some of your habits so that you have more time, energy, or a different attitude. Unnecessary burdens will not feel good, but carrying the obligations which add stability to your life is another story.

Jupiter, the energy of expansion and abundance, brings a series of positive opportunities to enhance your personal fulfillment. This is especially powerful from February through July, when you may feel that you're reaching a new plateau and when your confidence is at its highest. After that, you'll need to make a more careful assessment of anything arriving in the guise of opportunity, since you may be facing the temptation to get in over your head.

The slower moving transits provide the most disruptive and challenging elements in your life, but because they are slow-moving you'll be most affected when the cycle is in a close aspect to your Moon. If your Moon is between 0–2 degrees Cancer, you will be making preparations for the future and taking careful steps during the summer and early fall when Saturn transits in semi-sextile to your Moon. If your Moon is between 3–8 degrees Cancer, Neptune travels in quincunx aspect to your Moon all year and you feel a need to make more room for your spirituality. The challenge of this cycle is to determine the best way to maintain your personal boundaries, since it's easy to sacrifice yourself to the needs of others or give in to addictive tendencies. If your Moon is between 9–14 degrees Cancer, you will be trimming away the excesses while Pluto transits in quincunx to your Moon. This can be a powerful healing period, although you may also have to deal with losses you had not anticipated. Uranus brings unexpected disruptions into your life if your Moon is between 15–21 degrees Cancer. You may decide to move or renovate, or your family may be undergoing big changes. The manner in which you respond to these changes will determine whether or not you feel safe or threatened. If your Moon is between 22–30 degrees Cancer, you will be feeling powerful stability while Saturn transits in sextile aspect. This cycle helps you set important priorities.

All individuals with Cancer Moon experience a period of increasing awareness of inner motivations and drives. Understanding these triggers can make a huge difference in whether or not you feel in control of your life. It's time to know yourself and to honor your real needs.

Affirmation for the Year

I accept the essence of my soul and surrender my life to
fulfilling my real needs.

January

Measuring your involvement helps you keep pressure at bay. However, you can feel rushed, even if there's nothing to justify it. Turning inward to evaluate your priorities and reflect on your motivations, even just a few minutes each day, helps you maintain your center. You're eager to express your feelings, and you will receive a better response if you're coming from a place of pure intentions. On and after the New Moon on the 6th you feel that others need a bit of encouragement to see things your way, but if you become too manipulative you'll end with regrets. By the same token, you will resent anyone else who attempts to take charge of your life. Watch for power struggles from the 8th through the Lunar Eclipse on the 20th, when tolerance will be necessary to deal with differences of opinion.

February

Sharing your concerns and feelings with others in your inner circle can deepen your understanding and commitment from the 1st to the 12th. You may feel frustrated with attitudes of apathy or aloofness during the Solar Eclipse on the 5th, but there is little you can do to change things other than on a strictly personal level. Differences of opinion or situations involving value judgments can be upsetting from the 11th to 15th. Fortunately, your indirect approach buys you a little time, and by the Full Moon on the 19th you may finally make headway connecting with someone who understands and values your point of view. For the remainder of the month you may be reviewing a situation requiring adjustment or repair, and you will be getting to the core of long-standing problems.

March

Friction at work or at home can keep you on edge, especially if you feel that you're being pushed around by selfish attitudes. You may be a victim of a misunderstanding, and clarifying the situation can lead to progress during the New Moon on the 6th. Philosophical differences play a role in relationship disputes. You may discover your values stand in contrast to your partner's. Fortunately, your ideals are more grounded, though the temptation to fall into a judgmental attitude can overwhelm your tolerance. Feelings of love fill your heart after the 13th, and a romantic interlude near the Full Moon on the 19th can be gratifying. You are inspired to open your heart and surrender to the transformational flow of love.

April

Solidifying your commitments may be your intention, but you will turn to mistrust if a person's intentions seem vague to you. You may feel that it's safe to explore your concerns from the 1st to the 12th, when communication is easiest. Whether exchanging vows or keeping a promise to a friend, the significance of your actions can be very meaningful. Distractions enter the picture after the 13th, and during the Full Moon on the 18th you may feel upset if your preferences are either ignored or given too little credibility. It's easy to overreact, making a mountain out of the proverbial molehill, and you'll be happiest if you try not to read too much into your disappointments. Circumstances improve again after the 28th.

May

You're happiest when you can trust your security base, and this month you're in the position to firm up your foundations and take a stand about what you want. The New Moon on the 3rd can be a magical time, when everything seems to fall into place. It's time to determine a course of action that will promote growth in a stable framework. Whether you're moving, renovating, exchanging vows, or making other beginnings, putting your knowledge and experience to use while keeping an open mind about fresh possibilities allows you to make the greatest progress. Although you are tempted to cling to the past, changes beyond your control may arise near or on the Full Moon on the 18th. Make room for them!

June

With Mercury transiting over your Moon all month, you'll discover that it's impossible to separate your feelings from your decision-making. You're more likely to be very passionate about sharing your ideas, so this is the perfect time to exercise your creativity. After the 17th, with romance taking flight and passion ruling the day, you may feel like expressing your most intimate thoughts. During Mercury's retrograde from June 23rd to July 17th, an old love may enter the picture. This is not likely to complicate matters too much, though you will have idealize situations from the 21st to the 26th. Common sense may come to your rescue on the 27th.

July

You may be dealing with an truly amazing period of intuitive awareness and self-actualization during the Cancer Solar Eclipse on the 1st. It's time to break away from situations that are counterproductive to your growth and move into a more nourishing life experience. Your courage and confidence are strong, and loving energy takes top priority in all things through the 13th. During the Lunar Eclipse on the 16th you're confronting unfinished emotional business from the past and may finally feel you can say goodbye to a few bad habits, too. Trusting your feelings and listening to your inner voice, you're more likely to feel contentment about your choices. During the second Solar Eclipse on the 30th, it's time to "tweak" a few things; fine-tuning makes all the difference!

August

Although you are pushing to finish something before the 7th, you may not feel motivated unless there's a "prize" waiting. Expectations can be a problem now, and dealing with disappointments can be rather daunting. Clarify plans early in the month, and if you get iffy answers, try again after the 23rd when more reliable energy comes into play. Making room for distractions and unanticipated events during the Full Moon on the 15th gives you room to maneuver through a testy time. You may be learning what you do not like, not always a pleasant experience. Reassured by more realistic options after the 22nd, you're eager to move forward into safer territory during the New Moon on the 29th.

September

Maintaining your emotional center can be a challenge, since others around you seem stuck in a loop of trivial disagreements. It's a case of logic and propaganda running amuck! Though you like to make a difference, you may have to step aside and let the chips fall before you step into the picture. In other words, it may be time to choose your battles. There are likely to be some gratifying periods. Connecting with close friends feels great from the 1st to 7th, and getting involved in a project that makes a difference in the quality of life lifts your spirits near the Full Moon on the 13th. Create an escape after the 25th, when getting away with your sweetie to enjoy an entertaining or romantic interlude can be what you need.

October

Finally.... relief! Larger conflicts—the stuff permeating the news—may not be settled, but you're finding solutions to your personal quandaries. If you have to stand up for your ideals, you're likely to find others standing right beside you. It's a good time to be involved with a cause. You're more likely now to reach for situations which feel good and be involved with others who value the real you. Love relationships grow passionate, and you may allow yourself to open to the experience of receiving and giving love in a more profound way. New love grows from the 1st to the 13th, and after the 18th you may feel ready to move into fresh territory. Let your fervent expressions take charge during the New Moon on the 27th.

November

Unfinished business, some dating back to April/May of this year, needs special attention. Resolving old circumstances prior to the Full Moon on the 11th can be especially freeing, and may open the way for an easier connection in a close relationship. Balancing your needs with the concerns and needs of those who share your life can be confirming on many levels—but that means you have to honor your needs, too! It's easy to miss important details from the 12th to the 18th, particularly if you're distracted by your romantic inclinations. Keeping things light, simple, and unencumbered brings positive results. Try not to let your impatience get the best of you!

December

Although you may dislike the idea that you can be motivated by guilt, you discover a few of those guilt triggers from the 1st to the 8th. This is a good time to concentrate on practical matters. However, you may sense that there's something pretty fascinating waiting in the wings, and you're right! Romance, passion, and intrigue all blend together for an amazing experience after the 22nd. You may be wishing for a little magic, and sometimes, if you can allow it, wishing can be an invitation to the things you deeply desire. The Solar Eclipse on the 25th can lead you into realms you've not yet explored. Have your adventure gear nearby.

Leo Moon

L eo Moons need drama, intensity, and passion. You feel satisfied only when living on a grand scale. You know how to make the most of the spotlight and project a radiance. When challenged to meet expectations, you rarely let anyone down. Many times, others look to you for inspiration, and your enthusiasm and faith in their abilities lifts their spirits. Likewise, you feed on the respect and admiration of those you hold in high regard.

For those close to you, a bond of devotion and love is strengthened under your protective grace and nourished by your warm hugs. Unless you feel you can follow through, however, you're unlikely to make a promise. You can be generous to a fault, but if you're feeling insecure your emotional pendulum swings toward selfishness. Once you fall into the rut of self-absorption you can lose your objectivity, and it is then that your relationships suffer—particularly if you become willful or intractable. Enhancing your adaptability strengthens your ability to cope with life. Since you prefer to feel that you're in charge of your life, you'll be happier when you're the one directing changes. Of course, these extremes of emotion are part of your sensational approach to life.

From your first breath you may have felt that your soul was on fire with a passion for life and the potential to create. Your yearning to share an expression of love with the world prompts you to embrace your inner fire, and your connection with a higher power releases you from self-focus and makes you more eloquent. As the ultimate performer in the drama of life, you are truly never far from the hearts of those who adore you, even when you cannot hear the applause.

Famous Individuals with Leo Moon
Queen Elizabeth II, Queen Latifah, Barbra Streisand

The Year Ahead for Leo Moon
You're stepping into the new century with a desire to establish a foundation in a time of change. Your visionary abilities can be right on the mark, although executing the realization of your insightful dreams may get off to

a slow start. The greatest challenge you face stems from the need to keep a balance between breaking free and dealing with your responsibilities.

Saturn continues its cycle in Taurus, with the focus on maintaining the status quo standing in stark contrast to your impulse to move forward. You can use this cycle most effectively by making an honest assessment of your obligations and responsibilities. It may be time to hand over controls to someone else, and, in so doing, you may actually free yourself to devote more time to new dimensions of personal growth. However, if you try to dodge your own responsibilities the first emotion you'll experience is guilt. Let that be your signal to re-evaluate.

As an interesting compliment to Saturn's cycle, Jupiter also transits in conflicting aspect to your Moon from February through July, adding a feeling of impatience about what you do not have. Watch this impulse, since it can stimulate poor choices. From July through December, Jupiter's cycle moves into a helpful aspect to your Moon, enhancing your confidence and stimulating greater faith in yourself.

The slow-moving cycles have a significant impact on your emotional nature and can bring alterations which last for years. If your Moon is between 0–1 degrees Leo, you feel much more stability during the late summer and early fall with Saturn's sextile to your Moon. You feel a need to escape the bounds of the ordinary if your Moon is between 2–7 degrees Leo while Neptune transits in opposition to your Moon. You're much more sensitive now, and may feel an impulse to devote more time to your spirituality and creativity. If your Moon is between 8–14 degrees Leo you feel a sense of personal empowerment while Pluto transits in trine aspect; this is a healing cycle, and you're ready to embrace your talents and fulfill your needs. If your Moon is between 15–21 degrees Leo, you will experience a powerful testing period. Uranus is transiting in opposition to your Moon while Saturn is forming a tense square. You may feel you're between a rock and a hard place, since you may see that you need to make sweeping changes but have to wait for a while before you can complete them. If your Moon is between 22–30 degrees Leo, it's time to eliminate the things from your life you no longer need while Saturn transits in square aspect to your Moon.

Everyone with a Leo Moon has an opportunity to establish a more powerful connection between conscious awareness and inner self. It's time to feel the power which emanates from the center of your being and allow it to light the flame of self-acceptance and divine love.

Affirmation for the Year

Love is the fuel which keeps my soul alive with hope
and confidence.

January

You're initiating the year with hope, confidence, and personal renewal.
By expressing your gratitude and allowing your spirit of generosity to
shine forth, you can experience a revitalization of a significant relationship. Regenerating your life through love sets your soul afire with visions
of what can be, and this feeling can carry throughout the year in your
work and personal relationships. By releasing attitudes which stand in
contrast to your need to be hopeful and positive, you'll feel more alive.
Meeting the challenge of the Leo Lunar Eclipse on the 20th may require
that you step back and evaluate how you truly feel about your choices,
relationships, and life circumstances. It's time to listen to your inner voice
and to reflect on the best ways to put that guidance to use.

February

While you may feel eager to make a few changes, taking action requires
that you set your priorities first. Altering your life course may not be
immediately possible, but breaking out of a rut into the light of new
options is very promising under the stimulus of the Solar Eclipse on the
5th. Your courage and enthusiasm gain momentum after the 13th, though
you may have to pull back to deal with repairs, communication break-
downs, or other delays after the 21st when Mercury turns retrograde. It's
easy to be drawn into unrealistic attractions after the 20th as your vision
of love may turn out to have little substance. However, developing the
spiritual essence of a strong loving relationship can allow you to feel that
you've transcended the bounds of ordinary reality from the 19th to 23rd.

March

Your desire to experience something fresh and different through love can
be overwhelming from the 1st to the 10th and again after the 23rd. Someone can misread your signals, particularly if you're unsure about your own
feelings. You may be limited by circumstances or other obligations,
although the temptation to throw caution to the wind and give in to your
passions can preempt your common sense. If you determine to funnel this
energy into your personal environment, it's a good idea to keep your

budget in mind before you buy furniture, redecorate, or move, since you may be drawn to something now that you will not like later. It's the perfect time to buy slipcovers or paint, or to simply rearrange the room. Trying to make permanent changes can be a headache.

April

You're dealing with an interesting mix of responsibility and passion. Directing your enthusiasm toward an ongoing project or your career may be easier than trying to start something entirely new. Initiating changes in your habits and routine works to your advantage during and after the New Moon on the 4th. The only trap can be underestimating the time or energy required to fulfill obligations, which means you may have to say "no" to some opportunities knocking at your door. In all matters, you can rely on your judgment to tell you whether or not a situation will be beneficial to your self-esteem. If not, it's best to decline. Upholding your ethical values and reinforcing your loyalties leads you down the most fulfilling path after the 6th.

May

Remaining focused on your priorities saves you a lot of heartache and frustration. While some changes may arise beyond your control, your attempts to make sweeping modifications are likely to be resisted by those with a more conservative attitude. There is some flexibility when it comes to evaluating your long range plans, and if there's a shakeup in your life from the 12th through the Full Moon on the 18th, it may be due to changes others are experiencing. Taking into account these alterations, you may feel that it is time to work toward fulfilling obligations so you can move on. Planting seeds, making connections and reinforcing your relationships strengthens your self-confidence from the 7th to the 30th.

June

You've always preferred to think in terms of possibilities instead of limitations, and a spirit of hope stirs in your soul during the New Moon on the 2nd. Spending more time developing your special interests, hobbies, or artistic leanings can place you in touch with others who appreciate your talents from the 2nd to the 18th, and joining together to accomplish an important task gives you a chance to exercise your leadership abilities. Make time for fun and romance during the Full Moon on the 16th, when your spirit of adventure is awakened. Extending a helping hand feels like the right thing to do this month, although you're keeping your boundaries clearly in sight. You're not willing to let others see your vulnerability!

July

While others are mired in the trap of emotional attachment, you feel inclined to move away from loyalties which have outlived their usefulness. Letting go of the past is a feature of the Sun's Eclipse on the 1st, and as a new day dawns you sense that it's time to experience a shift in the natural order. After the 13th you will feel the stimulus of Venus transiting over your Moon, awakening love and aiding your ability to make choices which confirm your values and needs. The Moon's Eclipse on the 16th emphasizes letting go of the past. Taking advantage of a chance to let your altruistic nature emerge from the 14th to the 31st, your actions make a lasting impact. By the time the Solar Eclipse in Leo arises on the 30th, you're ready to answer the call of your heart and welcome good fortune.

August

Your soul is ablaze with passion, and expressions of love and creativity are underscored by a sense of conviction which is impossible to ignore. From the 1st through the Full Moon on the 15th, you're in the perfect position to move forward with your plans, and it's much easier to make important changes. This is your time to shine and to accept the admiration and love others feel for you. Essentially, you're experiencing a period of reward for your dedication and devotion. Whether you're moving, striking out in a new relationship, or beginning other significant life changes, you can now take action with confidence and more clearly see where you need to go.

September

You're feeling more alive as you lend your energy to larger concerns, such as community efforts. Your drive to fulfill your needs is strongest from the 1st to the 17th, and you may not experience resistance if you're on a steady course. Maintaining a balance between your desires and the needs and limitations of others flows fairly well, but it can be enhanced if you make an effort to talk about your feelings with others. Although you can be successful initiating an important action during the New Moon on the 27th, watch for the potential of value differences or philosophical disagreements with others. You might accidentally run roughshod over sensitive territory, or can say and do things which others take as an offense, even if that is not your intention.

October

You can feel frustrated dealing with misunderstandings or details which may be meaningless to you. Before you exhaust your patience, look for ways to delegate some of your tasks and responsibilities so you can give your full attention to the things that can thwart your progress. You may be appalled at the track some people seem to be taking, particularly if you cannot understand their motivations or attitudes. Rather than taking a defensive stand, look for common ground near the Full Moon on the 13th. You're not one to stick your head in the sand when things are difficult, but knowing when to get involved and when to step aside is the key to remaining emotionally unscathed.

November

On some levels, you may feel you're retracing your steps, although you're actually just dealing with the details which will free you to move forward. By keeping a loving and generous attitude from the 1st to the 13th, you'll feel much better handling some of the demands arising from old obligations. Watch for potentially difficult emotional issues arising near the Full Moon on the 11th, when you may finally get to the bottom of a long-standing problem. As each day of the month progresses, you may feel lighter and more free, until the stimulus of the New Moon on the 25th works to help you head toward the realization of your most cherished hopes. Your confidence is empowered, and you may also feel ready to step into a well-deserved spotlight.

December

Love can stimulate you to explore new ways to express your talents, and if you've been keeping your feelings under control you may be bursting to share what's in your heart. If the situation is appropriate and healthy, taking action to express your affections works to the benefit of all concerned. During the Full Moon on the 11th, furthermore, your sense of the romantic touches everything you do. Stepping outside the bounds of the ordinary, you may also decide that it's time to establish fresh traditions—although you can run into resistance from others if your approach is too controlling. Take care of health issues later in the month, and do something wonderful for your body-mind-spirit balance after the Solar Eclipse on the 25th.

Virgo Moon

You view life through a lens that magnifies details, and through the filter of your Virgo Moon there flows a highly proficient sense of discrimination. You definitely know what you like, and you are able to distinguish infinite details when you're evaluating anyone or thing. Emotionally, you're sensitive, although your conservative, down-to-earth nature may fool most people into thinking that you're rather invulnerable.

Your analytical abilities, in fact, are legendary, and when a situation requires serious deliberation—you're the person to call. This same ability can create havoc emotionally when you give in to your tendency to worry. Your tendency to want everything to fit into neat, sensible categories can sabotage your ability to accept what you truly feel, since emotions frequently defy logic or reason. To maintain your need to be emotionally grounded, you may be especially appreciative of natural surroundings or environments which reflect and incorporate the earth's beauty.

Productivity soothes your soul. You seek to do your best on a project or make a difference in someone's quality of life. Sometimes it's difficult to believe in yourself, since your critical eye tends to find your faults first. In relationships, you need a partner whose tolerance is well-developed and who will appreciate your perceptive insights. However, your tendency to be overly critical of yourself and others can be very difficult for others to handle unless you do it in a loving and gentle manner.

To find inner peace, search for positive outlets for your discerning nature while lightening your endless list of expectations for yourself and others. Remind yourself that though you may want perfection, you may actually just desire acceptance. Start by looking in the mirror and allowing the light from your soul to reflect through your eyes—therein lies perfection.

Famous Individuals with Virgo Moon

Florence Griffith Joyner, Dolly Parton, John Travolta

The Year at a Glance for Virgo Moon

Carving out the best possible pathways for change requires the kind of finesse you've exercised many times in the past. Now, as you step into the millennium, the paths lead into unknown territory, and you're opening your soul to release old hurts and experience healing on many levels. Fortunately, in this period of change your feet may be planted on terra firma.

Jupiter's transit adds a quality of confidence which supports your hopes and dreams from February through July, and during this time you may find it easier to nurture expansion in several arenas. You may decide to travel, move, or study—whichever seems to offer the most inner satisfaction. From July through December, setting limits can be almost impossible, and reminding yourself that you may have a tendency to underestimate resources required to complete a project or fulfill an obligation will keep you from getting in over your head.

All Virgo Moons are benefiting from Saturn's transit in Taurus this year—a cycle which enhances your practical nature and helps you build on what you've learned. Emotionally, this can be a stabilizing year. If your Moon is between 0–2 degrees Virgo, you may feel emotionally stressed from July through October as Saturn moves into a square to your Moon. If your Moon is between 3–7 degrees Virgo, you may be unrealistic about your needs while Neptune transits in quicunx to your Moon. The link between your soul, mind, and body is sensitive this year. If your Moon is between 8–14 degrees Virgo, you will feel the impact of Pluto transiting in square to your Moon, and as a result you may have to deal with a series of endings as you bid goodbye to things you've outgrown. Uranus adds its stimulus to your life if your Moon is between 14–21 degrees Virgo. Its transiting aspect can have an unsettling effect and leave you feeling restless and dissatisfied. Fortunately, Saturn's transit in trine to your Moon stimulates your ability to remain grounded, and helps you stay on top of this cycle from February through May. If your Moon is between 22–30 degrees Virgo, count your blessings! Saturn is transiting in trine to your Moon, and though you'll have your share of responsibilities to shoulder, you will also feel that you're capable of handling almost anything.

By carefully evaluating your priorities and making necessary alterations to your life in order to make the most of your resources, this can be an exceptionally powerful year. It's time for you to determine what you need to feel safe, secure, and confident, and to build the foundations which allow you to sustain those feelings.

Affirmation for the Year

I invite changes which bring hope, healing, and affirmation of my worth as an individual.

January

You may feel that you're starting the year with more on your plate than you can handle, and although delegating can be helpful, personal issues will simply require personal time to resolve. You're feeling more emotionally grounded during the New Moon on the 6th when outlining your ideas helps you feel that you're on a more secure platform. Frustrations from Mars diminish your patience, especially if you're continually picking up the pieces for somebody else. However, you will also see your relationships as they are, and as a result you can set reasonable limits and define realistic expectations. Until the Lunar Eclipse on the 20th you will make excellent headway getting rid of clutter and saying goodye to unhealthy situations. After the 25th your sense of stability increases.

February

Eager to make changes in your routine or your environment, you'll get on track by creating a series of fine adjustments which allow you to build on an established base while incorporating innovative changes. Health issues, particularly the balance between mind-body-spirit, require careful consideration during the Solar Eclipse on the 5th, when you may finally feel ready to abandon a long-held attitude in favor of true healing. The difficulty lies in avoiding the impulse to throw out everything simply because you've finally reached the end of your proverbial rope. The Virgo Full Moon on the 19th marks a time when you're more objective about your needs and motivations, though others may not appreciate your blatant honesty. Include positive critiques of yourself and others instead of simply making a list of complaints.

March

It's easy to fall into the "victim" trap, especially if your ideas or actions contrast others. Think of this as a time of housecleaning—emotionally, spiritually, and physically. Your impulse to let go of situations and habits which are draining your energy may be right on the mark, but actually doing it is another story. The amazing thing is that there is another Full Moon in Virgo this month (on the 19th), marking a period of endings. Allowing yourself to forgive—releasing old hurts and resentments— you're ready to empty your soul of the things that inhibit your ability to feel good about yourself. Surrendering to the guidance of your inner self, you move into the experience of rebirth and leave the pain behind.

April

On an inner level, a sense of stability and focus is emerging, making this the perfect time to lay foundations for long-term growth or to establish important routines. Working in concert with others may be easier, since grievances have been aired and you're ready to move ahead. After the New Moon on the 4th the planetary cycles lend stability to your emotions and you feel more inclined to initiate changes. Renewing your commitment to a relationship or endeavor helps to assure that you'll be happy with the outcome. Watch for potential disruptions from the 17th to 22nd, when an unanticipated event can require that you alter your course. After that time it may be easier to make more permanent changes.

May

It's time to move beyond your comfort zone and reach into more diverse ways to fulfill your needs and desires. Making alterations in your personal environment—like moving or redecorating—from the New Moon on the 3rd through the 20th can function as an extension of your creativity and will also feel stabilizing. There is a potential of getting into a situation only to realize that you've overlooked an important factor, although your intuitive sensibilities are very likely to clue you in before you've gone too far. Since most of your choices may be based on practical considerations, the changes themselves may involve breaking out of the rut of predictability and trying something completely different. For this reason paint and slipcovers may be a happier option than knocking down a wall! Relationship changes may follow the same formula.

June

If you've been bottling your feelings, the truth rises to the surface as anger or frustration. Dissipating pressure sooths you, and although you are tempted to work your fingers to the bone, recreation, physical activity, and play can bring healing. Setting emotional boundaries is necessary during the Full Moon on the 16th, or you may feel that you've been drawn into a situation which has little to do with your needs. After the 16th you'll feel more emotionally centered, and you may find wonderful solace in earth-oriented activities like gardening, sculpting, or cooking. Romantic inclinations blossom from the 19th to the 30th, but you may have to take the initiative to get exactly what you want.

July

Think of this as "eclipse month." Situations which have been brewing toward a crisis are almost certain to get there during the Solar Eclipse on the 1st, although this can be a good thing for you! Fortunately, you're ready to assert your desires and are feeling open to the prospects of actual fulfillment. If an old love has entered the picture you may not be ready to deal with your feelings until the Lunar Eclipse on the 16th, since this is the time when you're feeling more grounded. There's a potential trap, though, since you may be drawn into a purely fanciful experience, and you may abandon your good judgment in favor of a thrill. If you're in a safe situation, you might decide to go for it. Any repercussions will be evident during the second Solar Eclipse on the 30th.

August

Completing an ongoing project can be the perfect application for your keen observational abilities, though you may be a bit sensitive to any criticisms. Before you jump to the conclusion that you're not good enough, give yourself a break and take time to determine the actual situation! Your ability to express your feelings is strengthened from the 6th to the 31st when Venus transits in Virgo and enhances your connection to your value systems. You will project your most loving qualities and may finally feel that you're ready to make significant changes. If you need to strike out into fresh territory, make your move during or just after the Virgo New Moon on the 29th when everything falls into place.

September

While differences of opinion seem to be swirling around you early in the month, you can keep a cool head from the 1st to the 7th by using your analytical abilities to determine the best course of action. Unfortunately, you can be pulled into difficult situations during the Full Moon on the 13th, especially if you've left the wrong impression about what you need or want. If you feel you're about to reach your boiling point with some of the ignorant attitudes blowing about, take time out. Asserting yourself is much easier after the 17th, when Mars moves into Virgo, though you may take actions which others interpret as abrasive. Your desire to take the high road arises as you set limits and identify your priorities. Others may follow your lead after the 25th, so be sure your mountain-climbing equipment is in working order!

October

While your personal relationships may be undergoing improvements, you still feel that you're working too hard to keep others from jumping to wrong conclusions. Extra attention to communication saves you from headaches. Friction may run high from the 1st to the 7th, but afterward you'll know if you're on the right path. It's possible that you'll have a change of heart from the 8th through the Full Moon on the 13th, when your tolerance for an unhealthy emotional situation reaches its limits. For a while, you may feel more inclined to withdraw for intimacy, and a time of retreat can help you positively alter your perspectives. Concentrating on your personal priorities is important during the New Moon on the 27th.

November

As the rough waters are calmed, you're feeling more self-assured. A time of spiritual retreat can help soothe your soul—whether it's getting away on a vacation or simply spending time each day on contemplation or prayer. Proportioning time for fulfillment of your personal needs during the Full Moon on the 11th can lead to an expression of deep romantic love, but you may also be focused on a creative endeavor as a means of allowing loving energy to flow from your heart. Alterations to your schedule may be necessary if you are to incorporate everything on your personal agenda. Give this some thought on the 19th and 20th, when you're more inclined to be honest with yourself about what you actually want. Turning family obligations into an experience of celebration can heal old hurts during the New Moon on the 25th.

December

Start the month with a weekend (from the 1st to 3rd) dedicated to the people and things you love, and take the time to share your caring and compassion. From the 7th to the 21st you may feel stretched beyond your comfortable capacities, and it can be more difficult to take time to play. Knowing your limits is one thing, honoring them is another! After the 21st you feel much more at ease, and until the 31st your first priority is pure enjoyment. To avoid guilt remind yourself that taking care of your needs builds your capacity to care for the people and things you adore. The Solar Eclipse on the 25th marks a time of reflection, hope, and clear focus on the future. Taking the time to let your visions dance will lift your heart and can provide a sense of direction for the year to come.

Libra Moon

Your deep yearning to experience a life filled with peace, beauty and harmony reflects the essence of your Libra Moon, and the allure of beautiful people, breathtaking artistry, and charming places is an integral projection of your truest self. Your creativity is shown in the way you present yourself to the world—with a sense of style. Your artistry is an expression of your soul. You prefer refinement in everything, including people. In your ideal world, you would live a beautiful life with the perfect partner by your side. Your subconscious need to feel a connection may seem to drive your soul's search for a mate. While finding someone to share your life may be a top priority, you also fill the spaces of your heart with friends, family, and social contacts.

Your hunger for perfection can be seen in your mannerisms and attitudes, yet your sense of balance makes it possible to integrate the inevitable flaws into the rich tapestry of life woven through experience and time. While searching for the right partner, you may explore many possibilities, and finding the right ingredients for a healthy relationship can be quite a learning experience. However, even in relationships you realize that the ultimate goal may be to make room for the human imperfections and ever-changing balance of needs which are part of the ups and downs of life. There are probably times when you think that all the fairy tales need to be rewritten.

At home, in career choices, and in personal matters you strive for balance. Decision-making can sometimes be torturous for you and can also frustrate others. When you know what you want, you can take a passionate stand for your needs, but when logic dictates the need for reevaluation, you often change your mind. If your sense of peace is shattered by anger or hostility, you can feel fragile and uncertain, but once you've forged a solid connection to your inner self, you can handle any adversity. Learning to fight your own battles is part of the discovery of your inner partner, and as you reflect a quality of self-assurance you're more likely to manifest that perfect mate.

Famous Individuals with Libra Moon

Nicolas Cage, Julia Child, Mary Hart

The Year at a Glance for Libra Moon

Containing your excitement at the prospects of a new millennium may seem unnatural, since you're ready for new experiences and creative challenges. Opening your horizons to fresh opportunities may require that you take on new responsibilities in the process, and while getting into the rhythm of change may seem awkward at first, your graceful manner ultimately takes control and lets you step right into the flow.

The expansive energy of Jupiter provides a series of opportunities during the year 2000, although finding your limits can be trying from February through July while Jupiter transits in Taurus. During this time, your desires to grab hold of the golden ring can stimulate a tendency to overindulge, and the price you pay could be an uncomfortable series of restrictions in order to get things back into balance. After July, when Jupiter transits in Gemini for the remainder of the year, you're greeted with more palatable choices.

Saturn completes its cycle in Taurus this year, bringing limitations in the form of practical considerations. If you've taken on too many new projects or have yielded to the tendency to experiment with a situation before you're ready, you'll feel Saturn's pinch almost immediately. Moderation may be the key to contentment.

The slower moving planets bring the most intensive changes into your life this year. If your Moon is between 0–2 degrees Libra, you'll feel the supportive energy of Saturn transiting in a trine aspect from August through October. This is the time to put things in order! You will experience a profound enhancement of your imagination and spirituality if your Moon is between 3–7 degrees Libra, since Neptune is transiting in trine aspect to your Moon. If your Moon is between 8–14 degrees Libra, Pluto's sextile helps you fine-tune your needs during this cycle of positive transformation and rebirth. Uranus transits in trine to your Moon if your Moon is between 15–21 degrees Libra, marking a year of freedom, independent self-expression, and fresh beginnings. Saturnian restraints and tests will come into play, helping you determine if you can afford to make all the changes. If your Moon is between 22–30 degrees Libra, you experience a need to make a series of adjustments while Saturn travels in quincunx to your Moon. The late spring to early summer months and the months of November and December are the most challenging, though keeping your boundaries intact will help you measure your choices.

All individuals with Libra Moon are feeling a sense of renewal and hope, and despite any challenges or obstacles, you can find a way to rise above and make the most of your life. Your creativity shines but only if you remove the barriers!

Affirmation for the Year
My heart dances with joy! All of life is a celebration!

January

Your sense of possibility keeps you going, and you're feeling sufficient confirmation for your efforts to believe in yourself. If you feel there's a problem, it may be because time seems to be running short—a sensibility fueled by the stimulus of Jupiter's final six weeks in Aries. While you may gain increased recognition, you may also feel emotionally tired and ready for a break. The Lunar Eclipse on the 20th brings a series of titillating possibilities—dreams, hopes, illusions of what can be fill your mind and soul. Surrendering to your utmost creative and artistic drives may be your best option, particularly if you can feel the hunger driving you. At the core, you may be reaching an emotional breakthrough, ready to release the vestiges of pain in favor of a experience of redemption.

February

Impatience is your theme this month, as the world is lagging behind in fulfilling the promise of your dreams. The experience of awakening which accompanies the Solar Eclipse on the 5th can lead you to break away from your own inhibitions, which helps a great deal with the feeling that you'll never have exactly what you want or need. As your needs evolve into a complex of changes and fresh starts, you may look back only to realize that you've finally moved beyond an old prejudice or fear and that you're actually looking forward to the prospect of experimenting with something completely different. Artistic expression can be an outlet, but your relationships may be the primary target for these changes.

March

On some fronts, logic and reason may prevail, but you can be frustrated in situations where emotionality plays too big a part in decisions or policies. Expressions of love flow most easily from the 1st to 13th, and if there's a chase or potential conquest at stake, you may feel absolutely up for the game. However, if you're receiving little or no confirmation for your efforts, you may just as easily withdraw and focus your energy on a situation where you know you can win. Maintaining your objectivity works to your benefit, and this will be necessary in circumstances emerging after the 20th when you may be called upon to mediate a dispute. It's easy to be fooled by appearances, but your intuitive sensibilities will alert you if there's something more. All you have to do is pay attention to them!

April

With so many demands on your time and energy, you may find it difficult to maintain your balanced perspective. You may feel pressured to please everyone else first, especially if you sense that not satisfying their demands could jeopardize your relationship. The desire for others to like you can get in the way of standing up for your own needs. Eliciting from your partner what he or she wants may be easiest during the New Moon on the 4th, although keeping your personal boundaries can leave you feeling uncomfortable. Take an inventory of your needs and concerns during the Full Moon in Libra on the 18th, when you may have the greatest clarity about your situation. Meanwhile, leave room in your routine for unanticipated changes. Moving now can add excessive stress.

May

An emerging sense of courage helps you initiate positive changes, although progress can be slowed by situations you cannot control. The New Moon on the 3rd stimulates a period of stabilization, but the prevalence of stubborn attitudes can leave your muttering, "life is not fair." By remaining attentive to any person's hidden agenda, you'll move through the Full Moon unscathed on the 18th, though taking things at face value can get you into hot water! Fortunately, a more logical and reasonable climate ensues on the 21st, when your actions and attitudes reflect a more emotionally complete quality. Loving relationships can take on a more enchanting quality from the 26th to the 31st, when getting away from everyday pressures to enjoy romance will feed your soul.

June

Your enthusiasm grows, and whether you're funneling these positive feelings into a relationship, your job, or a creative project—you're likely to experience confirmation and support for your efforts. Initiate actions during the New Moon on the 2nd, when you're more in touch with your motivations and needs, but try to maintain an awareness of how your association with others can alter the course of your life. Allowing time for pure enjoyment keeps you smiling during the Full Moon on the 16th. Communication problems can inhibit your sense that everything is going smoothly, but by being especially attentive to signals from others you can avoid the pitfalls of these cycles as they intensify after the 22nd.

July

Argumentative situations can result from an emotionally charged experience on or near the Solar Eclipse on the 1st. While you may feel awash in excessive emotionality, your spiritual perspectives help you stay afloat. Giving in to pressure during the Lunar Eclipse on the 16th is tempting, although you're not likely to compromise on anything which will undermine your security. Withdrawing from intense circumstances may be your best option, and spending more time in retreat or involved with a creative or artistic venture is a constructive passtime. After the 23rd you feel safe to air your needs and concerns. The tide turns, and the Solar Eclipse on the 30th can stimulate a period of rebirth.

August

Making a dramatic gesture helps assure that others know how you feel and where you stand, and clarifying your position now can save you some grief. While negotiations may be bogged down from the 1st to 6th, you'll feel comfortable concentrating your energy on a project at home or dealing with personal matters. From the 7th to the 21st you may be called to provide mediation or share insights on a difficult matter. During the Full Moon on the 15th you may feel that you're one of the few people with a grasp on the larger picture, but before you pat yourself on the back, look closely at your personal situation. Taking advantage of this time to strengthen an emotional commitment or offer your support to someone you love can make all the difference in your relationship.

September

With life going much more smoothly while Venus transits in Libra through the 24th, you feel emotionally refreshed. Adding your special touches at home—redecorating, moving, or renovating—is gratifying, but you might prefer to funnel your increasing sense of artistry into your work. From the 7th through the 29th, your negotiating skills may be on-call in an important situation, and you're not likely to mince words when something significant is on the table for discussion. Political and philosophical disputes can run rampant, and although you have your priorities and preferences you may have to maintain a cool air of detachment to avoid being pulled into a heated conflict. Initiating an innovative plan works to your benefit on and after the Libra New Moon on the 27th.

October

Now that battle lines are drawn, you may want to fade into the woodwork and watch the proceedings. Dealing with other people's prejudices can try your patience. With your close associations, making every attempt to be truly fair in your attitudes goes a long way toward supporting trust and understanding, although you may be extrasensitive during the Full Moon on the 13th. Directing your emotional energy, nurturance, and care toward situations which show growth potential can be exceptionally beneficial after the 20th. You may feel ready to try something different, like an unusual cultural experience, from the 25th to the 31st.

November

Mars moves into Libra on the 3rd, adding zest to your emotional expression but also stirring the cauldron of your discontent. If you're unhappy with anyone or anything, your impatience can keep you on edge. Fortunately, your ability to move away from highly charged situations long enough to determine what you need and want can protect your emotional vulnerability. Once again, your stastes may stand in stark contrast to others from the 18th to 27th, although you may try to be nice and avoid excessively critical commentary in public. You're feeling more at ease after the 24th, and you may feel confident initiating contact with someone intriguing during the New Moon on the 25th. A getaway can lead to romance.

December

As you continue to experience the stimulus of Mars in Libra, you're ready to direct your energy toward your deep desires. An intriguing romantic possibility from the 8th to the 26th can stir your passions, and if you're sending clear signals you're likely to get a delightful response. Whether you're sharing an adventure with your sweetheart or a close friend, plan something fabulous for the Full Moon on the 11th. A weekend party could be nice, but you might prefer a more intimate exchange. The Solar Eclipse on the 25th can stimulate crisis in the family, although it may also be a time to venture away from tradition in favor of setting a new precedent. That, in itself, could be the source of crisis in some families!

Scorpio Moon

The Scorpio Moon hungers to experience life at its most profound levels. Your penetrating gaze offers only a hint of your trademark intensity. To the casual observer, you appear calmly in control, but you have the potential to be an emotional extremist. With an aura of mystery, you can confound those who are most comfortable with easy explanations or emotionally sterile experiences. While others may be fearful of probing beneath a polite exterior, you relish the rush of energy you feel when delving beneath the surface to uncover the mysteries of life.

Your subconscious domain is filled with emotion and an awareness of the true nature of healing and regeneration. Dishonesty is abhorrent to you, and anyone wishing to get close to you will have to prove themselves worthy of your trust. You're a natural detective and have a keen sixth sense. Uncovering secrets may be one of your strengths and very useful in science, healing arts, or detective work, but you are not likely to appreciate when someone exposes your secret self without your permission. That vault in your psyche containing unresolved anger, guilt, self-doubt, and shame is covered with layers of barriers to protect your emotional vulnerability. Releasing those feelings and inviting healing into your life requires that you forgive yourself every time negatives arise. Then, and only then, will you be able to rise to the heights of ecstasy you imagine.

Welcoming the processes of regeneration, you have a capacity for healing that few possess. As a result, the course of your life has probably seen phenomenal changes. Harnessing the power resting in the core of your being, you can become a true healer and have the capacity to maintain insights about the pure essence of life itself.

Famous Individuals with Scorpio Moon

Whoopi Goldberg, Bette Midler, Steven Spielberg

The Year at a Glance for Scorpio Moon

One thing is certain: you will not be bored this year! Your strongest drive may be to establish a solid foundation for growth, but circumstances around you may make that a frustrating task. Dealing with your internal responses to outside changes and your desire to remain safe, you may feel that you're dancing the cosmic three-step: two steps forward, one step back. This is your time to eliminate unnecessary restraints while embracing your responsibilities. While that may seem like a paradox, in truth, you are simply reprioritizing.

Jupiter's energy stimulates your sense that it's time to broaden your horizons, particularly from February through July when Jupiter transits in the sign opposite your Moon sign. You may be able to exercise reasonable restraint in this period of growth, however, since Saturn is traveling alongside Jupiter over this period. Educational pursuits may become a top priority, but you may also feel ready to make significant changes in your personal environment, taking steps to create a more stable home life.

The cycles represented by the slow-moving planets have the most noticeable impact on your life changes. If your Moon is between 0–2 degrees Scorpio, you may feel rather stable for most of the year, but from July through October you may need to adjust your habits or make improvements while Saturn transits in quincunx to your Moon. You will feel more emotionally sensitive if your Moon is between 3–7 degrees Scorpio, since Neptune's transit in square to your Moon opens your emotional boundaries. Dealing with your emotional needs realistically can be challenging, since your tendency may to be to escape. If your Moon is between 8–13 degrees Scorpio, you will fine-tune your habits and attitudes and may feel a true sense of renewal emerging in your life. Stability can be difficult to accomplish if your Moon is between 14–21 degrees Scorpio, since the transit of Uranus in square aspect to your Moon prompts you to make far-reaching changes. You may decide to move or make changes in your career path or relationships. If your Moon is between 22–30 degrees Scorpio, you feel Saturn's restraint this year as it transits in opposition to your Moon. While you may wish that someone would rescue you from some of the burdens you're carrying, you can rescue yourself by eliminating habits, attitudes, and situations that are counterproductive to your growth.

All individuals with Scorpio Moon are challenged to establish the priorities which will allow room for personal changes while assisting the requirements to remain responsible and stable. Your external changes can be a reflection of your need to work on your inner self, and as you become more aware of your transforming needs, you may gain better control over external events.

Affirmation for the Year

I embrace the soul of my desires and needs and have the
courage to fulfill them.

January

As you enter this new century, your ability to put the necessary energy
into satisfying your needs is strong. While you feel the rumblings of insta-
bility, the impulse to push beyond the barriers that have inhibited your
sense of fulfillment is powerful. Rather than focusing on far-reaching
goals, you can enhance your feeling of self-control by working on immediate concerns. Buckling down to finish a project and taking action to
complete course work or other ventures provides a feeling of confirmation
that adds joy to your life. Initiate plans, conversations, or changes with
the New Moon on the 6th, and work steadily toward your aims. The Lunar
Eclipse on the 20th can stimulate some confusion about your priorities,
especially if others are sending mixed signals about what they want.

February

Unsettling changes brought on by circumstances beyond your control can
alter your sense of security, though you always have options. During the
Solar Eclipse on the 5th, try to determine the pressure points. Your ability
to get to focus works like a charm, and instead of feeling off-guard when
dealing with change you may actually experience a growing sense of excitement. Romantic entanglements can be a positive challenge, although
there's a tendency to misread signals. The Full Moon on the 19th marks a
period when opening your heart to a trusted lover or friend fortifies your
affection. You may sense the dawning of all sorts of possibilities!

March

Despite the confusion surrounding you, you feel you're gaining an understanding of your inner drives. An intimate relationship gains momentum
during the New Moon on the 6th, and something you thought lost can be
rediscovered. The key is forgiveness—letting go of old hurts in favor of
moving ahead. While you may not heal overnight, relief from carrying
your unnecessary burden lifts your spirits. Loving energy flows freely after
the 12th, and by the Full Moon on the 19th you allow yourself to trust
again. The problem is not other people; you have been struggling against
yourself. Looking in the mirror reveals more than a shallow reflection.

April

Bringing everything into focus requires discipline, and though Jupiter and Mars make you rush from the 1st to the 13th, you can redirect this drive to increase your courage and confidence. In personal and professional arenas, you may feel that it's safe to express what's on your mind. Furthermore, the New Moon on the 4th indicates it is time to make changes. Altering your habits to take care of yourself becomes a higher priority, and you may respect the need to put your health first. Unanticipated changes can scatter your energy and alter your plans midmonth, and if you try to remain in control you might work against your own best interests. Following may be much easier than leading for the time-being.

May

Partnerships or cooperative ventures take most of your time. You may experience support and encouragement from your partner, and by offering your care and concern you can help bring balance into the situation. Unfortunately, you have little patience for purely selfish attitudes, and if anyone is pulling rank you may feel resentful if their actions bring harm to you in any way. Healthy relationships become more stable under these influences, but an unhealthy union can reach the breaking point. From the 16th through the Scorpio Full Moon on the 18th, search your heart for your true feelings. Coming to grips with the reality of your situation, you may feel it's time to make a serious commitment to accomplish measurable growth. Compromising your needs is no longer a workable option.

June

Forging a solid foundation is your highest priority, especially if your power base has been altered in any way. You may be distracted by others whose ideas sound good initially, but if you have concerns about their validity you're ultimately not likely to buy into them. Probing into a situation reveals all you need to know so that you can move on if necessary. Increasing your activity level feels best after the 15th, when burning off some of your frustrations in the gym can alleviate a great deal of emotional tension. Since you may have to juggle several priorities early in the month, you'll appreciate a chance to let go and focus more on your primary concerns after the 20th. Making improvements at home fares best after the 14th, although extensive remodeling is ill-advised since Mercury enters its retrograde cycle on the 23rd.

July

The Solar Eclipse on the 1st prompts a period of awakening and positive change. Many of the pressures you've been feeling are dropping away, and the planetary cycles stimulate your desire to express yourself more fully. Creative ventures can be a source of pure joy, and close ties to those you love provide true gifts. If you've been waiting for someone else to take the lead, you can send the go-ahead. Meanwhile, do something fabulously romantic during the Lunar Eclipse on the 16th, but take your sweetheart's concerns into consideration to avoid getting into hot water. That does not mean you should ignore your passions, and if you've been afraid to talk about your feelings, you're very likely to find other ways to communicate them that go beyond words. Career concerns may be emphasized after the second Solar Eclipse on the 30th.

August

While the outside world can be a bit distracting this month, you're also getting some important clues about your motivations. If you've been working toward a goal because it holds meaning for others and your heart is not really into it, then you can feel that you're carrying the weight of the world on your shoulders. But if your priorities and aims are an outreach of your true needs, any challenges you face will seem less daunting. Your support system is there, but you may not be utilizing it to its fullest. It's easier to keep a cool head when you have a plan of action. Formulate your strategies after the 7th, and target the New Moon on the 29th to put them into action. Be on the alert for power plays.

September

Agitations from the 1st to the 17th are a response to the energy of Mars in aspect to your Moon. Venting your concerns with a trusted friend is helpful from the 1st to the 6th, though you may have to deal with an uncomfortable situation. If the problem is an unlikeable person, keep your emotional boundaries intact since anyone trying to disrupt your plans or your life does not deserve to know your capacities. Of course, you'll let it leak a little by sending your subtle signals. From the Full Moon on the 13th through the 17th most situations will reach a climax, and after the 18th you may feel that the conflicts are under control. After the 23rd most of your attention goes to your intimate relationships or creative endeavors, leaving the pressures of life on the other side of the door.

October

The planetary cycles stimulate a true comfort zone, prompting you to express yourself more fully and to take actions which help to solidify your needs. From the 1st to the 17th, the influences of Venus and Mercury add the right ingredients for moving or enhancing your nest in some way. You may also be feeling rather amorous, hungry for adventure in the love department. A playful attitude lifts any relationship, but you may also be intrigued by someone who's not exactly available. Part of the fascination could be the titillation of what you cannot have, but you can get into trouble from the 7th to Full Moon on the 13th if you're really out of bounds. Dreams can be wonderful if they're in harmony with your needs. Look into your heart for truth during the Scorpio New Moon on the 27th.

November

While you may feel that you cannot trust many people, you're testing the validity of those feelings in your close relationships. You're ready to let go of people and situations which foster ill-will and a lack of trust. You're also feeling the need to eliminate your mistrust of yourself, since you're seeing good evidence that your ability to evaluate situations can be truly exceptional. During the Full Moon on the 11th, breaking away from the things you no longer need may be a part of moving into a gratifying situation. Staying out of arguments which do not involve you is perfectly fine with you, but you can be drawn in to help pick up the pieces. Those who want your opinion should be prepared to hear the truth.

December

You're tying up loose ends and dealing with unfinished business. If you've hoped to make a favorable impression, your carefully considered actions on the 1st to the 3rd can be significant. You may also feel it's important to get in touch with an old friend or relative, and you'll have the best success from the 1st to the 8th, and then after the 22nd. In the interim, you may have to juggle activities which arise from the demands and preferences of others. Once again, a ridiculous conflict can emerge, but you're likely to stay out of it. You're initiating positive changes after the 23rd and can feel positive confirmation during the Solar Eclipse on the 25th. During this last week of the year, you're filled with anticipation for the year ahead.

Sagittarius Moon

Your Sagittarius Moon drives an ever-present urge to search. Whether you're searching for answers, adventure, or interesting ideas, you love the quest and perform it best when you feel truly free. Eager to explore the boundless possibilities of life, you may love to learn. Travel or educational pursuits nourish your soul and help you keep an open mind. Wide open spaces are very appealing—philosophically, physically, and emotionally. You're likely to resist attempts from others to contain your boundless spirit, but you will easily warm to people and experiences which foster your higher ideals and driving sense of principle. You may enjoy writing or teaching, since your words and ideas can be on the mark.

Emotionally charged situations that require your continual presence or attention grow tiresome, since in all relationships you expect to feel independent. You can be loving and loyal, and will be most committed in situations which promote your higher values. When you're inspired, your generosity can be unmatched, and you have a flair for uplifting others. But you will resent being smothered and may feel your spirit withering under the influence of excessive restraints. You do not respond well to the word "no." Those who are emotionally insecure may feel that you're hard to pin down, particularly if you frequently yield to your needs to follow your own path or travel extensively. Others whose philosophical values are different from your own are welcome—as long as they do not try to force you to adopt ideals which contradict your sense of truth.

Ideally, you strive for a feeling that wherever you go, you're home. Place may be secondary to the feeling of acceptance, and sometimes you may move for the experience itself. Adapting to different surroundings can be exciting to you, and you may ultimately have homes scattered around the planet. Once you decide to nest, you need an inspirational environment and may enjoy living close to nature. In your soul, heart and mind you will be forever journeying, wondering, and questioning the vast possibilities while continuing your quest for the grail of Truth.

Famous Individuals with Sagittarius Moon
Larry King, Courtney Love, Donald Trump

The Year at a Glance for Sagittarius Moon

While you may be inspired to try something completely different this year, you will be restrained by increasing responsibility. By making accommodations for your obligations and managing the necessities, you can have a purely enjoyable year. Your primary challenge is to remain grounded in the face of continuing opportunities to expand your horizons.

The boundless energy of Jupiter stimulates a desire to reach into fresh territory, emotionally and spiritually, during January and then from July through December. During the rest of the year, it's easy to overdo it, as you feel that you're constantly required to keep your enthusiasm in check to avoid getting into something you beyond your emotional capacities.

Incorporating habits and attitudes that help you stay connected with your physical, mental, emotional, and spiritual needs may be a challenge. If you're on the go, making time for your health can be a headache, but if you fail to take care of your needs, you may feel under the weather emotionally and spiritually. Think of this as a time when honoring your limits and adopting an attitude of moderation can work to your advantage.

The slow-moving cycles have an impact on your emotional sensibilities this year. If your Moon is between 0–2 degrees Sagittarius, you will experience a year of inner stability, but you may face a challenge from July to October when Saturn transits in opposition to your Moon, forcing you to reprioritize your time. You will experience an inspiring year of expanded awareness and spiritual insight if your Moon is between 3–7 degrees Sagittarius. Neptune will transit in sextile to your Moon, indicating that you will find it necessary to forgive and to make a difference. If your Moon is between 8–14 degrees Sagittarius you feel the impact of Pluto's once-in-a-lifetime conjunction to your Moon, adding intensity to your emotions as you release things you no longer need and become open to true healing. It's easy to be obsessive, and remaining objective can be difficult unless you make a concerted effort to do so. If your Moon is between 15–21 degrees Sagittarius you may feel unfettered since Uranus is transiting in sextile aspect. However, from February through June your zeal may be quelled under the influence of Saturn's transit in quincunx to your Moon. Basically, this is a reality check. If you can afford it, break free. If you have duties, you must honor them. If your Moon is between 22–30 degrees Sagittarius, you feel the full impact of Saturn transiting in quincunx to your Moon. Think of this as a time when you need to finish whatever you've been putting off, since your obligations will definitely not go away.

Affirmation for the Year

I consciously choose to balance my desires for change
with my highest needs.

January

Your passions are stimulated by the energies of Venus and Mars, and while
you have powerful desires, determining whether or not you should pursue
them can produce anxiety. Although you might prefer to act and deal with
the consequences later, you may wonder, if you've been burned before,
whether you're not heading into a reality-check from the 6th to the 18th.
While your intuitive voice whispers helpful hints, study the body language
of others for clues. Love is promising during the Lunar Eclipse on the
20th, when playful romance can lead to good things!

February

You may think you know what you want, but your interest is difficult to
hold whenever something more intriguing comes along. Keeping your
options open keeps you out of trouble, but someone may question your
commitment. If you are not certain, say so, since leading anyone on will
complicate your life. A pure infatuation can take your breath away during
the Solar Eclipse on the 5th, and whether the object of your attention is a
person or circumstance, you may feel that you have to act quickly to avoid
losing your window of opportunity. Showing your intentions helps you determine whether or not you want to go further in your pursuit. From the
13th to 29th you may be unstoppable, although the most sensitive types
can overreact to your attentions during the Full Moon on the 19th.

March

You may not be the problem, but there is a high potential for misunderstanding and poor communication now. Your sense of humor can come to
the rescue just in time to salvage a sticky situation. Showering your affection on someone you adore from the 1st to the 13th can bring the response
you hope for, but you have to go the extra mile to be sure everything goes
off without a hitch. If you're in a relationship that's becoming complex,
you may pull away or have a change of heart after the 20th when you've
reached your tolerance levels. Once the pressure is off you may be willing
to reconsider, but meanwhile you may be off on a more exciting path.

April

Taking the initiative during the New Moon on the 4th can get things off to a good start, but you may lose interest if a situation gets predictable. Distractions can be a problem but are more troublesome if you're stuck in a boring or predictable circumstance. You are hungry for a change of pace, and all around you there may be signals that the pace is becoming more measured. Before you stage a revolution or decide to vanish from the scene, examine the intrinsic value of the things you're considering leaving behind. Opening your heart to the potential of a loving relationship seems quite natural after the 6th, and you're feeling a renewed sense of anticipation during the Full Moon on the 18th. It's a good time to clear away the clutter from your life so that you have more room to breathe.

May

Tip from the cosmic insider: The New Moon on the 3rd is accompanied by five other energies in Taurus. This signals a time when consistency and predictability are very important. If you're going too far against the stream of this energy, you may feel like a ton of rocks has just landed in your backpack. With Mars transiting in opposition to your Moon sign, your patience is at a minimum, and to maintain your sanity you simply must stay active. Actually incorporating distractions into your schedule or making room for spontaneity will help immensely. There is potential for a power struggle with someone whose actions are controlling, and the way you respond will determine whether or not you get to go anywhere!

June

Although you love grand gestures, this is a month to play your hand close to your chest: keep things simple. Taking your lead from the actions, invitations, or attitudes of others helps you determine the best direction during the New Moon on the 2nd. If a competitive situation arises, you may enjoy it but your enthusiasm may spark a more volatile response from your competitor than you intended. During the Sagittarius Full Moon on the 16th you're especially aware of your needs and feelings, and it's the perfect time to take stock of your situation. The adjustments you make for the remainder of the month can lead to a series of improvements, but if other people are involved things can definitely become complicated.

July

Emotional circumstances can arise from the time of the Solar Eclipse on the 1st through the Lunar Eclipse on the 16th. During this time period you may feel frustrated by situations you cannot change, although you can offer understanding and emotional support to those who need it. If you're too uncomfortable, you may stay away, but staying in touch can make a huge difference if a friend or family member is dealing with a crisis. Opening your heart is easier after the 14th, when you may feel more creatively inspired. From the 23rd to 31st you're feeling more alive, and some of those prickly circumstances may seem far behind you. Travel, writing, or making contact with others who share your ideals can be especially rewarding on or after the second Solar Eclipse of the month on the 30th.

August

Feeling playful, you're ready to set out on an adventure. Whether you're exploring the possibilities of love, loading up the SUV for a cross-country trek, or delving into an intriguing subject—it's time to expand horizons. Responsibilities are likely to limit the scope of your adventure, but the spirit bubbling inside makes all the difference. Sharing enthusiasm for your future plans brings others into the picture. Pleasurable relationships, including connections to family, can be especially enjoyable during the Full Moon on the 15th. You may be witness to the very intense philosophical diversity of the world, but before you become involved make sure you know where you stand. Practical considerations take top priority after the 23rd, although you may still be busy with your adventure.

September

With less emotional frustration, you feel you can finally get something done. You're even gaining cooperative support from others who share your enthusiasm. Disagreements over the best way to do things can be irritating from the 1st to the 7th, though your sense of humor will improve communication. Unfortunately, politically sensitive people may not like your jokes, so it's important to know your audience and judge your words and actions appropriately near the Full Moon on the 13th when emotions can get in the way of clear thinking. Philosophical disagreements continue, and if you're uncertain about a personal or professional situation, seeking counsel during the New Moon on the 27th about the best way to proceed can help you maintain your dignity!

October

Despite your best efforts to remain objective or avoid being pulled into conflicts, you may have no choice. Taking a stand on important issues is necessary, particularly if they affect you directly. While you may have some allies supporting you, it's entirely possible that you will still feel that you're dealing with some circumstances by yourself. If your intimate relationship is in trouble, the underlying cause may involve your ideals or beliefs. For you, getting close includes sharing ideals. While you may be open-minded, your tolerance for moral or ethical abuses is low. Allowing love to guide your actions works best after the 19th, although you may still bid goodbye to a person or situation. It's just easier to move forward with love in your heart.

November

The heated battles are subsiding in favor of open discussion. The bonds of friendship prove to be rewarding from the 1st to the 13th, and you can see the value of friendship in a relationship. You're feeling more generous after the 4th, though you're not willing to waste your efforts on futile things. Old issues can raise their heads during the Full Moon on the 11th, although you're not likely to feel regrets. Instead, you're ready to experience forgiveness and move forward, even if someone else wants to hang onto the past. After the 23rd you're feeling the stirring of new life, and the Sagittarius New Moon on the 25th initiates a period of hope and renewal. A journey to a place which inspires you can have immeasurable value.

December

You feel on fire with passion, and you're eager to express your feelings—particularly the good ones! Celebrating may be the first order of business, and whether you're attending parties or planning your own gatherings, you adore being around the people who have your affection. Your romantic desires grow stronger after the 8th, and during the weekend of the 8th to the 10th you'll do yourself a favor by taking advantage of opportunities to illustrate your desires. The period prior to the Full Moon on the 11th stimulates your ability to reach out to others. While you may be feeling especially generous, you're also eager to make positive changes in your home life to reflect your growing contentment and hope for the future. Take your cues from others during the Solar Eclipse on the 25th.

Capricorn Moon

There's little that can compare with the satisfaction you feel when you've overcome a challenge and stand tall in the light of your accomplishments. Your Capricorn Moon invites you to develop a true sense of self-respect by forging a sure and steady path in life. In some ways, the experience of climbing toward your goals recharges you. Even as you stand at the summit of achievement, your desire to fortify your security prompts you to reach toward new peaks. When times are tough, your marvelous dry wit comes to the rescue.

It's all too easy to place your primary emphasis on work, but your dedication to home and family can be your motivation to work diligently. You take your commitments seriously and respect others who have the dignity to honor their own. Relationships with friends and family are securely shaped through your desire to set a solid, reliable foundation which can withstand the ravages of time. While you may sometimes feel that you're the one whose shoulders carry the brunt of the responsibility, in some ways your controlling attitude and desire to get the job done invites others to rely on you to be the one who can carry the load. This is only a problem if you grow resentful. To avoid resentment, give others a chance to shoulder their own burdens when they're ready can make a significant difference in your personal and professional relationships.

Despite your apparently serious nature, when the job's done, you do love to play. Wasting precious time and effort is just not your style! Giving those who love you time to enjoy your company can provide a surprising level of personal contentment. Opening your heart, expressing your feelings and allowing yourself to experience the flow of love and tenderness can be impossible if you're hiding behind your walls. Your need to feel that you are in control of your emotions can actually block the love and support which you may deeply crave. Protecting yourself is one thing, but shutting out the joy of life diminishes the returns on your hard work.

Famous Individuals with Capricorn Moon
Lucille Ball, Al Gore, LeAnn Rimes

The Year at a Glance for Capricorn Moon

Strengthening your foundations may be your primary drive as you enter the millennium. You're inviting new responsibilities which afford you the opportunity to advance in your career or reinforce your family ties. After all, you'll feel more secure if you're following a well-defined path. Although circumstances around you are changing rapidly, your ability to remain emotionally grounded is enhanced.

Good opportunities to take the lead or carve a successful path abound from February through July, while Jupiter and Saturn are both transiting in Taurus. These months are the most significant times to make successful long-term commitments such as moving, marrying, changing careers, or adding to your family. While the other parts of the year offer their particular strengths and challenges, you may feel most emotionally in control during these cycles.

The longer lasting cycles of the slow-moving planets have the most significant impact on your life, since your response to these energies can have lasting effects. If your Moon is between 0–2 degrees Capricorn, you will feel fairly certain of your needs and directions for most of the year but will need to accommodate changes or make adjustments from July through October when Saturn transits in quincunx to your Moon. You may be tempted to see things only as you want to see them if your Moon is between 3–8 degrees Capricorn, since Neptune's transit in semi-sextile to your Moon can create a fog of unrealistic illusion. While this is a good time to better acquaint yourself with your spirituality, it's also a time when you can get lost in the fog. If your Moon is between 9–14 degrees Capricorn, you feel a need to become more clearly acquainted with your inner drives and needs while Pluto transits in semi-sextile to your Moon. You may make alterations to your personal environment or significantly change your habits under this influence. Meanwhile, if your Moon is between 15–21 degrees Capricorn, you are under the disruptive impulse of Uranus traveling in semi-sextile to your Moon. This can be an unsettling influence since you may have to deal with changes that are not of your own making, but finding a way to incorporate them into your life will help you catch up with the times. If your Moon is between 22–30 degrees Capricorn, you gain the greatest benefit from Saturn's transit in trine aspect to your Moon. This is a stabilizing cycle, and one of the best times to make long-term commitments or take on increased responsibilities. You definitely know what you want and what you're willing to give for it!

All individuals with Capricorn Moon may have their eyes on those shining peaks in the distance. This year, you will be drawing out the maps that will get you there.

Affirmation for the Year

I surrender my will to the wisdom and guidance of
the Source.

January

Although you may not make formal resolutions, you certainly feel the need to develop a clear plan for the year. Initiating changes—even small changes such as altering an old habit—has significant impact during the Capricorn New Moon on the 6th. This is also an excellent time to explore what you want, since you're feeling more in touch with your priorities and may be able to make the best determination about whether your wants and needs are compatible. Differences in taste or values can be at the core of disputes from the 9th to the 17th, but if a power struggle follows from the 16th to 20th it may be more trouble than it's worth. You're in a good position to do damage control after the 25th, although you'll be most inclined to focus on your personal desires.

February

In many respects, this can be a gratifying month, especially if you're paying attention to situations close to home. Disruptions in the status quo can throw you off during the Solar Eclipse on the 5th, but you're in a great position to handle these changes. Your love life gains momentum from the 1st to the 19th, when the benefits of Venus transiting in Capricorn promote your affections. Expressing your love and concern can have a positive impact and may be what's needed to solidify a relationship. The Full Moon on the 19th stimulates your desire to balance romantic and practical concerns, and an evening of simple pleasures may be especially gratifying. Follow through with consistent and open communication.

March

Although you may try to ignore the confusing circumstances around you, when it comes time to state your opinion or take a stand your insights into the long-term effects of certain actions can be helpful to others. During the New Moon on the 6th your confidence may be strong, although you may feel that more research needs to be done before you make a final decision about something. Since Mars is transiting in a tense aspect to your Moon you may feel impatient or easily agitated, especially if you're dealing with excessive restraints. Fortunately, you're likely to be focusing your attention on the things you can control during the Full Moon on the 19th and can see evidence that you're moving away from circumstances which do not fit your needs. Your frustrations calm after the 23rd.

April

Practical considerations take top priority, although convincing others to use common sense can be an interesting challenge. The little things can get on your nerves throughout the month, especially if you feel that you're not getting the cooperation you need to get everything done. However, large-scale projects may be going rather nicely, and you're seeing progress where it's most important. Holding onto the rules too intensely can lead to unnecessary frustration near the time of the Full Moon on the 18th, and circumstances beyond your (or another's) control can require a more flexible attitude. Making room for someone whose disposition is different from your own can be worthwhile if your goals are in harmony.

May

Everything seems to be coalescing, and it's time to focus your energy on top priorities. The New Moon on the 3rd marks a powerful time to concentrate on fulfilling your needs, and whether you're dealing with career or family matters you may have little doubt about what you hope to accomplish. Remaining open to the best ways to establish stability while leaving room for innovation can be a challenge, and you'll see the weak links in this plan around the time of the Full Moon on the 18th. Fortunately, you'll find it easier to attract others who understand your sensibilities and concerns. On a personal level, love relationships are strongly emphasized through the 25th. There's no room for waffling about your intentions or desires, and you may have to be the initiator.

June

It's time to diversify, as some people may not be as committed as you want. Once again, control issues can surface, indicating that you need a focus in your life. Let that be your choice; perhaps being attentive to your creative urges and desires to get back to nature can channel your drives. After the 16th, relationships can be trying if your aims are different from your partner's. You're also likely to run into situations which may require that you put someone else's needs before your own. If it's the right thing to do, you'll be okay with it, but if you feel forced into a circumstance which seems out of harmony you're likely to object.

July

You really don't like feeling pushed, and if you're being driven by something or someone your frustrations with the situation can be intensified during the Solar Eclipse on the 1st. By the time the Capricorn Lunar Eclipse rolls around on the 16th, you're setting stronger limits and looking for ways to assert your needs and opinions. However, you may also become embroiled in disputes. Healthy debate and reasonable conflict are inevitable, but if you're feeling unfairly victimized or accused it's time to take a stand for yourself and your needs. While you may feel that some situations are calming by the second Solar Eclipse on the 30th, you may still have a few loose ends to connect before you're satisfied.

August

Distractions can result from situations that are not quite as you expected them, and although the disruption will leave you feeling emotionally out of sorts you're willing to give someone the benefit of the doubt. Your feelings for someone may change from the 7th to the 15th, especially if you sense that you're moving along different pathways. If things are changing too rapidly during the Full Moon on the 15th you may need to pull back to make an assessment of how you feel before you determine your actions or responses. After the 23rd, the cycles allow for more emotional grounding, and you can take the initiative during the New Moon on the 29th.

September

While communication is working on some levels, you are left out of the loop if others are focusing on philosophical concerns and not real problems. Your patience wears thin with those who force their own agenda over the good of others , although you do have a way of effectively getting your point across near and during the Full Moon on the 13th. Dedicating your time and concern to matters close to home feels good, and after the 18th you may be most comfortable spending as much time as possible doing things to satisfy personal concerns. Pressure from relationships can be irritating after the 23rd unless you feel that you are truly appreciated.

October

You're sure of your opinions and values, and although someone may be trying hard to sway you to another way of thinking, you won't budge unless you see the value in making a change. In fact, if there's coercion involved, you are likely to dig in your heels just for the heck of it! Friction peaks from the 2nd through the Full Moon on the 13th, when you may see a contrast between your beliefs and those of others. Spending time on a constructive project takes the edge off tense emotional situations and may result in a wonderful product in the long run. In fact, your creativity is likely to surge and will need a positive outlet. You're ready to bid good-bye to an old, useless habit during the New Moon on the 27th.

November

While you may meet a few challenges, they're geared to help you break free of the things you no longer need. Examining your deepest feelings about your work may take top priority, although your relationships are also undergoing close scrutiny. If you sense that there are blocks to getting really close to someone, explore you own fears and try to make room in your life to let go of the demands on your time during the Full Moon on the 11th. Expressing your sentiments is easier after the 12th, when Venus transits in Capricorn for four weeks, aiding your ability to let love flow into everything you do and say. You may run into problems over details when it comes to shopping or entertainment late in the month, but you can quickly find solutions to quell any brewing storms.

December

Solidifying an emotional commitment can set you on sound emotional footing from the 1st to the 7th. These are also good days to finalize an agreement, especially if it's something you really want. You may feel a bit put off by situations which seem to be more pie in the sky than reality near the Full Moon on the 11th, but if it's important to someone you love, you may play along just to offer your support. Giving a false impression is not a good idea, though, since you could be trapped into having to repeat yourself later! Fortunately, everything is falling into place for a wonderfully rewarding holiday season, and the Capricorn Solar Eclipse on the 25th marks a highly significant time to explore your feelings and act from your heart—after all, your heart and mind are in close proximity now.

Aquarius Moon

Allowing your individuality to filter into everything you do is an expression of your Aquarius Moon. You're drawn to uniqueness, and you thrive on the leading edge of evolutionary change. Although you cherish your friendships, in some respects you're a loner. Your need for self-realization may take you along an untrodden path. You have the soul of a visionary humanitarian and often help liberate people from ignorance while illuminating the extraordinary qualities in the world around you.

You're comfortable when things make sense, which can be a problem since feelings frequently fall into a nonsensible category! Your natural reason gives you an air of objectivity, but over the years you've learned that you have to allow your feelings and emotions to be what they are. In relationships, you value unconditional acceptance, but you also find that you have to settle for less until you get beyond being merely human. You are a devoted friend but need your freedom. Only those who are at ease with your independent way of doing things will be able to get close to you.

Since you like doing things your own way, others can presume that you're close-minded. To avoid projecting an air of aloof disinterest, you will allow others the same individuality that you value for yourself. Your urge for freedom can seem like a cold shoulder when someone needs confirmation that he or she is special to you, and even though you may not mean to break hearts you sometimes will. Since your intuition can be as precise as a laser, surrendering to your intuitive guidance when you're in a quandary about personal issues can illuminate the best path. Your powerful urge to shatter the boundaries of ordinary reality always wins out.

Famous Individuals with Aquarius Moon
Glenn Close, Princess Diana, John Lennon

The Year Ahead for Aquarius Moon

The visions of what can be are coming fast and furious during this first year of the millennium, and you're eager to make the necessary changes in order to actualize them. There's only one problem: the physical plane moves much more slowly than you do! You're seeing evidence of the need to step back and allow things to end before you attempt to build a new structure. Pushing against the tide of conservatism or tradition may only waste your time and energy. You cannot avoid your responsibilities because there's something else you'd rather do. Completing your obligations are your first requirement. If you're in an unhealthy situation, avoiding things which are inhibiting your needs may require some patience.

Jupiter's transit marks two distinctive periods of challenge and opportunity. From February through the end of June, you may find it more difficult to set reasonable limits, especially if you are afraid to say "no" to something you think you want. Staying in touch with your deepest needs will help you set boundaries so that you can take advantage of the best options. From July through December, Jupiter moves into a supportive aspect to your Moon, opening the pathways to opportunity along lines which seem a better fit.

The actual degree of your Moon determines when certain cycles are most significant, and the long-term transits to your Moon this year will be apparent to you. If your Moon is between 0–2 degrees Aquarius, for instance, you will feel more stabilized, particularly from July through October while Saturn trines your Moon. The visions guiding you last year may now become reality. If your Moon is between 3–7 degrees Aquarius, you experience a cycle of transcendence and heightened imagination while Neptune transits in conjunction to your Moon. This once-in-a-lifetime cycle provides the perfect energy to forgive and let go of what you no longer need in order to improve the creative quality of life. You will experience a cycle of healing and rebirth if your Moon is between 8–14 degrees Aquarius while Pluto transits in sextile aspect to your Moon. You may make the most sweeping changes in your life from August through November, when moving, home improvements, or powerful relationship changes bring new vitality into your life. If your Moon is between 15–21 degrees Aquarius, you are ready to break free and move beyond unnecessary restraints. Uranus is transiting in conjunction to your Moon, prompting your urges for freedom, although from January through June you may still have to deal with completing old obligations before you're finally free as Saturn is squaring your Moon during those months. For the remainder of the year, however, your wings are ready to carry you. If your Moon is between 22–30 degrees Aquarius, you will experience a year of testing while Saturn transits in square aspect to your Moon. Although this cycle is most intense after May, you'll see increasing responsibilities earlier in the year.

Affirmation for the Year

I am ready to take responsibility for making significant
changes which help me fulfill my highest needs.

January

Creative ideas spawn new objectives. You're ready to remove barriers to
your happiness, although preparatory work is required. Talking about your
feelings brings transformation in a relationships and may be just what you
need. Rushing into changes early in the month will create unnecessary
havoc. Setting the stage works to your advantage. Meanwhile, by the Lunar Eclipse on the 20th you're ready to share your insights and vision.
From the 18th to the 26th, you're in an exceptional space spiritually and
emotionally, and you can share your insights with those you trust in ways
that will inspire them. Relationships take onspecial meaning, and the
spiritual quality of your connection can be exceptionally powerful.

February

By listening to your intuitive voice, you may determine that it's best to
allow a situation to mellow before making changes from the 1st to the
4th. You're ready to make a serious commitment to yourself about your
priorities and needs during the Aquarius Solar Eclipse on the 5th, when
breaking away from inhibitions become your primary drive. While you
feel that you can no longer tolerate certain circumstances, you also discover that some situations are changing beyond your control. All you can
do is respond in the most appropriate manner. Progress continues,
although you may feel that you need to withdraw from the action for a
while after the Full Moon on the 19th. Devoting more time to a creative
venture or the exploration of your inner self can become very important.

March

Developing a close relationship is rewarding from the 1st to the 13th,
although you may be distracted in the process by something intriguing.
This is an good time to indulge your artistry and to express your special
talents. If you're working with others who inspire your sense of vision, you
can be successful. Your drive and determination work to your best advantage through the 23rd, when actualizing your dreams and desires is easiest.
Unrealistic expectations can emerge from the 24th to the 31st, when you

can also run into contradictory attitudes or actions from others who may not understand your motivations or concerns.

April

With the necessity of keeping one foot on the ground, you're striving to maintain your stability while you're moving through a period of change. You may feel impatient with the process, especially if you're experiencing a sense of delayed gratification, but by reminding yourself that what you want is worth the effort you can stick around long enough to get what you want. Small steps open a fresh pathway during the New Moon on the 4th, when you may find a kindred spirit whose support offers some hope. Doing something wildly romantic may be on your agenda for the Full Moon on the 18th, although getting into an emotionally charged situation may leave you feeling trapped. Watch for emotional quicksand this month—a situation may not look dangerous until you've stepped into it!

May

You may feel you're being pulled back to an earlier state this month. While you may desire to break away from a stifling situation, if you've made a commitment, you are stuck until your obligation is satisfied. Frustrations are highest near the Full Moon on the 18th, although you may be feeling a weakening in the resistance to change and are likely to be initiating plans of your own, even if surreptitiously. Unless you must, this is not the best time to make promises, since you may be roped into something that's much more intensive than you realized. Progressive change is more hopeful after the 21st, and it may be through your artistic expression that you make headway during the remainder of the month.

June

Adding your own touches to your environment makes you feel more at home. Moving or renovation may be on your agenda, and from the New Moon on the 2nd through the Full Moon on the 16th, you are energized to break out of stale routines and try something different. You may not be able to change everything, but by concentrating on the things you can influence, you'll satisfy your deep desires for a change of pace. In your closest relationship, you may be ready to share your hopes and forge a stronger connection. But after the 18th you may feel the wheels of change are stuck in a rut, particularly if traditional values are holding your needs hostage. Fortunately, a few minor adjustments can make things tolerable.

July

The Solar Eclipse on the 1st may seem to affect people around you, but you are not likely to be pulled into an emotional crisis. Your ability to remain objective can be helpful to those who are uncertain about the best ways to handle things. During the Lunar Eclipse on the 16th your view of circumstances around you can be one of quiet discernment, and as your situations change, you're likely to see how simple innovations can be helpful. The most challenging elements of this eclipse cycle center around the way you handle your partnership and social commitments, since others may need your support. This is especially important during the second Solar Eclipse on the 30th, when your involvement can give someone else courage.

August

Your drive to fulfill your needs grows daily, and while you might prefer a policy of noninvolvement you're willing to get down and dirty if necessary to make things happen. You have a special way of handling conflicts which relies on opening lines of communication first and using the right words at the right time to create a wave of change. Emerging philosophical disagreements can lead to skirmishes, and seeing the potential for problems you may feel it's necessary to bring issues into the open. Even though you are emotionally vulnerable during the Aquarius Full Moon on the 15th, you also feel that this is the best time to bring an offer to the table. Rely on your higher principles to guide your decisions and actions.

September

Competitive situations may still remain a problem, although you're feeling a more supportive alignment of friends and associates when it counts the most. Maybe you just need a cheering section for a while—someone to be there as you push past the finish line. Keeping the lines of communication open allows you to deal with political issues at work, but this can infiltrate your family structure, too. You may be stepping away from the old family structure in favor of developing your own agenda in ways that others do not understand, but as long as you appear safe enogh for those who insist on protecting you, you're not likely to have much interference. It's time for you to forge your own way and to prove to yourself that you can satisfy your needs.

October

A sticky situation emerges early in the month, especially if you've been doing things which fly in the face of convention or have stirred the pot of controversy too intensely. Since you rarely do anything halfway, you may not realize the power you emanate, and you've been on fire lately. Anyone who feels threatened by your actions or attitudes may lash out against you, although if their responses are based on emotionality, you can respond with a logical alternative instead of getting into an emotionally intense struggle. Of course, that can anger those who want to fight! You can be blind-sided by another whose personal agenda revolves around control, so watch for signals that tell you when to retreat from the 25th to the 31st.

November

You may feel that your best approach is to withdraw from a heated situation, and if misunderstandings are running rampant, a quiet period from the 1st to 4th can give you a chance to reflect. In your desire to protect someone who may be stuck in an apparently hopeless situation, you can get in over your head, so knowing when to ask for assistance or when to pull in your boundaries can foster your ability to help. Working in concert with others who share your ideals or passions can be especially rewarding after the 5th. Watch for signals that tell you if you're stepping on the wrong toes during the Full Moon on the 11th, and then make quiet preparations for change. You're in the right place to initiate significant changes or bring others together to share a special time during and after the New Moon on the 25th.

December

You're feeling almost lighthearted, and finding ways to do things which bring smiles into the hearts of others refreshes your emotional vitality. Look for opportunities to laugh, and make room in your schedule to enjoy special times with your close friends or sweetheart near and during the Full Moon on the 11th. Making contact with others who stir your imagination and who tend to think futuristically can be like a magical elixir in your life from the 1st to the 23rd. Fortunately, though there is change in the wind during the Solar Eclipse on the 25th, you're still likely to be confident about your ability to illustrate to others how much you value their presence, talents, and pure essence. It's a good idea, however, to be sure that they're ready for that experience before you get things rolling!

Pisces Moon

Your emotional sensibilities are finely tuned, and through your Pisces Moon you have the capacity to connect to the most subtle realms of feeling. You feel at an intangible level in the realm of vibration, imparting to you the essence of what's happening. For this reason, you experience music, art, and life itself at a dramatic level, and your own approach to life is as an artist. Your sensibilities allow you to surrender to the realm of pure imagination; you can bring magic into any situation.

Your deepest urging may be to find ways to transcend the earthly plane. You are perched on the boundary between reality and fantasy and prefer to think of life as it should be—filled with serene beauty, forgiveness, and peace. You also often try to lend your help—when others show bigotry, you offer tolerance; where there is misery, you bring compassion. You're especially intuitive, and this sensibility can help you determine the underlying truth of a person or situation. Maintaining your emotional boundaries requires conscious effort, especially in your personal life, when you can sacrifice your own needs in favor of the more pressing demands you feel from others. To preserve your truly compassionate sensibilities without being taken in by unscrupulous characters, you need to create an emotional filter which allows you to discern the reality of a situation.

Falling victim can be your greatest potential trap—whether you're victimized by addictive behavior or deceitful people. Your urge to escape the ordinary is strong, and finding healthy ways to release the heaviness of everyday life is important. Meditation, especially movement-oriented practices such as tai chi or yoga, can be helpful, as well as a safe haven where you can feel free to exercise your creativity. Your need to fill your soul with tranquillity allows you ultimately to inspire the world.

Famous Individuals with Pisces Moon

Helen Keller, Martin Luther King, Jr., John Denver

The Year Ahead for Pisces Moon

Your sense of emotional stability allows you to yield to an increasing fascination with the prospects of change, and you may be able to take advantage of innovations without undermining your security base this year. Many of the options available to you will invite you to incorporate simple steps in order to make changes, but some changes will be happening outside your control. Your only option may be in the way you respond, though your creative resources are rich.

Jupiter's expansive cycle brings a mix of confidence and challenge. Through July, Jupiter's transit stimulates your optimism in positive ways, and you make changes which enhance the comforts of life. This can be a good period to alter your home environment. From July to December, you may feel that you're caught in a marathon. Though your sense of hope is strong, you will find yourself wondering if you'll ever get to the finish line.

The influence of the slower-moving cycles has the most significant impact, and determining the exact degree of your Moon will clue you into these cycles. If your Moon is between 0–2 degrees Pisces you may feel stable for most of the year, with the exception of the months from August to October when Saturn transits in square to your Moon. At this time, the bottom may drop out of your security base and you may feel you're being tested. If your Moon is between 3–7 degrees Pisces, you feel a need to forgive so that you can move into a more peaceful state of awareness while Neptune transits in semisextile to your Moon. You will experience profound changes if your Moon is between 8–14 degrees Pisces, since Pluto tests you as it transits in square to your Moon. During this year you may feel that you're under emotional reconstruction, while you are eliminating things you no longer need. If your Moon is between 15–21 degrees Pisces, you feel a strange mix of stability and desire to break free. Uranus transits in semisextile to your Moon all year, provoking you to try something completely different, but from February to June you have to focus on practical considerations while Saturn influences your Moon. You feel an enhanced sense of security if your Moon is between 22–30 degrees Pisces, since Saturn is traveling in sextile to your Moon this year. Most notably during May and June you may feel ready to forge lasting commitments.

Affirmation for the Year
I am a realistic idealist!

January

Putting energy behind your drives and needs helps you accomplish your aims, but you may feel somewhat hesitant if you sense that what you want or like does not meet with another's approval or tastes. Before you cave in and decide that your needs don't count, take another look. If you fail to deal with a genuine concern or need, you'll later have regrets. It's also conceivable that it's time examine why you're connected to certain people. If your feelings have changed or you're ready to end your involvement, your desire to take action can be powerful this month. Although outside frustrations can be troublesome from the 14th to 22nd, knowing what's in your heart helps you remain committed to your course. Invite the visions of what can be during the Lunar Eclipse on the 20th.

February

Changes happening around you can be unsettling near the time of the Solar Eclipse on the 5th, although they may not directly affect you. While others may be struggling to make sense of things, you can explore the emotional fallout and offer the right support at the right time. Spending extra time on your creative expressions can be especially satisfying, and you'll feel most comfortable sharing the results of your efforts from the 1st through the Full Moon on the 19th. In many ways, the full Moon cycle seems to bring things to a closure, especially if you've been laboring with a personal concern. You're ready to let go and allow nature to take its course, even if others may seem to want to fix things.

March

Going with the flow continues to feel best, since you're in no mood to stir up controversy at the moment. Nourishing a project or dealing with your obligations can be fulfilling, but by the Pisces New Moon on the 6th you may be ready to change things a bit. While you may not be ready to set the world on fire, this can be an excellent time to express what's in your heart or to let your talents shine. Loving relationships grow in intensity after the 13th, when, for the next four weeks, Venus travels in Pisces, emphasizing your need to express your emotions. Time spent with those you love can be remarkably satisfying, and during the Full Moon on the 19th you're experiencing rewards for your patience and compassion.

April

Blending your love of whimsy within the context of practical concerns can be especially gratifying from the 1st to the 14th, when some people may be taking things entirely too seriously. You're eager to allow your needs for romance to flourish, and while you're most comfortable allowing love to guide your course through the 6th, you may feel rather playful throughout the month. Disruptions in the status quo or alterations in your routine can leave things in a bit of a mess around the time of the Full Moon on the 18th, but you'll do yourself a favor by keeping things in perspective. You may not actually have to place yourself in jeopardy in order to rescue someone, and you gather support and resources from others to make the job easier.

May

Getting close to others, and enjoying hugs and laughter will feel great during the New Moon on the 3rd. You may feel the stirring of discontent, but its unlikely that you'll take rash action, since the idea of doing anything unsettling may seem rather distasteful. However, if you need to let someone know how you feel, you can get right to the point from the 1st to the 15th. Romantic encounters will be especially gratifying before and during the Full Moon on the 18th. But watch out, since argumentative circumstances can turn into tawdry mudslinging after the 19th, and if you've left yourself open to attack you may feel especially vulnerable. Calling on your friends or others who understand your motivations helps to shield you from unnecessary attacks.

June

Your feelings about someone may be changing, and even if you want to let go of a situation you may find the experience frustrating. If you're dealing with a hostile circumstance, you may want to withdraw, although it's unlikely that will resolve anything. Finding the peaceful place in your own heart before you speak or act will give you a feeling of safety as you make important changes. Boisterous attitudes or disruptive situations can get on your nerves during the Full Moon on the 16th, but after that you're experiencing a different stimulus. For the remainder of the month you're actively pursuing peaceful solutions and opportunities for creative expression. Romantic encounters can bring wonderful surprises, and you may even rediscover an old love!

July

Your impulse during the Solar Eclipse on the 1st may be to initiate a significant exchange of ideas and expressions, and it's a marvelous time to set things in motion which can have a lasting effect. While this can be an emotionally charged period, you may find that tapping into your feelings gives you what you need to bring a dream into reality. During the Lunar Eclipse on the 16th you may feel more reflective than usual and need time for solitude and illumination. While you may be eager to communicate what's in your heart, you may run into a few complications if someone else is not ready on the receiving end! Let your intuition guide you.

August

Maintaining your sense of emotional balance requires attentive, conscious effort, since stimulation from the planetary cycles can be unsettling. Unanticipated events or demands from others can leave you feeling uncomfortable during the Full Moon on the 15th, and you may fall into worry if outside circumstances create concern. Determining what you can and cannot do about a situation helps, since you may be unable to change anything right away. Your style of handling things contrasts with the words and actions of others. In shared experiences, look for the strengths you bring, and forge a path which allows you to utilize them. Philosophical conflicts can lead to divisive actions, so keep your boundaries intact if you don't want to play!

September

You may wish you could tell in advance where the battle lines will be drawn, but you may have to stay out of the line of fire in situations that are burning out of control. While the biggest problems revolve around moral concerns, you may quickly discover that the perceptions of right and wrong are unclear. Your ability to quietly retreat will serve you well, and during the Pisces Full Moon on the 13th you may want to spend time reflecting on your primary needs and the way you truly feel about what's happening. Propaganda can run rampant, but you're not likely to buy into it! Devoting time to bringing beauty and harmony into your home offers the perfect distraction after the 23rd.

October

Your desire to express the depths of your emotions is strong, and finding a dramatic outlet to share your feelings can be just what you need. It's easy to carry things a bit too far from the 1st to the 11th, although your motivation may be to try something completely different. If you're with people who love, you'll be fine, but those who don't know you well may not know how to interpret your actions. At the same time, you may wonder why others get away with their shenanigans! Engaging a positive emotional response from others flows best after the New Moon on the 27th, although you may be receiving mixed signals. If you're uncertain, clarify before you jump into the pool.

November

While everyone is sorting through the confusion of miscommunication, you may find creative inspiration in the experience of a path you might otherwise have never seen. You're not likely to feel the necessity to row against the tide, and certainly, if there are emotionally stormy circumstances early in the month you may find that the best way to deal with them is to shelter yourself in the experience of your creativity. You may be dancing, singing, cooking, painting, writing—it does not matter what you do—it's simply important that you allow the inspiration you're feeling to stimulate your creative flow. Your inner calm can radiate to others who may be in a bit of a frenzy. By taking someone under your wing to share the beauty of this time, you may make a great difference.

December

It's easy to be caught up in the pressure of getting everything right, and since your opinions may differ from those of others perfection can be an exceptional burden. Using your natural sensibilities to tune in to what someone needs or wants from you will put you on the right track from the 1st to the 8th. After that time, and especially near the Full Moon on the 11th, you may feel somewhat out of touch with others, particularly if they seem to be caught up in the externals. Pulling yourself back into your own emotional center is easiest after the Solstice on the 21st. You're ready for something truly enchanting and can share an exceptional experience which confirms the true nature of joy during the Solar Eclipse on the 25th. Clarify your thoughts, and project your hopes. This is your time to experience magic!

About the Authors

Gloria Star has been an internationally renowned astrologer, author, and teacher for over two decades. She has been a contributing author of the *Moon Sign Book* since 1995. Her most recent work, *Astrology: Woman to Woman*, was released by Llewellyn in April 1999. She is also the author of *Optimum Child: Developing Your Child's Fullest Potential through Astrology* (Llewellyn Publications, 1987), now translated into four languages. She edited and coauthored the book *Astrology for Women: Roles and Relationships* (Llewellyn, 1997) and contributed to two anthologies, *Houses: Power Places in the Horoscope* (Llewellyn, 1990), and *How to Manage the Astrology of Crisis* (Llewellyn, 1993). Her astrological computer software, *Woman to Woman*, was released by Matrix Software in 1997. Listed in *Who's Who of American Women*, and *Who's Who in the East*, Gloria is active in the astrological community and has been honored as a nominee for the prestigious Regulus Award. She has served on the faculty of the United Astrology Congress (UAC) since its inception in 1986, and lectures regularly throughout the United States and abroad. A member of the Advisory Board for the National Council for Geocosmic Research (NCGR), she also served on the Steering Committee for the Association for Astrological Networking (AFAN), was editor of the AFAN Newsletter from 1992–1997, and now serves on the advisory board. She currently resides in the shoreline township of Clinton, Connecticut.

Skye Alexander is the author of *Planets in Signs* (Whitford Press, 1988) and the astrological mystery *Hidden Agenda* (Mojo Publishing, 1997). She may be reached through Mojo Publishing, P.O. Box 7121, Gloucester, MA 01930, or mojo@shore.net. Her website is www.shore.net/~mojo.

Leeda Alleyn Pacotti embarked on metaphysical self-studies in astrology and numerology at age fourteen after a childhood of startling mystical experiences. After careers in antitrust law, international treaties, the humanities, and government management in legislation and budgeting, she now plies a gentle practice as a naturopathic physician, master herbalist, and certified nutritional counselor.

Chandra Moira Beal is a freelance writer. She has also been published in the magazine *Texas Beat*. The name Chandra means Moon.

Heyde Class-Garney is an astrologer, spiritual counselor, and certified tarot grandmaster. She has a B.S. in metaphysical ministry and is currently the vice president of the American Tarot Association (ATA). She has written her own column for the ATA newsletter entitled "Astro-Tarot" since 1995. Heyde has studied many new-age topics, including herbs and aromatheraphy. One of her specialities is baking, which she has done for thirty years.

Estelle Daniels has been a professional astrologer since 1972 and is the author of *Astrologickal Magick* (S. Weiser, 1995), and coauthor of *Pocket Guide to Wicca* (Crossing Press, 1998). Estelle has an astrological practice in Minnesota, and lectures around the U. S. Her work appears irregularly in *The Mountain Astrologer*, and she has been contributing to Llewellyn's annuals since 1997. She is an Initiate and High Priestess of Eclectic Wicca and teaches the Craft to students of varying interests and levels.

Alice DeVille is an internationally known astrologer and writer who has been practicing for twenty-five years. Alice conducts workshops and lectures on a variety of astrological, spiritual, and business topics, including a popular workshop called "Finding Your Soulmate." Her focus is on helping clients discover the psychological and spiritual attunement that supports their life purpose. You can reach Alice at: DeVilleAA@aol.com.

Ronnie Gale Dreyer is an internationally known astrological consultant and lecturer based in New York City. She is the author of *Vedic Astrology: A Guide to the Fundamentals of Jyotish* (S. Weiser, 1993) and an upcoming book about astrology and health. Ronnie is also a contributor to the anthology *Astrology for Women: Roles and Relationships* (Llewellyn, 1997).

Anna T. Duchon, age fourteen, is an eighth-grade student in Decatur, Georgia. She lives with her two parents, two sisters, two dogs, and two guinea pigs. This is her first published article.

Deborah Duchon is an anthropologist and ethnobotanist at Georgia State University in Atlanta. She often writes about plant-related issues.

Verna Gates teaches folklore classes at the University of Alabama at Birmingham, and has been featured on *NBC Nightside* as a folklorist. She was a writer for CNN and has been a freelance writer for fourteen years. Her specialties are wildflowers, moonlore, and storytelling.

Kenneth Johnson was born in southern California, where he obtained his degree in the study of comparative religions at California State University, Fullerton. He has been a practitioner of astrology since 1974 and is the author of six books published by Llewellyn, including *Mythic Astrology: Archetypal Powers in the Horoscope* (1993) and *Jaguar Wisdom: An Introduction to the Mayan Calendar* (1997). He has lived in London, Amsterdam, San Diego, Santa Fe, and currently resides in California.

Penny Kelly has earned a degree in naturopathic medicine and is working toward a Ph.D. in nutrition. She and her husband Jim own a fifty-seven-acre farm with two vineyards, which they are in the process of restoring using organic farming methods. Penny is the author of the book *The Elves of Lily Hill Farm* (Llewellyn, 1997).

Gretchen Lawlor combines twenty-five years as an astrologer with more than ten years as a naturopath in her astromedical consultations and teachings. Regularly traveling around the U. S., U. K., and New Zealand to visit hospitals and alternative medical practitioners, Gretchen teaches natural approaches to common health problems. Her work has been published in almanacs for the past twenty years. She can be reached for consultations at P.O. Box 753, Langley, WA 98260, email: light@whidbey.com.

Harry MacCormack is an adjunct assistant professor of theater arts (play writing, screen writing, and technical theater), and owner/operator of Sunbow Farm, which is celebrating a quarter century of organic farming.

Dorothy Oja has been an astrologer for twenty-eight years in a practice called Mindworks. Her specialties include electional work, and composite and Davison relationship analysis. An active writer for numerous magazines, her articles are available on request. Dorothy now acts as co-chair of AFAN's legal information committee, dedicated to protecting the constitutional rights of astrologers when faced with anti-astrology ordinances.

Louise Riotte is a lifetime gardener from Oklahoma and the author of several gardening books from Storey Publications, including: *Sleeping With a Sunflower* (1987), *Carrots Love Tomatoes* (1998), *Roses Love Garlic* (1998), *Planetary Planting* (1998), and *Catfish Ponds and Lilypads* (1997).

Kim Rogers-Gallagher is an astrologer and author of *Astrology for the Light Side of the Brain* (ACS Publications). She also writes columns for the magazines *Welcome to Planet Earth*, *Dell Horoscope*, and *Aspects*.

Kaye Shinker teaches financial astrology at the Online College of Astrology (www.astrocollege.com). She serves on the National Council of Geocosmic Research (NCGR) board of examiners and is active in the Chicago and New Orleans chapters. A former teacher, she and her husband own race horses and travel around the U. S. in an RV.

Nancy Soller has been writing weather and earthquake predictions for the Moon Sign Book since 1981. She is currently studying the effects of the Uranian planets and the four largest asteroids on the weather. She has lectured at the Central New York State Astrology Conference.

K. D. Spitzer is an accomplished astrologer and tarot reader. She teaches, consults, and writes about the planets and the Moon, contributing articles to various magazines. She gardens, cooks, and cuts her hair by the Moon, which hangs in the clear sky over seacoast New Hampshire.

Lynne Sturtevant has had a life-long fascination with ancient cultures, mystery cults, myths, fairy tales, and folk traditions. She is a freelance writer and lives in Artlington, Virginia with her husband and pets.

Carly Wall is author of *Flower Secrets Revealed: Using Flowers to Heal, Beautify and Energize Your Life!* (A.R.E. Press, 1993), *Naturally Healing Herbs* (Sterling, 1996), *Setting the Mood with Aromatherapy* (Sterling, 1998), and *The Little Encyclopedia of Olde Thyme Home Remedies* (Sterling 1998). A regular article contributor to Llewellyn, she holds a certificate in aromatherapy and lives on a farm with her husband, Ron, and cat, Missy; both of whom happily indulge in her natural healing methods.

Rowena Wall is a counseling astrologer in Georgia, and a writer and builder of web sites. She is webmistress to the ZodiacGazette Astrology Forum (http://www.starflash.com/zodiac), and she teaches and writes about astrology. She is an active member in NCGR, ISAR, AFAN, and other national astrological organizations, and she publishes "Star Source," an Astrology newsletter. If you would like a sample copy, or are interested in an astrological or tarot reading, contact her at P.O. Box 1405, Warner Robins, GA 31099, or starstuff@starflash.com.

Directory of Products and Services

Llewellyn's Computerized Astrological Services

Llewellyn has been a leading authority in astrological chart readings for more than 30 years. We feature a wide variety of readings with the intent to satisfy the needs of any astrological enthusiast. Our goal is to give you the best possible service so that you can achieve your goals and live your life successfully. **Be sure to give accurate and complete birth data on the order form. This includes exact time (a.m. or p.m.), date, year, city, county, and country of birth. Note: Noon will be used as your birth time if you don't provide an exact time. Check your birth certificate for this information! Llewellyn cannot be responsible for mistakes made by you.** An order form follows these listings.

Services Available

Simple Natal Chart
This is your best choice if you want a detailed birth chart and prefer to do your own interpretations. It is loaded with information, including a chart wheel, aspects, declinations, nodes, major asteroids, and more. (Tropical zodiac/Placidus houses, unless specified otherwise.)
APS03-119 . . . . **$5.00**

Astro*Talk Advanced Natal Report
One of the best interpretations of your birth chart you'll ever read. These no-nonsense descriptions of the unique effects of the planets on your character will amaze and enlighten you.
APS03-525 . . . . **$30.00**

TimeLine Transit/Progression Forecast
Love, money, health—everybody wants to know what lies ahead, and this report will keep you one-up on your future. The TimeLine forecast is invaluable for seizing opportunities and timing your moves. Reports begin the first day of the month you specify.
3-month TimeLine Forecast - APS03-526 . . . .**$12.00**
6-month TimeLine Forecast - APS03-527 . . . .**$20.00**
1-year TimeLine Forecast - APS03-528 . . . .**$30.00**

Friends and Lovers

Find out how you relate to others, and whether you are really compatible with your current or potential lover, spouse, friend, or business partner! This service includes planetary placements for both people, so send birth data for both and specify "friends" or "lovers."

APS03-529 . . . .**$20.00**

Child*Star

An astrological look at your child's inner world through a skillful interpretation of his or her unique birth chart. As relevant for teens as it is for newborns. Specify your child's sex.

APS03-530 . . . .**$20.00**

Woman to Woman

Finally, astrology from a feminine point of view! Gloria Star brings her special style and insight to this detailed look into the mind, soul, and spirit of contemporary women.

APS03-531 . . . .**$30.00**

Heaven Knows What

Discover who you are and where you're headed. This report contains a classic interpretation of your birth chart *and* a look at upcoming events, as presented by the time-honored master of the astrological arts, Grant Lewi.

APS03-532 . . . .**$30.00**

Biorhythm Report

Some days you have unlimited energy, then the next day you feel sluggish and awkward. These cycles are called biorhythms. This individual report accurately maps your daily biorhythms and thoroughly discusses each day. Now you can plan your days to the fullest!

APS03-515 – 3-month report **$12.00**
APS03-516 – 6-month report **$18.00**
APS03-517 – 1-year report **$25.00**

Tarot Reading

Find out what the cards have in store for you with this 12-page report that features a 10-card "Celtic Cross" spread shuffled and selected especially for you. For every card that turns up there is a detailed corresponding explanation of what each means for you. Order this tarot reading today! Indicate the number of shuffles you want.

APS03-120 . . . .**$10.00**

Lucky Lotto Report (State Lottery Report)

Do you play the state lotteries? This report will determine your luckiest sequence of numbers for each day based on specific planets, degrees, and other indicators in your own chart. Give your full birth data and middle name. Tell us how many numbers your state lottery requires in sequence, and the highest possible numeral. Indicate the month you want to start.

APS03-512 – 3-month report **$10.00**
APS03-513 – 6-month report **$15.00**
APS03-514 – 1-year report **$25.00**

Numerology Report

Find out which numbers are right for you with this insightful report. This report uses an ancient form of numerology invented by Pythagoras to determine the significant numbers in your life. Using both your name and date of birth, this report will calculate those numbers that stand out as yours. With these numbers, you can tell when the important periods of your life will occur. Please indicate your full birth name.

APS03-508 – 3-month report **$12.00**
APS03-509 – 6-month report **$18.00**
APS03-510 – 1-year report **$25.00**

Astrological Services Order Form

Report name & number_____

_____

Provide the following data on all persons receiving a report:

1st Person's Full Name, including current middle & last name(s)

_____

Birthplace (city, county, state, country) _____

_____

Birthtime _____ ❐ a.m. ❐ p.m. Month _____ Day _____ Year _____

2nd Person's Full Name (if ordering for more than one person)

_____

Birthplace (city, county, state, country) _____

_____

Birthtime _____ ❐ a.m. ❐ p.m. Month _____ Day _____ Year _____

Billing Information

Name_____

Address_____

_____

City _____ State _____ Zip _____

Country _____ Day phone:_____

Make check or money order payable to Llewellyn Publications, or charge it!
Check one: ❐ VISA ❐ MasterCard ❐ American Express

Acct. No. _____ Exp. Date _____

Cardholder Signature _____

❐ Yes! Send me my free copy of *New Worlds*!

Mail this form and payment to:

Llewellyn's Computerized Astrological Services
P.O. Box 64383, Dept. K-953 • St. Paul, MN 55164-0383

Allow 4-6 weeks for delivery.

Save $$ on Llewellyn Annuals

Llewellyn has two ways for you to save money on our annuals. With a four-year subscription, you receive your books as soon as they are published—and your price stays the same every year, even if there's an increase in the cover price! Llewellyn pays postage and handling for subscriptions. Buy any 2 subscriptions and take $2 off; buy 3 and take $3 off; buy 4 subscriptions and take an additional $5 off the cost!

Please check boxes below and send this form along with the order form on the next page.

Subscriptions (4 years, 2001–2004):

☐ Astrological Calendar...............$51.80	☐ Astrological Pocket Planner......$27.80
☐ Witches Calendar......................$51.80	☐ Sun Sign Book............................$27.80
☐ Tarot Calendar.........................$51.80	☐ Moon Sign Book.........................$27.80
☐ Crop Circle Calendar...............$51.80	☐ Herbal Almanac..........................$27.80
☐ Daily Planetary Guide...............$39.80	☐ Magical Almanac.........................$27.80
☐ Witches Datebook.....................$39.80	

Order a Dozen and Save 40%: Sell them to your friends or give them as gifts. Llewellyn pays all postage and handling when you order annuals by the dozen.

2000	2001		
☐	☐	Astrological Calendar.....................................	$93.24
☐	☐	Witches Calendar...	$93.24
☐	☐	Tarot Calendar..	$93.24
☐	☐	Crop Circle Calendar.....................................	$93.24
☐	☐	Daily Planetary Guide....................................	$71.64
☐	☐	Witches Datebook...	$71.64
☐	☐	Astrological Pocket Planner...........................	$50.04
☐	☐	Sun Sign Book...	$50.04
☐	☐	Moon Sign Book..	$50.04
☐	☐	Herbal Almanac...	$50.04
☐	☐	Magical Almanac..	$50.04

Individual Copies of Annuals: Include $4 postage for order $15 and under and $5 for orders over $15. Llewellyn pays postage for all orders over $100.

2000	2001		
☐	☐	Astrological Calendar.....................................	$12.95
☐	☐	Witches Calendar...	$12.95
☐	☐	Tarot Calendar..	$12.95
☐	☐	Crop Circle Calendar.....................................	$12.95
☐	☐	Daily Planetary Guide....................................	$9.95
☐	☐	Witches Datebook...	$9.95
☐	☐	Astrological Pocket Planner...........................	$6.95
☐	☐	Sun Sign Book...	$6.95
☐	☐	Moon Sign Book..	$6.95
☐	☐	Herbal Almanac...	$6.95
☐	☐	Magical Almanac..	$6.95

Llewellyn Order Form

Call 1-877-NEW-WRLD or use this form to order any of the
Llewellyn books or services listed in this publication.

SEND TO: **Llewellyn Publications, P.O. Box 64383,
Dept. K-953, St. Paul, MN 55164-0383**

Qty	Order #	Title/Author	Total Price

Postage/handling:
ORDERS $15 AND UNDER: **$4.00**
ORDERS OVER $15:   **$5.00**
Subscription orders, dozen orders,
 or orders over $100:   **FREE SHIPPING**
2ND DAY AIR: **$8.00 for one book**
(add $1 for each additional book)
*We cannot deliver to P.O. Boxes; please supply a street
address. Please allow 4-6 weeks for delivery.*

Total price	
MN residents add 7% sales tax	
Postage/handling (see left)	
Total enclosed	

☐ VISA   ☐ MasterCard   ☐ American Express
☐ Check or money order – U.S. funds, payable to Llewellyn Publications

Account #   Expiration Date

Cardholder Signature

Name   Phone (   )

Address

City   State   Zip/PC

Questions? Call Customer Service at 1-877-NEW-WRLD

What Does Your Future Hold?

Sept. 8, 2001

Feb. 14, 2001

July 17, 2002